Life-Cycle of Structural Systems

This book aims to promote the study, research and applications in the design, assessment, prediction and optimal management of life-cycle performance, safety, reliability and risk of civil structures and infrastructure systems. The contribution in each chapter presents state-of-the-art as well as emerging applications related to key aspects of the life-cycle civil engineering field.

The chapters in this book were originally published as a special issue of *Structure and Infrastructure Engineering*.

Hitoshi Furuta is Professor at the Department of Informatics, Kansai University, Japan. He is the author or co-author of over 300 refereed publications, including books, book chapters, journal articles, and papers in conference proceedings. He is editorial board member of several international journals and was President of ICOSSAR, Chair of WC1 of IABSE, and currently Vice President of IALCCE.

Mitsuyoshi Akiyama is Professor at the Department of Civil and Environmental Engineering, Waseda University, Japan. He published many research papers on earthquake engineering, life-cycle structural performance, safety and reliability in structural engineering, and application of probabilistic concepts and methods to the design of civil structures. He is the recipient of several national and international awards.

Dan M. Frangopol is Professor and the inaugural holder of the Fazlur R. Khan Endowed Chair of Structural Engineering and Architecture at Lehigh University, USA. He has authored/co-authored over 350 articles in archival journals including 9 prize-winning papers. He is the Founding Editor of *Structure and Infrastructure Engineering* and of the book series *Structures and Infrastructures*. He is a foreign member of the Academia Europaea and of the Royal Academy of Belgium, an Honorary Member of the Romanian Academy, and a Distinguished Member of American Society of Civil Engineers (ASCE).

T0239613

Life-Cycle of Structural Systems

Design, Assessment, Maintenance and Management

Editor in Chief: Dan M. Frangopol

Edited by
Hitoshi Furuta, Mitsuyoshi Akiyama and Dan M. Frangopol

Routledge
Taylor & Francis Group

LONDON AND NEW YORK

First published 2018
by Routledge
2 Park Square, Milton Park, Abingdon, Oxon, OX14 4RN, UK

and by Routledge
52 Vanderbilt Avenue, New York, NY 10017

First issued in paperback 2020

Routledge is an imprint of the Taylor & Francis Group, an informa business

British Library Cataloguing in Publication Data
A catalogue record for this book is available from the British Library

ISBN 13: 978-0-367-57215-0 (pbk)
ISBN 13: 978-0-8153-8428-1 (hbk)

Typeset in MinionPro
by diacriTech, Chennai

Publisher's Note
The publisher accepts responsibility for any inconsistencies that may have arisen during the conversion of this book from journal articles to book chapters, namely the possible inclusion of journal terminology.

Disclaimer
Every effort has been made to contact copyright holders for their permission to reprint material in this book. The publishers would be grateful to hear from any copyright holder who is not here acknowledged and will undertake to rectify any errors or omissions in future editions of this book.

Contents

v

CONTENTS

Citation Information

The chapters in this book were originally published in *Structure and Infrastructure Engineering*, volume 13, issue 1 (January 2017). When citing this material, please use the original page numbering for each article, as follows:

For any permission-related enquiries please visit:
http://www.tandfonline.com/page/help/permissions

Notes on Contributors

Mitsuyoshi Akiyama is Professor at the Department of Civil and Environmental Engineering, Waseda University, Japan.

Gojko Balabanić is Professor at the Faculty of Civil Engineering, University of Rijeka, Croatia.

Fabio Biondini is Professor at the Department of Civil and Environmental Engineering, Politecnico di Milano, Italy.

Eleni N. Chatzi is Associate Professor and Chair of Structural Mechanics at the Department of Civil, Environmental and Geomatic Engineering, ETH Zurich, Switzerland.

Dimitris Diamantidis is Professor at the Faculty of Civil Engineering, Ostbayerische Technische Hochschule Regensburg, Germany.

You Dong is Assistant Professor at the Department of Civil and Environmental Engineering, The Hong Kong Polytechnic University, Hong Kong.

Dan M. Frangopol is Professor and the inaugural holder of the Fazlur R. Khan Endowed Chair of Structural Engineering and Architecture at Lehigh University, USA.

Hitoshi Furuta is Professor at the Department of Informatics, Kansai University, Japan.

Haijie Ge is a Senior Engineer at Shanghai Land Group, China.

Milan Holicky is Professor at the Klokner Institute, Czech Technical University in Prague, Czech Republic.

Haitao Jiang is Research Associate at the Department of Civil and Environmental Engineering, Waseda University, Japan.

Karel Jung is Researcher at the Klokner Institute, Czech Technical University in Prague, Czech Republic.

Kiyoyuki Kaito is Associate Professor at the Department of Civil Engineering, Osaka University, Japan.

Hideyuki Kasano is Assistant Professor at the Department of Civil and Environmental Engineering, Nihon University, Japan.

Hyun-Joong Kim is Post-doctoral Researcher at Engineering Risk Analysis (ERA) group, Technische Universität München, Germany.

Sunyong Kim is Assistant Professor at the Department of Civil and Environmental Engineering, Wonkwang University, South Korea.

Anne S. Kiremidjian is Professor at the Department of Civil and Environmental Engineering, Stanford University, USA.

Roman Klis is Enterprise Data Scientist at Philip Morris International, Poland.

Kiyoshi Kobayashi is Professor at the Research Center for Business Administration, Graduate School of Management, Kyoto University, Japan.

Hyun-Moo Koh is Professor Emeritus at the Department of Civil and Environmental Engineering, Seoul National University, Republic of Korea.

Atsushi Koizumi is Professor at the Department of Civil and Environmental Engineering, Waseda University, Japan.

Heang Lam is Research Associate at the Department of Civil and Environmental Engineering, Waseda University, Japan.

Michael D. Lepech is Associate Professor at the Department of Civil and Environmental Engineering, Stanford University, USA.

NOTES ON CONTRIBUTORS

Abbie B. Liel is Associate Professor at the Department of Civil, Environmental and Architectural Engineering, University of Colorado at Boulder, USA.

Jeong-Hyun Lim is a freelance civil engineer.

Sopokhem Lim is Assistant Professor at the Department of Civil and Environmental Engineering, Waseda University, Japan.

Weiwei Lin is Associate Professor at the Department of Civil and Environmental Engineering, Waseda University, Japan.

Jun Murakoshi is Professor at the Department of Civil and Environmental Engineering, Tokyo Metropolitan University, Japan.

Kuniei Nogami is Emeritus Professor at the Department of Civil and Environmental Engineering, Tokyo Metropolitan University, Japan.

Filip Oršanić is Product Manager at SOFiSTiK, Germany.

Joško Ožbolt is Professor at the Faculty of Civil and Environmental Engineering, University of Stuttgart, Germany.

Wonsuk Park is Professor at the Department of Civil Engineering, Mokpo National University, Republic of Korea.

Anirudh S. Rao is Seismic Risk Engineer at the Global Earthquake Model Consortium, University of Pavia, Italy.

Mark P. Sarkisian is Structural and Seismic Engineering Partner at the Skidmore, Owings & Merrill LLP, San Francisco, USA.

Junho Song is Professor at the Department of Civil and Environmental Engineering, Seoul National University, Republic of Korea.

Xiao-Yan Sun is Professor at the Department of Agricultural Engineering, Zhejiang University, P.R. China.

Miroslav Sykora is Associated Professor at the Klokner Institute, Czech Technical University in Prague, Czech Republic.

Hiroshi Tanaka is Chief Technological Engineer of Technology Headquarters at Nippon Koei Co. Ltd., Japan.

Jianhong Wang is Senior Research Fellow of the R&D Centre at Nippon Koei Co. Ltd., Japan.

Sarah J. Welsh-Huggins is a former PhD student at the Department of Civil, Environmental and Architectural Engineering, University of Colorado at Boulder, USA.

Ying Xu is Management Trainee at Fosun International Limited, China.

Jiwoon Yi is Senior Researcher at Korean Institute of Bridge and Structural Engineers, Republic of Korea.

Teruhiko Yoda is Emeritus Professor at the Department of Civil and Environmental Engineering, Waseda University, Japan.

Hiroshi Yokota is Professor at the Faculty of Engineering, Hokkaido University, Japan.

Life-cycle of structural systems: design, assessment, maintenance and management

The International Association for Life-Cycle Civil Engineering (IALCCE) was founded in 2006 to support this challenge and to create a fertile ground for promoting the study, research, and applications in the design, assessment, prediction, and optimal management of life-cycle performance, safety, reliability and risk of civil structures and infrastructure systems (http://www.ialcce.org). To accomplish this mission and following a series of International Workshops on Life-Cycle Cost Analysis and Design of Civil Infrastructure Systems held in Honolulu, Hawaii, USA (LCC1, 2000), Ube, Yamaguchi, Japan (LCC2, 2001), Lausanne, Switzerland (LCC3, 2003), Cocoa Beach, Florida, USA (LCC4, 2005), and Seoul, Korea (LCC5, 2006), it was decided to bring together the main advances on life-cycle civil engineering and related topics at the First International Symposium on Life-Cycle Civil Engineering (IALCCE'08), held in Varenna, Lake Como, Italy, 10–14 June 2008 (http://www.ialcce08.org), and afterwards at the Second International Symposium on Life-Cycle Civil Engineering (IALCCE 2010), held in Taipei, Taiwan, 27–31 October, 2010 (http://www.ialcce2010.ntust.edu.tw), the Third International Symposium on Life-Cycle Civil Engineering (IALCCE 2012), held at the Hofburg Palace in Vienna, Austria, 3–6 October, 2012 (http://www.ialcce2012.org), and the Fourth International Symposium on Life-Cycle Civil Engineering (IALCCE 2014), held in Tokyo Japan, 16–19 November, 2014 (http://www.ialcce2014.org). IALCCE 2014 has been organised on behalf of the IALCCE under the auspices of Waseda University, Tokyo, Japan. The interest of the international civil engineering community in areas covered by the IALCCE has been confirmed by the significant response to the IALCCE2014 call for papers. In fact, 436 abstracts from 44 countries were received by the Symposium Secretariat, and approximately 75% of them were selected for publication. Contributions presented at IALCCE2014 include state-of-the-art as well as emerging applications related to key aspects of the life-cycle civil engineering field.

The extended versions of selected papers presented at IALCCE 2014 and invited contributions are included in this special issue of *Structure and Infrastructure Engineering*. Sarkisian introduces new ideas to improve the seismic performance of buildings. Welsh-Huggins and Liel present a framework to assess the building life-cycle in terms of social, environmental, and economic impacts using probabilistic approaches. Yokota discusses the framework of the life-cycle management system for shore protection facilities. Wang et al. propose a maintenance framework for a shield tunnel. Rao et al. in two sequential papers

propose a simplified structural deterioration model for reinforced concrete bridge piers under cyclic loading and develop time-dependent fragility functions for deteriorating structures. Kobayashi and Kaito present a big data-based deterioration prediction model and infrastructure management. Biondini and Frangopol investigate failure loads and failure times of concrete structures exposed to corrosion and propose life-cycle performance indicators related to redundancy and elapsed times between sequential failures. Lin et al. investigate after-fracture redundancy of truss bridges through a case study of a five-span continuous truss bridge in Japan. Lim et al. investigate the spatial variability of the steel weight loss and corrosion cracking of RC members using X-ray photograms. Ožbolt et al. present the coupled 3D chemo-hygro-thermo-mechanical model for numerical analysis of non-mechanical and mechanical processes related to the corrosion of steel reinforcement in concrete. Kim and Frangopol present a novel approach to multi-objective optimisation process of probabilistic service life management. Dong and Frangopol present an integrated probabilistic framework for the healthcare - bridge network system performance analysis considering spatial seismic hazard, vulnerability of bridges and links in the network, and damage condition of a hospital at component and system levels. Koh et al. present a reliability-based design framework for accidental loads based on findings and outcomes of recent research efforts on bridge designs under ship-collision risk. Sykora et al. discuss the target reliability for existing structures considering economic and societal aspects. Finally, Klis and Chatzi propose a novel data compression framework for wireless sensor networks and present a case study to demonstrate the ability of the proposed scheme to efficiently detect structural modes.

The guest editors thank the authors and the reviewers for contributing to this special issue and hope that this collection of papers will represent a useful reference for researchers, students, and practitioners to promote and advance research and applications in the field of life-cycle civil engineering.

Hitoshi Furuta

Mitsuyoshi Akiyama

Non-prescriptive approaches to enhanced life cycle seismic performance of buildings

Mark P. Sarkisian

ABSTRACT
Safety is the primary focus of structural design of buildings and infrastructures for expected operational life. However, with the growing importance of performance and the relationship of structures to the environment, additional considerations must be made to achieve successful design. These considerations must assess the structure's impact on natural resources, long-term economic success and continuous operation following a significant abnormal loading event. Non-prescriptive approaches to design are becoming more important and perhaps will be common as we progress into the twenty-first century. These design approaches impart an improved understanding of behaviour, aides with economic predictions and increases efficiency of the building. Examples of non-prescriptive, performance-based approach to seismic design in both the United States and China will be presented as both regions represent major building markets utilising different methods for analysis and design. Specific building project examples will be described.

1. Introduction

Understanding building behaviour when subjected to extreme loading is the first and most important step in enhancing life cycle performance. Computational methods including finite element programs such as Etabs and Perform 3D have assisted designers in modelling behaviour in both linear and non-linear states. Building codes in major development regions around the world including the United States and China have been frequently updated in response to research and building performance observed after abnormal loading events including strong ground motion. Optimal performance will not be achieved however until all systems and components are designed to remain elastic during and after major loading events including extreme seismicity.

This paper was motivated by the need to consider performance-based design (PBD) beyond prescriptive requirements of building codes. In addition, guidelines and standards are compared between the United States and China where many of the world's tallest and most complex buildings exist or are being planned. The paper focuses on seismicity as it relates to PBD with specific references made to code sections that address analysis and design.

2. Considering U.S. codes for performance-based seismic design

Contrary to common perception, PBD does not necessarily lead to structures with better performance. This method of design specifically addresses seismic performance of tall buildings, including structures with long fundamental periods of vibrations, significant mass participation and lateral load-response in higher modes of vibration and a relatively high aspect ratio (slender profile). The process typically results in a better understanding of structural behaviour, but does not lead to enhanced performance unless specific higher performance objectives (minimum objectives are defined in the building code) are used, including ground motion input, components or systems. In addition, building codes have one primary goal and that is to protect the public and create structures that are life-safe. Many people have the incorrect perception that buildings designed to current code are earthquake-proof, but in fact they can sustain significant, and in some cases even irreparable, damage even when they are designed to be life-safe.

PBD was originally developed for structures that are an exception to the building code through an alternative, non-prescriptive approach to design with the methodology confirming code equivalence. These structures typically complied with the general requirements of the building code, with the exception of height limits described for particular seismic force-resisting systems, or for seismic force-resisting systems not specifically described in the code. For instance, concrete structures of various mass and stiffness characteristics over 49 m (160 ft) tall located in areas of high seismicity require a dual structural system composed of a shear wall core and moment-resisting frame to meet the prescriptive intent of the building code. Because of increased cost, increased construction time, and

Figure 1. Rendering (structure – core only lateral system) 500 Folsom Street, San Francisco, CA.

architectural impacts, many have designed and successfully built these structures without the frame by proving code equivalency through PBD methods. Figure 1 illustrates a 122 m (400 ft) tall residential tower that is being designed using performance-based methods.

2.1. Objectives

There are two primary objectives for performance when considering a non-prescriptive analysis and design. The first and required is defined as a minimum performance objective where code equivalency is met, and the second is an enhanced performance objective where systems and components are designed to standards beyond those required by the building code.

2.1.1. Minimum performance objectives

2.1.1.1. Maximum considered earthquake (MCE). The structure must withstand a MCE with a low probability (approximately 10%) of collapse without loss of gravity load-carrying capacity, without inelastic straining of important lateral load-carrying elements that would result in severe strength degradation, without excessive permanent lateral drift, and without the development of global structural instability. In addition, all elements of the structure must be designed for compatibility with the anticipated deformations of the seismic-force-resisting system. The MCE level typically has a probability of exceedance of 2% in 50 years or a return period of 2500 years.

2.1.1.2. Design earthquake (DE). The structure must withstand a DE also known as a design basis earthquake (DBE) having an intensity 2/3 of the MCE without creating significant hazards to individual lives while assuring that non-structural components and systems remain anchored in place and building drifts are limited so that undue hazards are not created. The DE level typically has an equivalent probability of exceedance of 10% in 50 years or a return period of 475 years.

2.1.1.3. Frequent or service level earthquake. The structure must withstand a relatively frequent, more moderate-intensity earthquake shaking with minimal damage. The structure must be designed to remain essentially elastic considering service-level earthquake ground shaking of a 50% probability of exceedance in 30 years or a return period of 43 years.

2.1.2. Enhanced performance objectives

It may be desired, and it is encouraged, to design structures to achieve a performance hiring than the life-safety requirements of the code. The following are examples of enhanced performance objectives.

2.1.2.1. Probability of exceedance. A lower probability of exceedance for service-level or MCE-level ground shaking or both can be considered in selecting ground motions. The intensity of shaking will be high resulting in designing the structure for greater seismic demand.

2.1.2.2. Drift and residual displacement. Establishing lower limits for lateral drift and/or reduced levels of acceptable levels of cyclic straining of ductile elements will lead to enhanced performance. Limiting residual displacements will lead to a structure that can be more easily repaired after an earthquake (most primary structural components will have remained elastic).

2.1.2.3. Non-structural components. Non-structural components and systems could be designed for accelerations based on higher ground motion intensities or storey drifts that are larger than those required by the building code.

2.1.2.4. Damage-tolerant or response modification devices. Damage-tolerant structural elements such as pin-fused seismic systems that are capable of remaining elastic during strong ground motions, or other devices that can withstand cyclic inelastic deformation or limit permanent distortion, will lead to enhanced performance. Response modification devices such as seismic isolation, energy dissipation systems such as viscous dampers, or passive and active control systems can be used to enhance performance and limit damage. Figure 2 shows a photograph of the first lift of reinforcing for the shear wall core of the 30-storey, 350 Mission Building.

2.2. Design approach

Since the design approach takes exception to the building code, it is important to obtain concurrence that this approach is acceptable to the building official reviewing the project. Once that is established, the following procedure should be used:

Figure 2. Under construction (core only lateral system), 350 Mission Street, San Francisco, CA.

(1) Performance objectives – either minimum or enhanced.

(2) Seismicity – two levels of ground motions must be considered. The Service-Level Earthquake must consider a 2.5%-damped acceleration response spectrum having a 43-year return period and the MCE shaking with a 5%-damped acceleration response spectrum having a 2475-year return period.

(3) Conceptual design – structural systems and materials, as well as intended elements that will be subjected to inelastic or pseudo inelastic beforehand, must be selected.

(4) Design criteria – a design criterion that establishes the design and analysis approach with all performance objectives, system types, codes / references and materials must be developed and submitted for review by the building official and a third-party peer panel.

(5) Preliminary design – a dynamic structural analysis must be used to confirm that the design is capable of meeting the established performance objectives. To perform this design, the structure must be developed to a level of detail where stiffness, strength and mass are defined, as well as hysteretic properties of elements that must undergo inelastic straining due to strong ground motion. To the extent possible, the structure should be configured to include simple arrangements of structural elements with clearly defined load paths and regular structural systems. Large changes in building stiffness, building mass as well as repositioning of bracing elements over the height of the tower, column transfers and system eccentricities should be avoided. This will limit complexity and uncertainty in the final design.

(6) Service level evaluation – this consideration must be used to demonstrate that the structure is capable of withstanding a frequent seismic event with limited damage.

(7) MCE evaluation – a non-linear dynamic analysis must be used to demonstrate that the structure will not collapse during this level of ground shaking.

(8) Final design – since the final design described in building codes is based on the Design or DBE, 2/3 times the intensity of the MCE earthquake must be considered with all load combinations and strength/ response modification factors applied.

(9) Peer review – because the design process is non-prescriptive and takes exception to one or more requirements of the building code, an independent third-party peer review is required. The peer review panel typically consists of experienced engineers: an expert in seismology, an expert in the practice of engineering and an accomplished academic.

3. Considering the Chinese code for seismic design

3.1. Basic seismic design philosophy and parameters

The Seismic Fortification Category of a structure is a classification used in seismic design that is based on the impact of potential damage, occupant life safety, economic loss and the effect on society, buildings are categorised as Special Fortification (Category A), Important Fortification (Category B), Standard Fortification (Category C) or Appropriate Fortification (Category D) (NSPRC-08 & NSPRC-10). Residential buildings and typical office buildings are typically at least Category C, while high-rise office buildings having more than 8000 occupants normally or gross areas of 80,000 m² or more fall under Category B (NSPRC-08, 2008).

Table 1. Building damage and inter-storey drift ratio reference indexes for performance objective levels.

Performance objective	1 = Frequent ($T = 50$ years)	2 = Design Intensity ($T = 475$ years)	3 = Rare ($T = 1600$–2475 years)
Performance level 1	Good condition $\Delta << [\Delta ue]$	Good condition $\Delta < [\Delta ue]$	Basically intact $\Delta \geq [\Delta ue]$
performance level 2	Good condition $\Delta << [\Delta ue]$	Basically intact $\Delta \geq [\Delta ue]$	Slight damage $\Delta < 2[\Delta ue]$
Performance level 3	Good condition $\Delta < [\Delta ue]$	Slight damage $\Delta < 2[\Delta ue]$	Moderate damage $\Delta < 4[\Delta ue]$
Performance level 4	Good condition $\Delta < [\Delta ue]$	Slight to moderate damage $\Delta < 3[\Delta ue]$	Almost serious damage $\Delta < .9[\Delta up]$

Note: $[\Delta ue]$ is the elastic drift ratio limit per code requirements and $[\Delta up]$ is the plastic drift ratio limit per code requirements.

The seismic design philosophy of the Chinese codes intends building structures to not be damaged in the frequent earthquake, reparable after the moderate earthquake and to not collapse in the rare earthquake (NSPRC-10, 2010). The frequent earthquake is defined to have a 63% probability of exceedance in 50 years (50 year return), the moderate earthquake is defined to have a 10% probability of exceedance in 50 years (475 year return) and the rare earthquake is defined to have a 2–3% exceedance in 50 years (1600–2475 year return). The moderate earthquake is also referred to as the 'design earthquake' since the seismic design parameters for a region such as the Seismic Precautionary Intensity, and the Seismic Design Basic Acceleration are defined based on a 10% probability of exceedance in 50 years.

For ordinary code-prescriptive building structures, individual members of the structure are required to remain elastic under frequent earthquake load combinations, and the inter-storey drifts of the structure are recommended to remain below the serviceability drift limit in the frequent earthquake (NSPRC-10). Building performance is assumed to satisfy the code intent in moderate and rare earthquakes based on reserve capacity and ductility considered to be built into the prescriptive code requirements, but is not typically required to be explicitly checked.

3.2. Seismic resistance grade and member design criteria

The codes define maximum building height limits, and the Seismic Resistance Grades of members for Category C buildings (NSPRC-10, ISPRC-98, ISPRC-10 and NCSSR) based on Seismic Precautionary Intensity and structural system type. For Seismic Design Category A or Category B buildings, height limits and Seismic Resistance Grades corresponding to one seismic intensity level higher are mandated.

Seismic Resistance Grade of a lateral system component is used by the codes as a primary device to stipulate applicable design requirements. For example, vertical structural members are recommended to have different axial stress ratio limits under seismic load combinations or under gravity combinations, moment frame beams and columns are required to have different internal force amplification factors, and members are required to have different minimum longitudinal and transverse reinforcement ratios based on their respective Seismic Resistance Grades. Appropriate levels of over strength and resilience of members, and thus robustness of structures are thus effectively ensured in seismic design through the use of Seismic Resistance Grades.

3.3. PBD in China

The requirements for PBD are contained in section 3.10 of GB 50011-2010 (NSPRC-10) and section 3.11 of JGJ 3-2010 (ISPRC-98, 1998). Additional guidelines are provided in the Technical Notes (NCSSR, 2010) and in CECS 160-2004 (CECS, 2004).

PBD in China follows a three-level seismic hazard and performance-objective approach consistent with the seismic design philosophy Table 1 identifies, from the seismic code (NSPRC-10), four performance objective levels with associated damage conditions at the three seismic hazard levels. Table 2 provides definitions of the damage conditions categories used in Table 3.

In these checks, desired performance is associated with levels of inter-storey drift and assured by implication by not exceeding the stipulated limits. Unless special performance requirements are mandated or desired, building structures are typically designed and checked to satisfy Performance Level 4.

Table 2. Building damage condition description and probablility of continuous use.

Damage condition	Damage description	Probability of continuous use of structure	Drift reference value
Good condition (including 'perfect' cond.)	Structural members are in good condition, a few non-structural members slightly damaged; ornament elements are damaged at different levels	Can be continuously used without repair	$<[\Delta ue]$
Slight damage	A few of structural members have slight cracks, a few of non-structural members significantly damaged; ornament elements are damaged at different levels	Can be continuously used without repair or slight rehabilitation could be needed	1.5–$2 [\Delta ue]$
Moderate damage	Most of structural members have slight cracks, some have significant cracks, a few non-structural members are seriously damaged	Regular rehabilitation required after safety measures are undertaken. Can be used appropriately	3–$4 [\Delta ue]$
Serious damage	Most of structural members are seriously damaged or some of them collapse	Needs a great amount of rehabilitation to eliminate risk. Partial demolition required	$<.9 [\Delta up]$
Collapse	Most of structural members collapse	Demolition required!	$>[\Delta up]$

In addition to inter-storey drift checks, typically performed using non-linear analysis at the moderate and rare earthquake hazard levels, specific strength performance checks corresponding to the four performance objective levels are also stipulated in the code to be performed (ISPRC-10, 2010). For instance, a member may be required to remain 'elastic' in flexure at a particular seismic hazard level or to 'not yield' in flexure in the same event; the difference being that in the former case, typical design procedures are used with seismic forces from the designated event; with factors of safety to elastic limits as built-in by the design codes, whereas in the latter case, load factors and strength-reduction factors are generally taken to be 1.0 and the response considered to be at the limits of elasticity.

Typical strength performance requirements are defined in the high-rise building code (ISPRC-10), and are designated for Performance Levels A to D instead of corresponding Levels 1 to 4 as in the case of the seismic code (NSPRC-10). Table 3 provides an example of the enhanced strength performance requirements for a tall frame-core wall reinforced concrete structure.

4. Creative solutions to enhancing structural performance

4.1. Using infill frames to protect primary structure

The Gemdale Plaza project in Beijing consists of two high-rise reinforced concrete office towers, one 100 m tall and the other 150 m tall, and a retail podium along Chang-an Boulevard, Beijing's principal thoroughfare.

The design for the towers envisioned the use of sun screens for shading purposes on the south and west facades of the towers. The sun screen expression was inspired by traditional oriental screens. Initially, the structural lateral system of the towers was conceived of as a core-perimeter moment frame system, with columns spaced at 6 m at centre and girders at every floor. The sun screens, however, yielded an interesting possibility: to serve dual functions as the architectural aesthetic and shading devices as well as structure; thereby eliminating redundancy and adding rentable space to the plans. The concept of 'screen frames' was thus conceived. For the conceptual drawing of the screen frames see Figure 3. Figure 4 presents Tower A, the taller of the two towers.

Table 3. Strength requirements for example structure developed using JGJ 3-2010 Performance Level D.

	Core walls	Link beams	Frame columns	Frame beams
Elastic in frequent EQ	X	X	X	X
Shear no yielding in moderate EQ				
No yielding in moderate EQ			X	
Shear elastic in moderate EQ				
Elastic in moderate EQ				
Shear no yielding in rare EQ	X		X	
No buckling in rare EQ			X	

Figure 3. Conceptual drawing of the screen frames.

Figure 4. Screen and conventional frames, Gemdale Plaza.

The first challenge was to rationalise seemingly random screen members into a pattern that could serve as moment frames. This was achieved by creating a regular, multi-storey mega-frame of primary members that carry the gravity loads of the tower as well as provide lateral resistance, and then using secondary 'infill' members within the mega-grid to provide lateral stiffness only. This resulted in a screen frame system with mega columns at 9 m at centre and mega girders every three storeys. Floor framing was laid out so that gravity loads from the floor framing were directed only to the mega columns of the screen frames, allowing the frames to be pulled away from the edge of the floors and creating a vertical gap between themselves and the exterior glazing. The frames on the north and east facades, subject to less direct sunlight, had conventional moment frames with columns at 6 m at centre and girders at every floor. The infill screen frames were designed to become inelastic before the other lateral load-resisting elements, protecting the mega-columns and interconnecting beams allowing them to remain elastic. The shear wall core and screen frame are shown in Figure 5.

Once the screen frames were rationalised, the next step was to 'tune' their lateral stiffness to match those of the conventional frames on the opposite faces of each tower to minimise torsional irregularity. The lateral stiffness matching displacement curve is shown in Figure 6.

In order to keep gravity loads in the mega columns from straying into the secondary lateral stiffening infill members of the screen frames, delayed pour joints were introduced where the infill members connected to the mega-frame members. See Figure 7 for a diagram indicating delayed pour joints and Figure 8 showing the areas where rebar is placed without concrete at the joints.

This consideration significantly reduced shear force demands on the mega-frame girders reducing reinforcement quantities and also reduced forces in general in the infill members allowing them to be as slender as possible (Figure 9).

4.2. External bracing used to minimise environmental impacts

The 56-storey, 232 m tall Jinao Tower in Nanjing, China, is a next-generation tower which maximises performance, efficiency and occupant experience. Its faceted form is derived from the juxtaposition of an innovative double-skin façade and an external lateral braced steel frame that wraps the tower from crown to base and defines the dimensions and folds of the building envelope (Figure 10).

The structural system reduces the impact on the environment by minimising the need for raw materials and by forming the space between two exterior wall systems that are used to control the tower's internal temperature. Introducing the diagonal steel brace system on each side of the structure (outside of the tube-in-tube structure and between the double-skin facade) resulted in additional stiffening of the structure and a 40% design reduction in concrete and rebar in the concrete lateral load resisting system and a 20% design reduction in concrete and rebar for the overall building structure. The integrated double-skin façade provides solar shading and creates a climatic chamber of air, offering improved insulation in both the hot summer and cool winter months. Vented openings in the outer exterior wall allow wind pressure to extract built-up heat out of the cavity, lowering temperatures along the inner exterior wall (Figure 11). Together, the external steel bracing system, integrated into the high performance double-skin façade, provides an easily identifiable iconic tower with cultural references to 'lantern' forms.

The initial structural system for the Jinao Tower was conceived considering an interior tubular reinforced concrete shear wall core and an exterior tubular frame. The perimeter folded planes of the architectural expression provided an opportunity to incorporate a perimeter bracing system along the fold lines to achieve greater structural efficiency. The additional lateral stiffness provided by a perimeter bracing system allowed the central shear wall core to be progressively 'punched' into an interior tubular frame, thereby providing greater flexibility to meet programmatic needs as well as allowing it to be designed for strength and not stiffness which, given the buildings height, would have controlled material economy. Together, the punched core, perimeter frame and perimeter braces provided the stiffness necessary to satisfy core drift limits.

The slender elegance of the bracing exoskeleton, with single 500 mm diameter diagonal steel pipe braces along each facade was achieved by largely keeping gravity loads out of them and having them function primarily as lateral system stiffening components. Delayed construction techniques were employed to accomplish this objective. The steel pipe braces connect to the base building reinforced concrete structure at the four corner composite columns every 16 m in height (four storeys at office levels and five storeys at hotel levels) with cast steel pin assemblies.

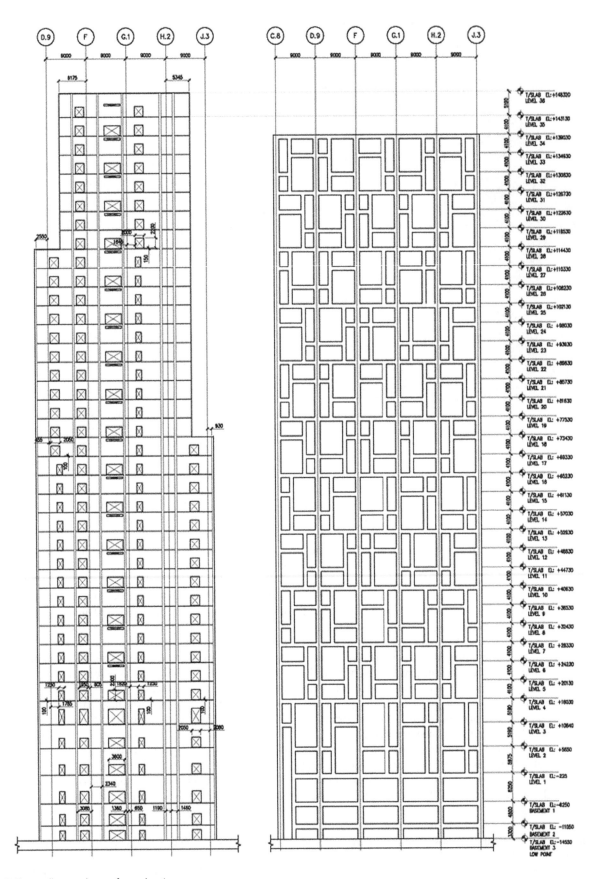

Figure 5. Shear wall core and screen frame elevations.

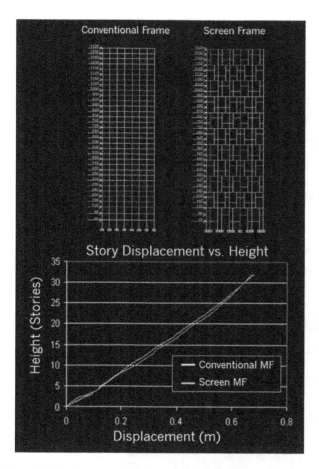

Figure 6. Relative displacement of frames due to rare seismic event.

4.3. Thin steel plates used for optimal performance

The 75-storey, 336.9 m tall Jinta Tower (Figure 12), has an elliptical plan footprint approximately 42 m by 81 m at the base which changes with height widening towards the middle and then tapering above to create an 'entasis' effect. The pleated exterior wall recalls the structure and lightness of Chinese paper arts and allows for maximum views for occupants. The premise of the original design was to use locally supplied materials, fabricated and constructed by local labour. The underlying idea was to construct this tower entirely of thin steel plate. The structure is the tallest steel-plated core system ever built.

The main lateral-force-resisting system for the tower comprises a perimeter ductile moment-resisting frame, and an interior steel plate shear wall (SPSW) core linked together with outrigger and belt trusses. The perimeter ductile moment-resisting frame consists of concrete filled tube (CFT) columns and structural steel wide flange beams. The interior shear wall core consists of CFT columns and structural steel wide flange beam ductile moment-resisting frames infilled with structural steel plates to create SPSWs. Four sets of outrigger trusses and belt trusses, located at mechanical floors have been used to link the core and perimeter frame and stiffen the structure in the transverse (narrow) direction (Figures 13 and 14).

As only eight perimeter columns are connected directly to the outrigger trusses, belt trusses are utilised to link the remaining columns to work with the outrigger trusses and increase their

effect. In the process of behaviour and material optimisation, steel plates were found to be necessary only over two-thirds of the building height in the core, transitioning to braces above and eventually to moment frames. The gravity system for the tower consists of conventional rolled structural steel wide-flange composite framing and composite metal deck slabs. The perimeter columns bend to follow the shape of the tower exterior.

The structural system was selected after considering various alternatives to efficiently deal with the structural load demands given the slenderness of its form (1:8 aspect ratio).

A major challenge for the designers was to reconcile the ductile tension field based load resisting action of SPSW's with the serviceability requirements of the Chinese codes that mandated that SPSWs not buckle under 50 year return lateral load events. Analysis showed that given the high gravity axial stresses induced in the steel plates because of the building's height, buckling was likely to occur even without lateral loads. While this would not materially impact the lateral capacity of the structure, it would be unacceptable considering serviceability.

Vertical stiffeners were added to the steel plates between the boundary elements to increase their buckling capacity and detailed in a manner to permit ductile load resistance through tensile field action in 475 and 2500 year return lateral load events. The detailing method involved introducing gaps between the ends of the vertical stiffeners and the horizontal boundary elements. These gaps reduced the ability of gravity loads to seek paths through the stiffeners and thus load the plates.

The size and spacing of the stiffeners, as well as the dimensions of the gaps at their ends were determined by non-linear pushover analysis trial and error methods: for each grouped condition of plate thickness, boundary element stiffness and gravity load, a lateral drift was introduced on the stiffened SPSW panel and the optimum stiffener spacing and gap dimension were thus determined. See Figure 15 for key aspects of the analysis model. The as-constructed plated shear wall system is shown in Figure 16. The concepts adopted were verified using scaled laboratory testing of joints and multiple bays and storeys of SPSW panels and non-linear dynamic analysis of the entire structure.

5. Robust structural design in the future

Robust structural design combines the incorporation of over-strength and resilience to assure reliable function and eventual deterioration of function when design loading condition ranges are exceeded. Evolution of structural design knowledge and computational techniques coupled with the pursuit of sustainability has led to the emphasis of resilience in robust design; systems that are not only ductile, but are capable of returning to their original un-damaged state after load conditions that exceeded the planned ranges are removed – with minimal, if any, need for replacing damaged members. Such systems include base isolation systems, pre-stressed self-centring brace systems, friction mechanism-based pin-fuse® systems and systems with replaceable dampers or buckling resistant braces that limit damage to selected easily replaced members. These systems at marginal initial cost premiums provide significantly improved performance and life cycle cost benefits and are expected to become the robust and sustainable structural systems of the future particularly in areas of high seismicity.

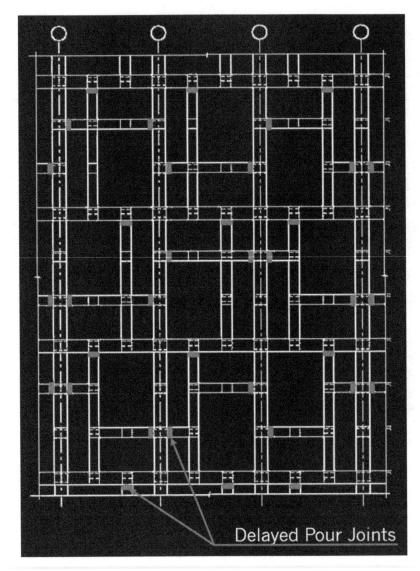

Delayed Pour Joints

Figure 7. Diagram indicating delayed pour joints.

The Pin-Fuse Frame™ (Figure 17) system comprises a pre-fabricated structural steel beam-column moment frame assembly combined with concentric braces in the bay. The braces are each divided into two parts connected across a gap that allows each part to slip towards or away from the other during a major seismic event. The Pin-Fuse Frame™ SMF Joint consists of side plates welded to both sides of the beam that are clamped to corresponding projecting plates welded to the column flanges using high-strength bolts in a circular pattern. Standard holes are provided in the beam side plates, and slotted holes in the corresponding column plates.

These beam-column joints at each end of the horizontal beam frame elements are designed to slip when subjected to a preset rotational load. Based on earlier work done by E. P. Popov, T. S. Yang and C. E., Grigorian, half hard cartridge brass shim plates (UNS 26000) with standard circular holes are used as friction shim material between the two clamped slip surfaces of every joint. A solid steel pin is provided at the centre of rotation of the joint to transmit the shear and axial forces from the beam to the column through the side plates.

Welded gusset plates with standard holes, and stitch plates with slotted holes are clamped together with interposing brass shims and pretensioned high-strength bolts to create the friction slip joints in the braces. In a major seismic event, the force in the brace exceeds a preset threshold, and the two connected parts slip relative to one another within the designed slots.

In a severe seismic event, the brace-slip joint fuses first, and as the drift levels continue to increase, the SMF joints start to slip; both dissipating energy in friction. In a very severe earthquake that causes drift levels to exceed the frictional slip limits of the slots (over 4%), the bolts in the SMF joints reach the end of their travel in the provided slots and bottom out. At drifts that exceed this condition, the forces in the frame members start to increase significantly and yielding starts to occur in the side plates of the SMF connections.

5.1. Pin-Fuse Frame™ design philosophy

A Pin-Fuse Frame™ bay, as shown in Figure 17, incorporates four Pin-Fuse Frame™ SMF Joints and a brace-slip joint. The beam, column and brace sections must be seismically compact

Figure 8. Tower A – under construction indicating Pour Joints where concrete is temporary not placed (rebar only).

and must be designed to be adequate for all load combinations specified by the applicable building code. The system for the purposes of selecting a response modification factor is considered equivalent to a special moment resisting frame (IBC, Response Modification Factor (R) = 8). In addition, the design objectives for Pin-Fuse Frame™ SMF Joints are summarised as follows:

(1) Plastic hinge formation is intended to occur primarily in the joint, which is located at a specific distance from the face of the column.

(2) Slippage is initiated when the moment demand of the beam at the joint reaches 80% of the flexural plastic moment capacity (M_p) of the beam, thus limiting the moment at the face of the column to 85–100% of M_p of the beam.

(3) The grade for the high-strength bolts and the number of bolts are determined such that the product of minimum bolt pre-tension and effective coefficient of friction (COF) will set the initiation of slipping in the joint. Also, the bolts should be capable of resisting the expected plastic moment capacity of the beams in shear and bearing.

(4) Slots in the side plates welded to the column flange allow for 4.5% rotation in either direction before locking.

(5) When the bolts reach the end of their travel in the slots, they lock and continued increase in moment demand will result in a condition similar to conventional beam-column joint.

(6) Shear is resisted by the centre 'pin' alone

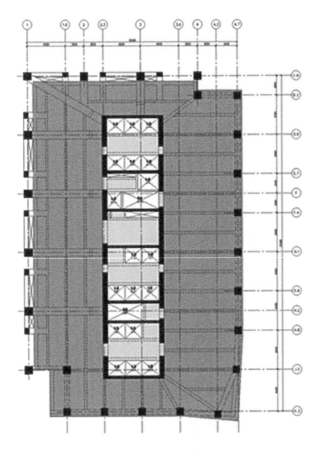

Figure 9. Typical floor framing plan for Gemdale Plaza.

Figure 11. Braces in double-wall cavity, Jinao Tower.

Figure 10 Jinao Tower, Nanjing.

Figure 12. Jinta Tower, Tianjin.

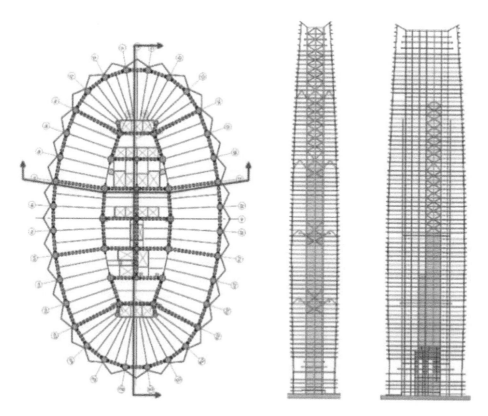

Figure 13. Jinta Tower structure: plan, sections.

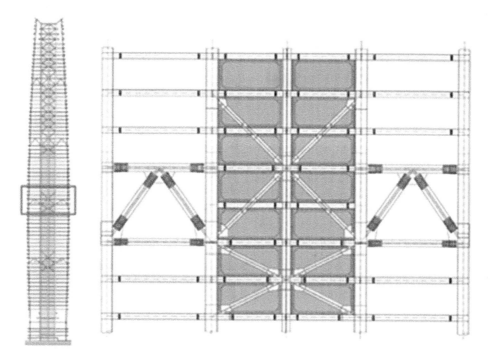

Figure 14. Enlarged view of outrigger system.

(7) The side plates to be designed for the probable maximum plastic moment capacity of the beam section at the face of the column.

(8) The strong-column weak-girder relationship shall be satisfied for the selected beam and column section combination.

In addition to the AISC 341-05 concentric brace design requirements, the design objectives for brace-slip joint are summarised as follows:

(1) The brace-slip joint is to be designed to slip at 85% of the design buckling capacity of the brace.

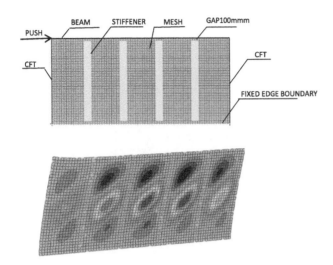

Figure 15. Pushover analysis of individual panel.

Figure 16. Web stiffeners with clear gap at bottom of each one.

(2) Slots in the brace-slip joint are to be designed to accommodate drifts of 5–7.5%.

(3) Bolts in the brace-slip joint are to be designed for maximum shear and bearing that is required to develop the expected tensile capacity of the brace section.

(4) The pins connecting the brace to the side plate (gusset plate) and the side plate are to be designed to develop the expected tensile capacity of the brace.

Combining the design philosophies of the two types of joints, the complete Pin-Fuse Frame™ is designed such that the following sequence of events occurs when it is subjected to a continuously increasing load:

(1) Brace-slip joint slips.

(2) Pin-Fuse Frame™ SMF Joints slip causing them to rotate.

(3) Bolts in the Pin-Fuse Frame™ SMF Joints reach the end of their slots and yielding initiates in the side plates at the beam-column joint.

(4) Bolts in the brace-slip joint reach the end of the slot and the buckling of the brace occurs. At this point, the force carrying capacity of the system considerably reduces. This condition occurs at drift levels exceeding 5–7.5%.

5.2. Details and performance of the pin-fuse frame

The system incorporates two different types of energy dissipating connections:

(1) The pin-fuse rotational joint (Figure 18) is an evolution of the pin-fuse joint which replaces the curved socket joint of the original design with a circular pattern of bolts to make the connection more economical.

(2) The brace-fuse joint (Figure 19) is used with an axial element and will slip under friction before allowing the brace to buckle. The connection consists of a symmetric arrangement of plates and friction surfaces located along the length of a diagonal brace.

Both joints of the pin-fuse frame are tuned to slip and protect the components they connect from damaging force levels. By preventing beams from yielding and braces from buckling, the system ensures the structural integrity of the entire structure and minimises costly post-earthquake repair.

The friction joints developed use arrays of pre-tensioned bolts that clamp two parts of a brace together and also clamp beam-column joints together with interposed friction shims until severe seismic activity causes the force in the joint to exceed a pre-set threshold. This threshold is the product of the COF and the normal clamping force at joint sliding surfaces. At this threshold, the two connected members are allowed to slip relative to one another within carefully designed slots. If the lateral load imposed on the structure continues to increase representing a severe seismic event, the bolts pass through the full length of the slots and re-engage with the connected members. At this point only do the elements continue to attract additional force and eventually create inelastic yielding behaviour in the frame members.

Testing of the pin-fuse frame was performed at the University of California, San Diego, once again using brass shims. See Figure 20 for an image of this test.

Figure 21 illustrate plots of the cyclic tests for the full frame. The Brace-Fuse Joint slipped first followed by the rotational slip of the rotational joints. No significant yielding or deformation was observed until 5% drift when the bolts bottomed out in the Brace-Fuse Joint and the brace eventually experienced in-plane buckling at 6% drift in compression. Buckling occurred at the long side of the brace just beyond the plate connections. A significant strength increase at 6% drift was observed due to the bearing action of the bolts in the brace. Note that slots in the brace were designed to accommodate drift up to 5% and the slots within the Pin-Fuse Frame™ rotational Joints were designed for 4.5% drift (with an additional 1/16th inch for tolerance). To observe the post-buckling behaviour, a decision was made to continue to load the system monotonically with the brace in compression. At drifts of 6.9, 8.3 and 9.7%, testing was paused for observations. At 6.9% drift, some yielding of the beam compression flanges just beyond the side plates was observed as evidenced by flaking of the whitewash.

At 9.7% drift, the frame was still capable of withstanding additional loads; but the test was stopped to ensure safety. It is important to note that conventional moment frames start losing

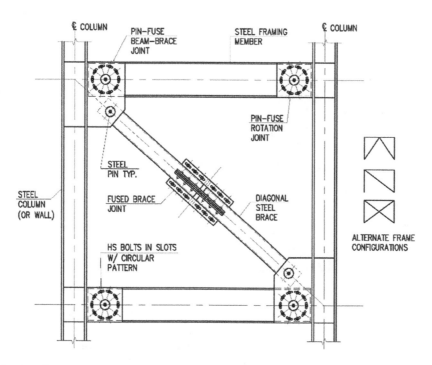

Figure 17. Elevation of Pin-Fuse Frame™.

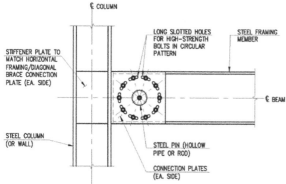

Figure 18. Pin-fuse rotational joint.

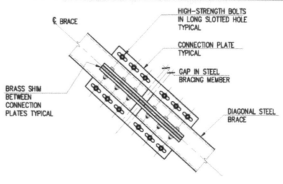

Figure 19. Brace-fuse joint.

4%, bolts are still slipping in slots under friction with no material yielding.

5.3. The Link-Fuse Joint™

The link elements that connect relatively stiff earthquake load-resisting components such as shear walls or steel braced frames usually experience significant ductility demand in severe

strength at drifts exceeding 4% accompanied by significant material yielding. But, the Pin-Fuse Frame™ proved that it is capable of resisting very high drifts without strength loss, thus providing additional ductility to the lateral load-resisting system. It is additionally pointed out that until drifts exceeding

earthquakes and undergo damage while dissipating energy. Shear is typically the critical force component in link elements. The Link-Fuse Joint™ was developed to withstand severe seismic

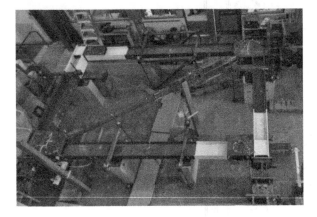

Figure 20. Pin-fuse frame test assembly.

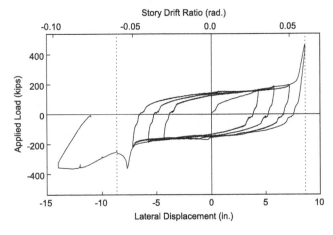

Figure 21. Pin-fuse frame: applied load vs. lateral displacement.

events without experiencing significant beam or joint yielding damage by dissipating energy through friction slip at a pre-set shear force level. The details of a typical Link-Fuse Joint™ are provided in Figures 22 and 23.

By incorporating a joint that 'slips' in friction at force levels just shy of potentially damaging shear force levels, the Link-Fuse Joint™ protects the integrity of the link beam components it connects, preventing yielding and damage in the components and, ultimately, costly post-earthquake replacement of the damaged elements.

The Link-Fuse Joint™ involves a pre-tensioned 'pin' that clamps two halves of a link (coupling) beam together until the shear force in the beam exceeds a certain threshold. At this point, the two beam halves slip relative to one another within engineered slots with minimal increase in shear force. As the lateral load continues to increase, the pin traverses the full length of the slots and eventually in a severe earthquake re-engages the two halves of the beams. Only then do the beam halves beyond the fuse attract additional force and eventually yield in flexure. The goal of the Link-Fuse Joint™ is to postpone inception of this flexural yielding by introducing mechanical energy-dissipating friction 'slip' prior to plastic hinging of the beam at its ends. Yielding and damage thus occur at drift levels significantly greater than in structures with traditional link elements.

The Link-Fuse beam can be a valuable tool for performance-based design of coupled shear-wall systems of tall buildings. If used properly, it allows for the proportioning and tuning of the lateral system to behave elastically (no slip) in a frequent seismic event (e.g. 43-year event). In a severe seismic event (e.g. 2475-year event), the Link-Fuse Joint™ can slip to accommodate larger displacements, with significant energy dissipation due to friction. This provides a high performance level with little or no structural damage and with minimal requirement for post-earthquake repairs.

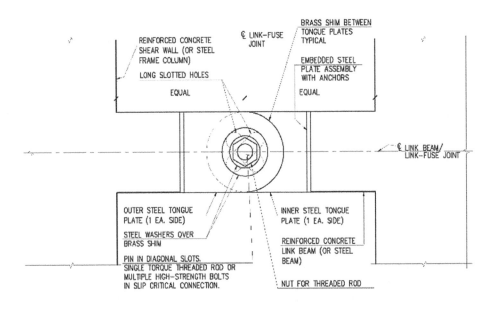

Figure 22. Link fuse assembled joint.

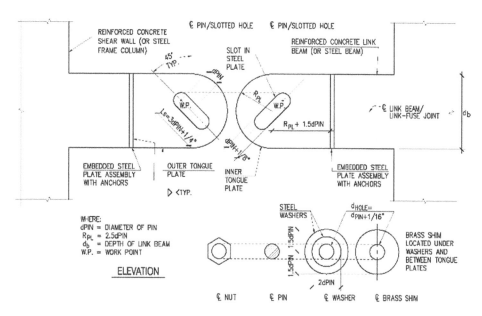

Figure 23. link fuse part piece elevation.

Figure 24. Dynamic friction test set-up at Ecole Polytechnique.

5.4. Friction tests

Prototype testing at Stanford University and the University of California, San Diego, as indicated, utilised brass as the friction shim material. Concerns about electrolytic corrosion and cold welding impacting friction coefficients have led to the search for non-metallic composite friction materials. After an exhaustive search for appropriate materials, a promising material used in vehicular brake linings has been identified as a suitable candidate material. Testing will be carried out to determine the COF of identified materials, their creep behaviour as well as the impact of fireproofing installation on their properties including coefficients of friction. Currently, the tests of candidate friction materials are being done at Ecole Polytechnique in Montreal, Quebec, Canada (Figure 24).

5.4.1. Test objectives

The main goal of the current series of tests is to find a friction material that meets all code requirements and is financially viable. The friction material is required to exhibit stable short-term and long-term friction coefficient, be corrosion-resistant and possess minimum cold welding/adhesion property, and exhibit minimal stick-slip behaviour. In addition, the friction material is required to be simple to apply to the steel plates, be capable of multiple re-use and demonstrate stable property even at higher temperature. In order to meet this goal, a specific test protocol has been developed as described below:

Static COF tests:

(1) Determine reliable static COF values
(2) Determine the factor to account for the secondary friction resulting from sliding of washer plates at the long-slotted holes

Cyclic COF tests:

(1) Determine reliable dynamic COF values at three different velocities
(2) Determine impact of loading frequency on COF values

The test velocities selected for the tests bound the rates of dynamic movement that the joint will experience typically when a structure (with fundamental period from .67 to 10 s) is subjected to seismic movement.

Fire-proofing impact tests:

(1) Determine impact of fire-proofing installation on friction material properties

Tension creep tests:

(1) Determine creep behaviour of the friction material and impact on COF values

5.5. Test results

Figure 25 shows a typical hysteretic plot of the ongoing friction tests. As seen in the figure, the composite material has shown robust hysteretic behaviour. After initial load peaks due to static break-away of the multiple friction surfaces, a constant coefficient

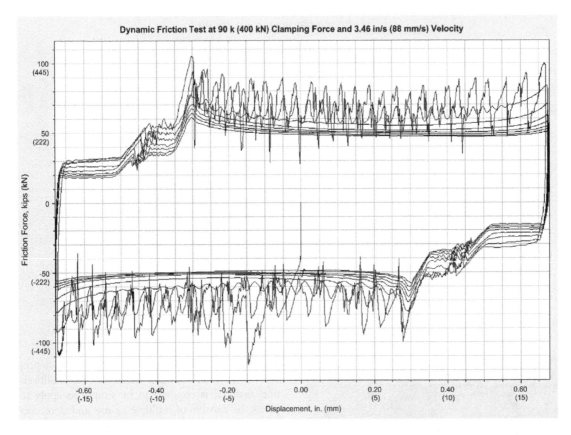

Figure 25. Dynamic friction test at Ecole Polytechnique: friction force vs. displacement.

is observed and the sliding force converges after multiple cycles. A stable COF under cyclic loading is essential to demonstrate the reliability of the material during a seismic event.

The results of these ongoing tests suggest the promise of the selected friction material for providing reliable friction performance. It is expected that the composite brake material will be able to be implemented into all three of the pin-fuse system.

6. Conclusions

Major code efforts have been developed to address performance-based engineering. When designing for enhanced life cycle performance, optimal designs go far beyond just modelling behaviour. The structural systems need to be conceived to minimise material use and provide natural behaviour in extreme loading events such as seismicity. Ideas like the pin-fuse seismic systems require further development. These systems will lead to best performance and increased life of buildings.

Disclosure statement

No potential conflict of interest was reported by the author.

References

CECS. (2004). *CECS 160:2004 (draft) general rule for performance-based seismic design of buildings*. Chinese Engineering Construction Standard, China Planning Press.

ISPRC-98. (1998). *JGJ 99-98 technical specification for steel structures of tall building*. Industry Standard of the People's Republic of China, China Architecture & Building Press, Sections 5.5.1, 4.2.4, 5.5.1.2, 5.5.1.3, 1.0.2, 3.11.

ISPRC-10. (2010). *JGJ 3-2010 technical specification for concrete structures of tall building*. Industry Standard of the People's Republic of China, China Architecture & Building Press, Sections 3.1.3., 3.12, 4.2.7, 3.7.3, 3.7.6, 3.3.1, 3.9.3, 11.1.2, 11.1.4, 5.1.13.

NCSSR. (2010). *Technical notes on special seismic review of code-exceeding high-rise building structures, National Committee for Special Seismic Review of code-exceeding high-rise building structures ministry of housing and urban-rural development of the People's Republic of China, Building Quality* (Document #164).

NSPRC-08. (2008). *GB50223-2008 Standard for classification of seismic protection of building construction*. China Architecture & Building Press, Sections 3.0.2, 6.0.11, 6.0.12, National Standard of People's Republic of China.

NSPRC-10. (2010). *GB50011-2010 code of seismic design of buildings*. China Architecture & Building Press. Sections 3.1.1, 5.1.6, 5.5.1, 6.1.1, 6.1.2, 8.1.1, 8.1.3, 5.1.1, 5.1.2, 5.5.2, M.1.1-2, 3.1.3.2C, National Standard of People's Republic of China.

A life-cycle framework for integrating green building and hazard-resistant design: examining the seismic impacts of buildings with green roofs

Sarah J. Welsh-Huggins and Abbie B. Liel

ABSTRACT

Building design and performance are increasingly being scrutinised from perspectives of both sustainability and hazard resistance. However, the approaches taken to consider these perspectives are disconnected; green building rating systems do not consider hazard resistance in their assessments, while performance-based engineering methods have tended to neglect consideration of environmental impacts. This study presents a framework to assess a building's life-cycle performance in terms of social, environmental and economic impacts using probabilistic approaches, considering the possible occurrence of an earthquake or other extreme event. The framework is illustrated through a case study of an office building in Los Angeles, designed with and without different types of vegetated (green) roofs, and at risk from varying earthquake hazard scenarios. The case study results demonstrate trade-offs between upfront building costs, material choices, hazard resistance and environmental impact.

1. Introduction

This paper conducts a life-cycle assessment evaluating impacts of green and hazard-resistant design decisions for buildings within an integrated green-resilience framework. Investigation of the synergies between green building design and hazard-resistant building design is a developing field of research. The building sector accounts for over 40% of primary energy consumption in the United States (DOE, 2012a). Activities in each building life-cycle stage, i.e. material extraction, construction, operation, etc., release CO_2, CO, NO_x, SO_2 emissions, volatile organic compounds and particulates, which can harm human health and the natural environment. Green building rating systems such as LEED (United States Green Building Council [USGBC], 2015), Green Globes (Green Building Initiative, 2014, http://www.thegbi.org/) and BREEAM (BRE Global, 2015) have become popular tools to credit buildings where owners and operators have taken steps towards sustainable design and operations. At this time, programmes such as the *Resilience-Based Earthquake Design Initiative* and the U.S. Resiliency Council (USRC) are still-developing systems to rank building resilience and hazard performance (Almufti & Willford, 2013; USRC, 2015, http://www.usrc.org/)

However, the development of resilience rating systems programmes such as REDi or the USRC demonstrates the increasing interest from both policy-makers and communities to design buildings for beyond code minimum requirements, in order to achieve targeted performance goals for resilient post-hazard functionality. In addition, several pilot credits to incorporate resilience into LEED certification have recently been introduced, to encourage developers to consider disaster preparedness and resilience in community development and structural design, including the suggested incorporation of the REDi credits into their planning initiatives (USGBC, 2015).

Goals for green, sustainable design and hazard-resilient buildings cannot be seen as separate or competing objectives. Modern structural engineers must balance structural integrity and resilience with sustainability (Feese, Li, & Bulleit, 2014). Academic research is needed to understand how a greener 21st century building stock will perform in the context of being placed near or alongside hazards such as floodplains and fault lines (Menna, Asprone, Jalayer, Protoa, & Manfredi, 2012; Comber, Poland, & Sinclair 2013). Probabilistic life-cycle performance assessment methodologies can provide a means to account for potential occurrences of hazard events in the context of green/sustainable design and vice versa.

This article opens with the motivation for the proposed green-resilience framework. We illustrate the framework with a case study of a Los Angeles office building, designed with and without different vegetated or green roof systems, and at risk from varying seismic ground shaking hazard scenarios. The objective of this case study is to explore how inclusion of green roofs in structural design affects both building environmental impact and hazard losses. We close with a discussion of case study findings and future applications for the proposed holistic life-cycle framework for building design and assessment.

For this purpose, this paper expands on initial work by the authors, which presented our ideas for a holistic framework that combines green building design and hazard performance (Welsh-Huggins & Liel, 2014a, 2014b). In this study, we significantly improve the method employed for computation of environmental impact and add an assessment of the uncertainty and variability of the results. We also include a more comprehensive qualitative and quantitative discussion of the trade-offs between building design alternatives.

2. Points of departure

Green building design strives to minimise resource use in building construction and operations and to provide healthier and more environmentally friendly living and working spaces, offering benefits such as reduced operating energy, lower water consumption, improved indoor environmental quality and enhanced occupant productivity (USGBC, 2009). Rating systems such as LEED are intended to evaluate the 'greenness' of new and existing buildings (USGBC, 2015). The green building initiative grew alongside ideas about sustainable development in the urban planning community. The concept of sustainable development stemmed from the broader sustainability movement, which recognises the need to reduce present-day negative environmental impacts to sustain resources for future generations (Berke, 1995; Bokalders & Block, 2010; Schwab & Brower, 1999).

Life-cycle analysis (LCA) can be used to systematically quantify economic and environmental impacts of a product or system over its lifespan (Hendrickson, Horvath, Joshi, & Lave, 1998; Lawson, Bersani, Fahim-nader, & Guo, 2002; Sharrard, Matthews, & Ries, 2008). LCA can be used to demonstrate the benefits of different green building design features. Process-based environmental LCA describes the inflow of material processes and outflow of environmental impacts during a product's lifespan. By comparison, input–output LCA defines life-cycle costing and environmental impacts based on U.S. Department of Commerce data for goods and services production and related energy and material consumption. For both approaches, a typical system boundary encompasses impacts associated with extraction of raw materials, product manufacturing, construction, operations/maintenance, repairs and demolition/end-of-life.

A seminal study by Junnila and Horvath (2003) quantifies life-cycle impacts of reinforced concrete office buildings. The study found that steel production dominates manufacturing impacts, given the high volume required for the structural reinforcement, external envelope, HVAC and water piping systems and other internal non-structural components. Compared to the manufacturing phase, the impact of construction activities was found to be relatively low. In the operations and use stage, lighting, heating and cooling have the largest impact on electricity consumption. During maintenance, replacement of steel components and re-painting of building interiors contributed greatly to life-cycle heavy metal and summer smog impacts. Impact from the end-of-life and decommissioning of buildings depends largely on whether building components are recycled, reused or disposed of in a landfill (Carbon Leadership Forum, 2012; Thormark, 2007).

In parallel to environmental impacts studies, engineers have developed the concept of performance-based engineering (PBE) to formalise hazard-resistant design practices that aim to satisfy societal and owner's performance objectives (Moehle & Deierlein, 2004). PBE was originally envisioned as a method to quantify building seismic performance in terms of death, dollars and downtime. Recently, engineering philosophy has moved towards a more holistic thinking, encompassing societal, structural and economic concerns through resilient design, which seeks robust, redundant structures that incur low levels of hazard-induced damage and limited loss of post-disaster functionality (Bruneau et al., 2003).

The interplay between green design and hazard resilience has the potential to change life-cycle environmental and post-disaster impacts and to change how we assess these impacts. From one perspective, green building practices can impact how a building resists hazard events, a potentially negative influence (Gromala et al., 2010). At the same time, poor hazard performance of a building may contribute to greater environmental impacts and economic costs over the building lifespan. Yet, current established green building rating systems have not yet adopted considerations of building hazard performance (BREEAM, 2015; USGBC, 2009) into their rankings, and PBE methodologies have not been developed to fully account for environmental impacts.

This paper joins a small, but growing body of the literature aiming to examine both building environmental impact and hazard performance through the addition of a 'hazard event' stage in building LCA. For example, researchers have proposed framework improvements to LCA that can account for life-cycle stages associated with hazard events (Hossain & Gencturk, 2014; Padgett & Tapia, 2013) and structural deterioration. Each of these frameworks produces a life-cycle impact score as final output, where the results from each life-cycle stage are normalised and summed either in terms of dollars or environmental impact indicators. In this vein, the ATC-86 initiative (Court, Simonen, Webster, Trusty, & Morris, 2012) proposed the inclusion of environmental impacts from seismic hazard damage within the PBE paradigm.

Nevertheless, a holistic framework that assesses the building life-cycle in terms of social, environmental and economic impacts without default to a common, dollar-based, denominator and using probabilistic approaches is lacking. Thus, this study expands assessment of 'green' buildings to include consideration of the environmental consequences of hazard resilience. The proposed framework combines environmental impact assessment for green building and hazard performance design objectives, which have, up to this point, largely been quantified separately.

3. Proposed framework for green-resilient building design and assessment

Figure 1 presents a graphical representation of the so-called 'green-resilience' framework developed and applied in this study. The *conceptual design* stage involves development of an initial building, designed to meet building codes, hazard performance objectives and green building rating system goals selected by the owner and/or building professional as appropriate for the function and location.

The second stage involves three consecutive, interrelated analyses of the initial design: (1) *structural analysis* to predict building response to hazard events, (2) *loss analysis* to quantify

Figure 1. Framework for 'green-resilient' building design and assessment.

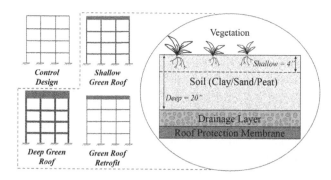

Figure 2. 2D elevation representation of the four study buildings, with diagram of the different green roof systems (based on Sailor, 2008).

potential economic losses and material quantities associated with hazard-related repair needs and (3) *environmental analysis* to evaluate the life-cycle impact of the design decisions, including environmental impacts from potential hazard events. The third stage of the framework, *integration of analysis results*, combines the results from the analyses in terms of the multi-objective social, economic and environmental losses and benefits to consider potential outcomes probabilistically. In the final stage, *implementation of green-resilient building design*, decision-makers evaluate the results to develop more holistic building designs. The case study presented here illustrates the framework and demonstrates how it can improve understanding of potential trade-offs between designing for green building objectives, such as a green roof, and designing for hazard resistance goals.

4. Life-cycle assessment of buildings with green roofs in high seismic areas

4.1. Green roofs

Engineering for green buildings encompasses a wide range of design choices, from decisions about structural features, to energy-saving equipment, to building orientation. Some of these design options may also impact building performance during hazard events. The construction of vegetated (green) roofs in high seismic areas offers both potential benefits and trade-offs for building life-cycle impact. In dense urban environments, plants on these roofs can lower ambient building air temperatures, reducing heat island effect (Saiz, Kennedy, Bass, & Pressnail, 2006). Green roofs also can help to manage storm water, reducing run-off and slowing release time of precipitation. Other benefits include improved air quality, external noise control and provision of wildlife habitat; green roofs may also improve psychological

health for building occupants (General Services Administration [GSA], 2011). Although U.S. adoption of green roofs lags behind Europe, the construction of buildings with green roofs is on the rise: Chicago leads the U.S. in terms of square footage of green roof coverage (DOE, 2012b, http://www.energycodes.gov), while Seattle's 2009 'Green Factor' legislation requires 30% vegetated roof cover for all new development in residential and commercial districts (Weiler & Scholz-Barth, 2009).

The basic structure of all green roofs is a plant or grass layer, supported by an engineered soil mixture. Protective, insulating and structural materials placed beneath the soil include filter fabrics, foam insulation and root barriers, waterproofing membrane and structural decking (National Roofing Contractors Association [NRCA], 2007). In this study, green roof types are distinguished by their soil depth following classifications from GSA (2011) and other green roof professionals. 'Shallow' green roofs, which are used predominately for storm water management and to mitigate heat island effect, are covered by sedum or other succulent grasses and defined as those with a soil depth of less than six inches. 'Deep' green roofs have 8–24 in of soil and provide the same benefits as shallow roofs, along with wildlife habitat and occupant access to green space (GSA, 2011). For both types, granite pavers or light aggregate acts as drainage perimeters around green roofs to store excess storm water before its release into the ground and provide roof access (NRCA, 2007). All green roofs require supplemental irrigation for the first three years after construction, but roofs with deeper growing media and larger plants may need permanent irrigation systems (Weiler & Scholz-Barth, 2009), especially in arid places.

4.2. Building designs

The proposed framework is illustrated through a case study of four commercial buildings in Los Angeles, as shown in Figure 2. The basic design of this modern code-designed four-storey office building is adopted from Haselton and Deierlein (2007) and Haselton, Liel, Deierlein, Dean, and Chou (2011). The special reinforced concrete space frame has a floor area of 120 ft by 180 ft with six frame lines resisting lateral loads in each direction. The storey height at the first storey is 15 ft, all others are 13 ft, and column spacing is 30 ft. The case study site in Los Angeles places the building in seismic design category D (American Society of Civil Engineers [ASCE], 2010). This site has a design spectral acceleration for short periods (S_{DS}) of 1.0 g and a design spectral acceleration at 1s (S_{D1}) of .6 g.

4.2.1. Green roof design

The control building does not have a green roof. The other three buildings consider three different green roof cases. The roof system designs resulted from correspondence with practicing engineers and a literature review of common material, structural and environmental requirements for green roofs. The shallow green roof building has a green roof with 4 in of soil. The deep green roof building uses 20 in of soil, with around 1500 ft² of paved area for occupant roof access. The shallow and deep green roof buildings have larger member sizes and greater amounts of reinforcing steel (larger reinforcement ratios) than the control building to account for the added weight and seismic mass. The fourth building is a 'green retrofit' case, in which a shallow green roof is placed on the building without modification, using the same structural system as the control building.

4.2.2. Structural design

The soil, with a saturated density of 120 pcf, is the heaviest component of green roof systems. The weight of small plants is less than 10 psf, providing only a small contribution to the total superimposed dead load (Shook, 2013). In the shallow green roof building, the vegetated green roof system covers a roof area of 21,000 ft². The drainage pavers add 1.3 psf along the roof perimeter and are supported by one inch of soil. Therefore, considering the entire roof area, the average weight of the shallow green roof system is 47.9 psf. The average roof dead load for the deep green roof building is 200 psf, due to the deeper soil. In addition to the calculated dead loads for each building, the International Building Code designates that 'where roofs are to be landscaped, the uniform design live load in the landscaped area shall be 20 psf' (ICC, 2009). The shallow green roof building is therefore designed for this recommended 20 psf of live load, because occupant access is expected to be sparse on these systems. By comparison, the deep green roof live load increases to 100 psf across the entire roof area, because it is designed to support more frequent roof access by occupants.

Seismic design utilises the equivalent lateral force method (ASCE, 2010). The special moment-resisting space frame system of these building has a response modification coefficient (R) of 8, indicating that stringent seismic capacity design and detailing requirements were met in accordance with ASCE 7-10. The control building has a design base shear of 193 kips resisted by each frame. The heavier load for the green roofs increases the lateral load, producing a design base shear of 212 and 258 kips for the shallow and deep green roof buildings, respectively. The retrofit green roof building is modelled with the same member sizes and ratios of reinforcement steel as the control building, because we assume that the additional roof mass is not accounted for in the original design. For each building, the size of columns

and beams was assumed to be the same at each storey, although reinforcement ratios varied (smaller areas of steel were placed in the exterior column lines for the control and retrofit buildings, and the shallow and deep green roof buildings have larger areas of reinforcement at upper stories to support the heavier roof masses). Key design variables are summarised in Figure 3. The member sizes and reinforcement ratios were designed to ensure that overstrength and ductility were as similar as possible between the different buildings.

4.3. Non-linear modelling and analysis

4.3.1. Ground motion selection and scaling

The seismic response of each of the buildings is simulated using a multi-stripe dynamic analysis (MSA) procedure. In MSA, a structure is subjected to a set of ground motion records at each of several hazard levels, creating multiple observations of structural response at each hazard level (Jalayer & Cornell, 2009). For the purposes of this study, the MSA approach is considered superior to incremental dynamic analysis because MSA permits the selection of ground motions that more appropriately represent the expected spectral shape at each hazard level (Baker, 2015).

This study employs the conditional mean spectrum (CMS) method for selecting ground motion records. A CMS curve represents the expected response spectra shape, conditioned on occurrence of a target spectral acceleration value at a period of interest for a particular site, corresponding to the hazard level of interest (Baker, 2011). Ground motion sets are selected to match (in terms of mean and standard deviation) the expected shape and amplitude of the CMS following a procedure defined by Lin, Harmsen, Baker, and Luco (2013). For this study, nine levels of seismic hazard, ranging from 50% in 50 years to 1% in 50 years, were chosen (see Table 1). Since the CMS changes shape at each ground motion intensity level, we selected a suite of twenty ground motions at each hazard level of analysis. The ground motion sets differ between buildings due to the different fundamental periods of each building model. Relatively small-scale factors (between .4 and 1.9) were applied to ground motion records in this process.

4.3.2. Non-linear building models

The OpenSEES seismic analysis programme (PEER, 2014) was used to conduct a non-linear analysis of two-dimensional, three-bay models of each of the four buildings. Beams and columns are modelled with elastic elements and concentrated hinge springs, i.e. a lumped plasticity approach. These hinges were assigned a material model developed by Ibarra, Medina, and Krawinkler

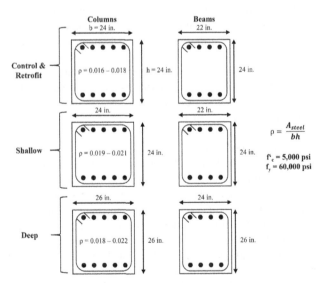

Figure 3. Column and beam designs for all four buildings (same size member dimensions at each floor for each building, but reinforcement ratios vary depending on position in the building). Concrete and steel material strengths are the same for all buildings.

Table 1. Hazard levels considered for Los Angeles site.

Hazard level	Sa(T_1 = 1.00 s) (g)	Return period (years)
1	.22	72 [50% in 50 years]
2	.40	224
3	.48	336
4	.55	475
5	.65	712
6	.74	975
7	.86	1462
8	1.05	2475
9	1.32	4975 [1% in 50 years]

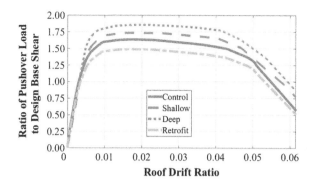

Figure 4. Results of static pushover analysis for all buildings.

Table 2. Pushover results for case study buildings.

Buildings	Overstrength ratio	Fundamental period (s)
Control	1.64	$T_1 = 1.24$
Shallow green roof	1.74	$T_1 = 1.31$
Deep green roof	1.85	$T_1 = 1.35$
Retrofit green roof	1.49	$T_1 = 1.33$

(2005), which is capable of capturing the effect of strain softening at large deformations associated with concrete spalling and rebar buckling. The hinge model can also capture cyclic deterioration. The properties of the hinge are calibrated to experimental results of more than 250 concrete columns, such that modelling of different components represents differences in design and detailing. For dynamic analysis, the buildings were assumed to have 5% damping, using Rayleigh damping in the first and third modes and assigned only to the models' elastic elements. More details about the structural modelling approach are available in Haselton et al. (2008, 2011).

In the OpenSEES models, the green roof is considered only in terms of the added mass and its column, beam and reinforcement sizes. Recent experimental studies of green roof seismic performance suggest that water content in the soil layer may provide greater damping during seismic excitation (Carmody, Jasarevic, Omenzetter, Clifton, & Fassman, 2009), but, to date, this effect is poorly quantified and we did not consider it here.

Eigenvalue analysis was used to calculate the first-mode period of the building models, which varies between 1.24 and 1.35 s. These values account for cracked section properties. Static pushover analysis results illustrate the as-modelled properties of the buildings in Figure 4, including ductility capacity and overstrength (defined as the ratio of peak strength from pushover to design base shear). Table 2 presents the overstrength ratio and fundamental period of each building.

The retrofit green roof building was not intended to carry the heavier load of the shallow green roof, which gives this building the lowest maximum base shear and overstrength values. The building follows a similar ductility trend to the control building, however, and experiences a loss of strength at a roof drift ratio of about 5%. The designs for the shallow and deep green roof buildings aimed to achieve the same ductility and overstrength of the control building as closely as possible. The control model and the deep and shallow green roof models show similar trends in ductility, although the deep and shallow green roof models experience a slightly earlier loss of lateral load capacity at a roof drift

ratio of around 4% due to the more significant p-delta effects. Despite this, the shallow and deep green roof buildings exhibit greater overstrength due to the larger gravity loads dominating the design.

4.3.3. Dynamic analysis results

Following the design and preliminary static analysis of the buildings, the framework moves into the *analysis* stage, beginning with a dynamic structural analysis. The four non-linear building models were analysed under 20 different ground motion records at each of the nine levels of seismic intensity. The results of the analysis are compared at each intensity level in terms of interstorey drift ratios (IDR), maximum floor accelerations and collapse fragilities.

As shown in Figure 5, all four buildings experience higher IDR and floor acceleration values at increasing levels of shaking intensity, with similar results (across all four buildings) at the lower levels of shaking. At the two highest levels of ground shaking, however, the IDR results were greatest for the shallow and retrofit green roof buildings, due to the extra mass at the top of the building, which impacted the response in the non-linear range. Similar trends were observed for the maximum floor accelerations: at all but the highest intensity levels, the three green roof buildings have maximum floor accelerations similar to those of the control building. Although the shallow green roof building has a stronger gravity and lateral system compared to the retrofit building, at a shaking intensity of Sa($T_1 = 1.00$ s) of 1.05 and 1.32 g, the retrofit building actually showed slightly lower interstorey drifts and floor accelerations than the shallow green roof, likely due to relative differences in mass and stiffness and the related non-linear effects.

Collapse fragility curves, which show collapse probability as a function of ground motion intensity, are provided in Figure 6. The median collapse capacity of all four buildings is similar, ranging from 1.1 to 1.25 g. However, the retrofit model experienced the greatest instances of collapse, while the deep roof building had the fewest instances of collapse. The difference in collapse risk between all four buildings is greatest at higher levels of excitation. At low levels of excitation, the shallow green roof building has a lower collapse probability than the retrofit green roof building, but this performance is reversed at higher levels of shaking, where the shallow green roof building demonstrates a slightly higher probability of collapse.

Overall, the MSA results demonstrate the complexities of adding green roofs to buildings. The 'up-sizing' of structural members to support heavier green roof mass appears to compensate for the larger mass of the green roof buildings. These buildings could have been designed without larger member sizes, but this would not have satisfied building code requirements for lateral force resistance. Conversely, the retrofit green roof building was not designed for the green roof. The performance of this building demonstrates a potential trade-off in decreased collapse capacity if additional roof support is not provided for a green roof system placed after building service life has begun.

4.4. Loss analysis

The next step in the analysis stage of the framework is loss analysis. Loss analysis refers to the probabilistic analysis of building

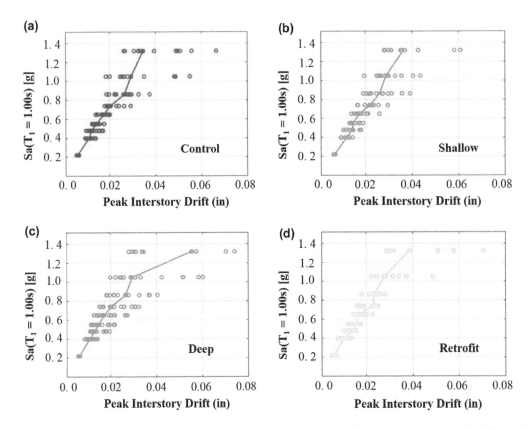

Figure 5. Peak IDRs as a function of ground motion intensity, for (a) control building, (b) shallow green roof building, (c) deep green roof building and (d) retrofit green roof building (showing only non-collapse results in all cases).

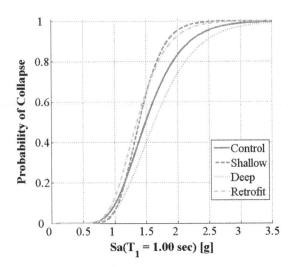

Figure 6. Collapse fragility curves for the four buildings.

response quantity a component is most sensitive (Mitrani-Reiser et al., 2006). Each damage state for a specific type of component is associated with a specific repair action. The consequences of incurring this state are quantified in terms of repair costs and repair time (Beck et al., 2002).

The basis for our loss analysis approach comes from advancements in seismic performance and probabilistic loss estimation developed by the FEMA P-58 project (Applied Technology Council [ATC], 2012a). The FEMA P-58 software, *Performance Assessment Calculation Tool* (PACT), provides a tool for incorporating building occupancy schedules, inventories of structural and non-structural components and results of structural dynamic analysis into predictions of seismic repairs and associated costs (ATC, 2012b). PACT takes an assembly-based approach, wherein total building repair costs and repair times are sums of losses associated with different building components (Porter, Kiremidjian, & LeGrue, 2001). Given our interest in forecasting building performance for a specific site and at specific hazard levels, we conducted an intensity-based assessment.

4.4.1. Inputs to loss analysis

Loss estimation requires data for structural and non-structural quantities at each floor. The structural components considered here are the following: beams, columns, concrete roof and concrete floor slabs. The non-structural components are the following: water piping and HVAC systems, gypsum wall partitions, raised access office flooring, suspended ceiling systems and concrete roof tiles (for the control building). The inventory of components within the building was calculated from typical

performance in terms of damage and repair costs. This analysis considers the different possible damage states for structural and non-structural components, taking dynamic structural analysis results as input. In loss estimation, damage analysis for each building component is based on empirically built fragility curves that express the probability that a component is in or exceeds a specified damage state as a function of the engineering demands on a building. Engineering demand parameters (EDPs) include peak IDRs and peak floor accelerations, depending on to which

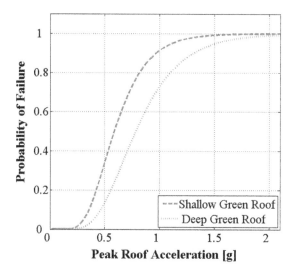

Figure 7. Fragility functions developed for shallow and deep green roof systems (based on data from Carmody et al., 2009).

quantities provided by PACT that depend on building occupancy type and gross building area. Previous studies suggest that repair costs after an earthquake are dominated by damage to the structural framing system, interior partition walls and interior paint (Goulet et al., 2007), making it particularly important to estimate accurately the quantity of and possible damage to these components.

At present, the PACT fragility library does not contain damage information for any green building features, including green roofs. We created user-defined components for the shallow and deep green roofs based on the limited body of experimental data pertaining to the seismic response of green roofs. Carmody et al. (2009) used a shaking table to test green roof scale models with and without plants under varying levels of excitation. This study quantified the peak roof acceleration that led to detrimental soil displacement, for different fundamental periods of hypothetical buildings. For the purposes of our study, the data for the Carmody et al. (2009) roof without the plants are assumed to represent the response of a shallow green roof, while the data for the Carmody et al. (2009) roof with plants represent the response of a deep green roof. These data were used to create fragility functions following recommendations from Porter, Kennedy, and Bachman (2007), and produced median detrimental roof accelerations of .59 and .78 g and dispersions of .39 and .40 for the shallow and green roof systems, respectively, as shown in Figure 7.

The user-defined green roof fragility function in PACT was completed by defining repair costs for damage to green roof components, based on industry standard minimum and maximum repair costs per square foot. The fragility functions are based on economies of scale. Repairs for roof areas 100 ft^2 or less were assumed to cost \$12.50/ft^2 for the shallow roof system and \$19.90/ft^2 for the deep roof system. Repairs for a roof areas greater than 600 ft^2 were assumed to cost \$10.30/ft^2 for the shallow roof system and \$16.20/ft^2 for the deep roof system (GSA, 2011; Peck & Kuhn, 2003). The roof pavers around the green roof systems were represented in PACT by the existing component for unsecured clay roof tiles.

The estimated replacement cost for the buildings is \$12.5 million for the control building, \$12.75 million for the shallow green roof building and \$12.9 million for the deep green roof building. The replacement costs come from a study of the same four-storey control building, which used RS means to estimate the construction costs (Ramirez et al., 2012). It is assumed that total collapse of a building would require the total replacement cost, equal to the capital cost.

4.4.2. Repair outcomes and bill of required materials

PACT uses a Monte Carlo simulation procedure to estimate a lognormal distribution of possible repair outcomes at each of selected hazard levels. At each realisation for each hazard level, PACT estimates the EDPs in the building from the structural analysis results, and then simulates the damage states DS $\in \{1, 2, 3, \dots, m\}$ experienced by a component, based on its fragility functions. Damage states in PACT are sequential, meaning that each subsequent damage state cannot be reached until the previous one has been exceeded. Some of the sequential damage states are also divided into mutually exclusive damage states, representing two or more distinct sub-states with different probabilities of occurrence, where only one state can exist at a time and each represents a distinct state of progressive damage. For example:

$$P(ME_{1,2}) = P(ME_{1,2}|S_2) * P(S_2) \tag{1}$$

quantifies the occurrence of the mutually exclusive damage state ($ME_{i,k}$), which is a subset of sequential damage state of $S \in \{1, 2, \dots, k\}$. Given a predicted damage state, PACT also reports the expected value of the repair cost for the component.

The loss analysis methodology can be adapted to assess environmental consequences associated with seismic damage. A challenge when conducting this integrated environmental loss analysis is that the repair estimates in PACT and other similar software (e.g. SP3) do not provide bills of materials for repair actions (ATC, 2012b). This study remedies this gap by calculating specific material quantities needed to conduct repairs, in order to assess the added environmental impacts from each possible repair scenario.

To calculate the material quantities needed for repairs of the damaged components, the first step is to determine the number of units of each component type damaged in each realisation. This information is not directly provided in the PACT output. In this study, calculation of the number of damaged units begins with Equation (2), which states that, for each realisation, j, the cost of repairing a given type of component, c, from PACT is equal to:

$$\begin{aligned} C_{c,j} = TU_{c,j} * (UC_{c,i} * P(i) \\ + UC_{c,i+1} * P(i+1) + \cdots UC_{c,n} * P(m)) \end{aligned} \tag{2}$$

where $C_{c,j}$ is the expected total repair cost for component c. $TU_{c,j}$ is the total number of units damaged for component c, and m is the total number of damage states for this component. $UC_{c,i}$ is the unit cost for repairs at (DS$_i$), and $P(i)$ could be either $P(S_i)$ or $P(ME_i)$, depending on the damage states defined for component c. Equation (2) can be rearranged to calculate the total number of damaged units, $TU_{c,j}$ for component type c, as well as the number of units within each damage state, based on the repair cost, probabilities of the different damage states and mean unit costs.

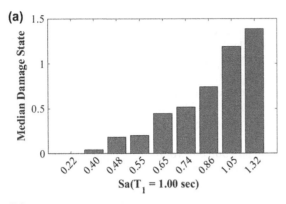

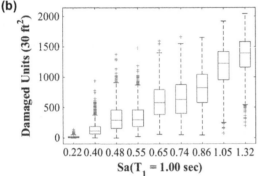

Figure 8. Damage to curtain wall units for shallow green roof building over all realisations at each hazard level, showing (a) median damage state and (b) box plot of number of damaged units (25th, 50th and 75th percentiles).

This process is illustrated in Figure 8(a) with damage to the curtain walls in the shallow green roof building. The first sequential curtain wall damage state (DS$_1$) describes how the glass will crack at a median storey drift ratio of .021. The second damage state (DS$_2$) is more severe, involving glass falling from the curtain wall frame occurring at a median storey drift ratio of .024. Following the trends of the structural analysis, the curtain walls move into more severe damage states with larger floor displacements. The Monte Carlo simulation enables probabilistic representation of the results, as in Figure 8(b) which shows box plots (25th, 50th and 75th percentiles) of the number of damaged curtain walls as a function of the hazard level. As for most components, the lower hazard levels have larger coefficients of variation. Similar results are observed for all component types.

PACT also lists qualitative descriptions of repairs for each damage state of each component. We used these descriptions along with industry data sheets for non-structural components and architectural drawings provided in *Building Construction Illustrated* (Ching, 2014) to convert the qualitative repair actions into numerical material quantities. Table 3 shows example calculations as applied to exterior curtain walls. The PACT repair action description suggests that the damage for DS$_1$ of the curtain

walls is light enough to warrant only replacement of the glass, while DS$_2$ requires new glass, as well as plywood to cover the area during replacement (ATC, 2012a). Therefore, repairs for a single 30 ft^2 unit (the PACT normative unit for this component) of curtain wall were converted into material volumes of 2610 in^3 of glass (assuming double glazed window thickness of .5 in, from standard industry specifications) for DS$_1$ and the same volume of glass plus a volume of plywood equal to 1080 in^3 (assuming a thickness of .25 in) for DS$_2$.

The total volume of required materials to repair the curtain walls at each hazard level for the shallow green roof building is shown in Figure 9. The results show how damage at the lower hazard levels mainly involves repair of the glass, but damage at higher hazard levels also requires plywood, to cover damaged openings during more intensive repairs. Similar repair descriptions were translated into material quantities at each damage state for each component.

Material quantities for repair came from combining repair costs and multi-state damage probabilities from PACT with the estimated repair action quantities. We then multiplied the estimated damaged unit quantities for each damage state in each realisation by the material quantities required to repair damaged units of a given component. The total number of materials – such as concrete, rebar and grout for the beam–column components – for a single realisation of one hazard level is the sum of materials for repair of a given component.

The median repair costs presented in Table 4 are the sum of all component repair and replacements costs for each building at each hazard level. At the lower hazard levels, the damage is concentrated in the non-structural curtain walls and interior wall partitions, while the beams, columns and floor slabs dominate the losses at higher hazard levels. The retrofit green roof building has higher potential median losses than the other three buildings. Similarly, while the shallow and deep green roof building have fewer losses at less intense hazard losses, they incur greater potential losses than the control building at hazard levels of 1.05 and 1.32 g.

4.5. Environmental impact analysis

4.5.1. Life-cycle boundary and scope

Following structural and loss analysis, the framework enters the final analysis stage, environmental analysis. LCA, in general, begins with definition of the functional unit for comparison, the system boundaries and the impact methodology (ISO, 2006). In our case, the functional units (Ciroth & Srocka, 2008) for comparison are the four buildings. This definition introduces a series of challenges due to the inherent complexity of buildings, compared with a simpler product such as a plastic cup (Vieira & Horvath, 2008). Here, the functional unit of each building is broken into separate structural and non-structural components, providing the basis for the assessment of the total

Table 3. Example calculation of material quantities needed for repair actions for curtain wall units.

Damage state description	Repair action description	Material quantities needed per damaged 30 ft^2 unit (in^3)
DS 1: Glass cracking	Replace glass	Glass = 2160
DS 2: Glass falls from frame	Replace glass; cover exposure in meantime	Glass = 2160 Plywood = 1080

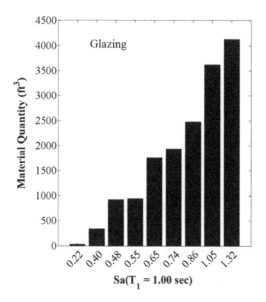
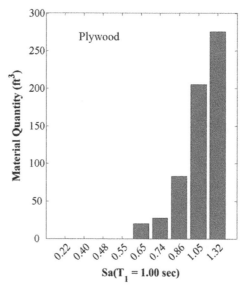

Figure 9. Median material quantities needed for curtain wall repair for shallow green roof building (including collapse cases).

Table 4. Median repair costs for total building post-earthquake losses from Monte Carlo simulation of all potential component repair costs.

Hazard levels	Control	Shallow green roof	Deep green roof	Retrofit green roof
1	$70,200	$680,000	$571,000	$901,000
2	$198,400	$1,484,000	$1,434,000	$2,429,000
3	$2,403,000	$2,402,000	$2,055,000	$3,434,000
4	$2,724,000	$2,531,000	$2,669,000	$4,116,000
5	$3,392,000	$3,533,000	$3,005,000	$4,775,000
6	$4,108,000	$3,807,000	$3,842,000	$5,623,000
7	$4,838,000	$4,535,000	$4,352,000	$6,818,000
8	$5,596,000	$6,126,000	$6,387,000	$7,788,000
9	$6,146,000	$7,163,000	$7,858,000	$9,105,000

material quantities associated with manufacturing or repair of these components.

The study takes a 'cradle-to-gate' approach in which building performance is compared in terms of the impacts produced by the extraction and manufacturing of new products and the operating energy, but not routine maintenance activities or end-of-life disposal. The system boundary was determined by identifying those life-cycle stages that differ between the buildings. In the product manufacturing stage, the impacts of only structural members (beams, columns and floor slabs) and roofing systems are considered, since the structural, but not the non-structural, design varies between the buildings. Previous studies (Bilec, Ries, & Matthews, 2010; Junnila & Horvath, 2003) have demonstrated that environmental impacts from a building's construction stage, i.e. impacts associated with on-site erection of the building, are minimal compared with manufacturing and production of raw materials or annual operating energy of a building. As such, construction impacts are not considered.

The impact of the transportation stage for initial construction and post-earthquake repair activity depends on the distance of manufacturing plants to the building site, and the weight and volume of material being transported. The impact of transportation emissions also depends on the type of freight vehicles, because larger vehicles (Class 6) used to transport

materials such as steel and concrete typically have more stringent emissions standards than smaller vehicles (Class 4), which would be used to transport materials such as the green roof soil (EIA, 2012). Transport effects are not considered in the current study.

The operating energy of the buildings is presented in order to quantify the differences between the green roof and non-green roof buildings in this context. The midlife addition of a green roof system on the fourth building requires creation of an additional life-cycle stage, entitled midlife roof retrofits, which accounts for only impacts from the manufacturing and production of materials for the new green roof system. In this stage, we assume that the shallow green roof was added five years after initial construction and that the computed impact from this stage comes only from manufacturing the additional materials needed to build the roof.

The final stage included in the system boundary is post-hazard repairs. Based on the loss analysis results, this stage considers the impact from the most heavily damaged building components: the structural members, wall partitions, exterior curtain walls, suspended ceiling tiles and green roof systems. The raised access flooring, water piping systems and HVAC system units are excluded, because the loss analysis suggested that most damage for these components could be repaired by manual labour (i.e. no material inputs required). Discussion of end-of-life building disposal is out of the scope of this study, because we chose to focus on in-service building performance, not differences between material disposal options and impacts.

4.5.2. Environmental impact assessment methods

The environmental impact of a functional unit and its sub-components comes from analysis of life-cycle inventory (LCI) data. Process-based LCI describes the main flows of energy and materials in and out (e.g. emissions) of the functional unit (EPA, 2008). Multiple databases estimate energy/material flows, as well as their associated emissions to land, water and soil (Cook, 2014). Here, inventory data were gathered using Ecoinvent, because

Table 5. Allocation unit environmental impacts for curtain wall repair materials.

Impact	Units	Glazing (per 1 ft^3)	Plywood (per 1 ft^3)
Ozone depletion	ton CFC-11 eq.	1.57E−10	2.71E−09
Carbon dioxide emissions	ton CO_2 eq.	1.14E−03	1.98E−02
Smog	ton O_3 eq.	1.03E−04	2.01E−03
Acidification	ton SO_2 eq.	9.89E−06	1.64E−04
Eutrophication	ton N eq.	2.81E−06	1.04E−04
Carcinogenics	CTUh	1.66E−10	9.21E−10
Non-carcinogenics	CTUh	2.64E−10	1.20E−08
Fine particle emissions	ton $PM_{2.5}$ eq.	9.36E−07	4.31E−05
Ecotoxicity	CTUe	1.12E−02	.197
Fossil fuel depletion	kWh surplus	13.42	310.77

it is a well-established and comprehensive source, with over 10,000 different processes for analysis (Goedkoop, Oele, Leijting, Ponsioen, & Meijer, 2013). Ecoinvent is a database that provides life-cycle inventories for all economic activities at a unit process level (Weidema et al., 2013).

We allocated the impact of the inventory processes using the EPA's *Tool for the Reduction and Assessment of Chemical and other Environmental Impacts* (TRACI). The TRACI methodology transforms a selected process activity into potential impacts associated with raw material production, as well as associated chemical releases (EPA, 2008). While the LCI presents the raw emissions of each process, the TRACI methodology is needed to quantify how these emissions combine together into environmental impacts, such as fossil fuel depletion or eutrophication, which are more typically presented as LCA results.

We used the process-based life-cycle software SimaPro (Goedkoop et al., 2013) to organise environmental impact calculations for all building materials. Product manufacturing impacts were determined based on unit impacts for the amount of concrete, reinforcing steel and roof materials needed for the initial production of building components. In the post-hazard repairs phase, calculations began with collection of the material quantities for earthquake-related repairs; here, we use these material quantities as inputs for the environmental impact analysis. Table 5 shows an example of manufacturing environmental impacts, for 1 ft^3 each of curtain wall glazing and plywood in the post-hazard repair stage. The table presents results for all ten different SimaPro environmental impact categories. The remainder of this study focuses on only the impact from carbon dioxide (CO_2) equivalents at the life-cycle stages of interest. Carbon dioxide equivalents compare emissions from different greenhouse gases with respect to their contribution to climate change. Embodied carbon is considered here as the total amount of greenhouse gas emissions, converted to CO_2 equivalents, required to produce a given material or building product (Werner & Burns, 2012). We refer throughout this study to the embodied carbon from manufacturing products or the CO_2 emissions from operating energy consumption and demand as climate change potential (CCP), considered in terms of tons of CO_2 equivalents.

For each realisation of each hazard level, we multiplied the unit environmental impacts by the respective material quantities needed to conduct repairs. The final output for each building was a lognormal distribution of the environmental impacts over the 1000 realisations of each hazard level. For realisations causing building collapse, it is assumed that all of the components need to be replaced, which therefore considers the environmental impact

from restoring the building to full functionality for the remainder of its service life.

4.5.3. Energy use assumptions

The energy portfolio impacts from electricity and natural gas production for building operation and the expected annual operating energy at our site are determined from assumptions drawn from survey data compiled by the California Energy Commission (CEC). In Los Angeles, the typical energy supply mix is 33% from coal, 21% from natural gas, 13% from wind power, 10% from nuclear power, 5% from biomass and natural waste, 6% from hydroelectric plants and 12% from other energy sources (CEC, 2006).

Controlling for state-wide differences in climate, the CEC survey suggests that a typical mid-sized commercial building consumes annually 17.72 kWh/ft^2 in electrical energy and 21.80 kBtu/ft^2 in natural gas (CEC, 2006). From these data, we assume that the control building would consume 1.53 million kWh/year and 1.88 million kBtu/year. The potential operating energy reduction from the green roof buildings is based on the LEEDv4 new building + construction prerequisite for minimum energy performance (USGBC, 2015). One suggested path in LEED to meet this goal is to follow ASHRAE design guides for advanced energy savings (ASHRAE, 2014). The shallow green roof building was assumed to meet ASHRAE 30% energy reduction standards for mid-sized commercial buildings, consuming 1 million kWh/year and 1.23 million kBtu/year. The deep green roof building, representing design changes for 50% annual energy savings, is assumed to use 765,000 kWh/year and 941,700 kBtu/year goals that would be reasonable to achieve with the chosen green roof systems. Here, the authors choose not to account for future changes in energy efficiency or green building technologies.

4.5.4. Environmental impact results

First, the LCA results at a building component level for the damaged curtain walls are presented. Figure 10 illustrates how the impacts of earthquake damage influence the climate change potential, or CCP, (in terms of CO_2 equivalents) for curtain walls in the shallow green roof building. The results show, at all hazard levels, that the impacts of the glass replacement in the curtain wall dominate the overall component environmental impact. At higher hazard levels, where the second, more extensive, damage state is more likely to occur, the use of plywood in repair activities begins to make a much greater contribution to the CO_2 emissions compared with lower hazard levels.

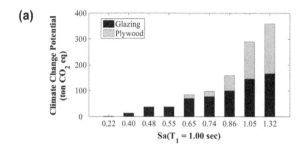

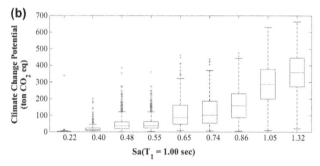

Figure 10. Tons of CO_2 emissions or equivalent associated with post-earthquake repairs of curtain walls in the shallow green roof building: (a) deaggregated by material requirement and (b) showing uncertainty in the estimate (box plot with 25th, 50th and 75th percentiles).

To study the uncertainty of the results, we also computed the median, 25th percentile, 75th percentile and coefficient of variation (COV) for the total number of units damaged and the total environmental impact of each component. For all components, we saw relatively small COV values, almost all less than 1, at lower hazard levels. However, in some cases, at higher hazard levels, the CO_2 emissions COV value for the curtain walls was larger

than 2, indicating large uncertainties in the projected results. In this study, the computation of uncertainty considered variability only in the damaged components and material quantities, not in sources of uncertainty from the emissions data, so the true uncertainties are likely substantially higher.

Subsequently, the combined life-cycle environmental impacts for all four buildings are compared. Figure 11(a)–(c) presents the CO_2 emissions for each building, in terms of the life-cycle phases of greatest interest: product manufacturing, operating energy and post-hazard repairs. (The results of the roof retrofit stage are not shown because its impact is minimal compared to the other stages and occurs only for the retrofit green roof building.) We can draw relative comparisons between each of the three life-cycle stages presented. For all four buildings, the life-cycle impacts are dominated by the operating energy, because the results are shown over the total 50 year lifespan (without any discounting). Among all buildings, the shallow and deep green roof buildings have the highest manufacturing impact for raw materials, due to the larger volumes of concrete and reinforcing steel for the initial construction of the beams, columns and floor slabs required to support the extra green roof loads under gravity and seismic load cases. Therefore, a decision-making trade-off appears with respect to the upfront environmental impact of the green roof building and the significantly lower annual operating energy impacts.

The impacts of post-hazard repairs assume that at most one earthquake occurs sometime during the building's service life, after addition of the retrofit green roof. Note that the results below are presented in lognormal space, due to the challenge of comparing the scale of the impacts between the lower intensity shaking and higher intensity shaking levels. At all hazard levels, the median impacts are largest for the retrofit building impacts.

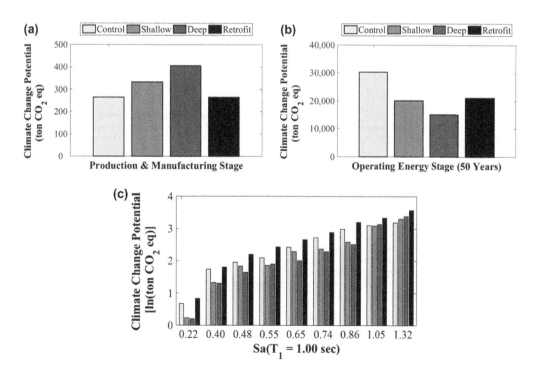

Figure 11. Climate change potential (CO_2 eq.) from life-cycle stage of (a) product manufacturing, (b) operating energy and (c) post-hazard repair (mean log values from lognormal distribution) for all four buildings after each of the nine ground shaking intensity scenarios.

If ground shaking event with an intensity ranging from $Sa(T_1 = 1.00 \text{ s}) = 0.2-0.86$ g was to occur, the control building is likely to have the next largest possible CO_2 emissions. Nonetheless, at the highest level of shaking, the non-linear effects and weak performance of the green roof models result in much higher environmental impact from post-hazard repairs. For example, at a hazard level of .74 g, both the shallow and deep green roof buildings have almost 20% lower CO_2 repair emissions than they do at the highest level of shaking. The values presented in Figure 11(c) are the median values, but we note that the uncertainty and variability in the results increase with greater level of shaking. However, similar trends between buildings are observed in the other percentiles.

At low levels of ground shaking, the results suggest positive trade-offs between the upfront material choices for the green roof buildings: these buildings provide both lower annual operating energy demand, costs and emissions and fewer post-earthquake losses and environmental impacts. Although the control and retrofit green roof buildings have lower upfront material production impacts, they trade this low upfront impact for greater operating energy impacts and higher earthquake losses. The retrofit green roof building in particular demonstrates a decision-making dilemma, in that the building can achieve a midlife reduction in operating energy with the addition of the green roof, but the building would perform relatively poorly during any significant seismic event. Design decisions must weigh the costly investments in green design practices, which may not benefit hazard performance in high consequence, low probability events, but may provide smaller annual operating benefits over the total lifespan of a building.

These trade-offs are demonstrated through an environmental impact exceedance curve, by presenting the results in terms of a fragility function for CCP (CO_2 eq.) at varying hazard levels. Figure 12 presents the CCP fragility function for the probability that post-earthquake repair actions for four buildings exceed 100 tons of CO_2 equivalents based on the analysis conducted here (without fitting to a probability distribution).

5. Limitations

This study has several limitations. First, there exists a significant challenge in comparing environmental and economic impacts at different points in time. Discounting could have been used to place all impacts in the same temporal boundary, but this introduces an ethical debate with respect to selection of appropriate discounting factors (Vieira & Horvath, 2008). Due to the social and ethical complexities of discounting environmental impacts, we chose to exclude discounting for all results and instead deaggregated the impacts in the different stages.

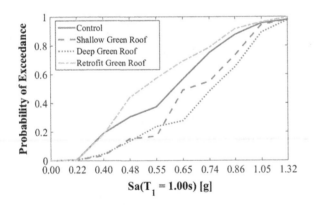

Figure 12. Environmental impact exceedance curves (representing CO_2 emissions fragility function) for exceedance of 100 tons of CO_2 during post-earthquake repairs.

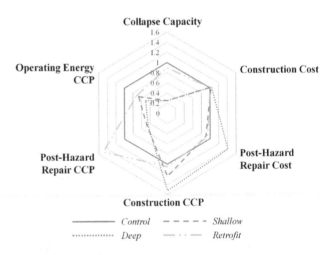

Figure 13. Radar plot showing trade-offs in building performance (where repairs are considered at a ground shaking intensity of $Sa(T_1 = 1.00 \text{ s}) = .86$ g) and environmental impact is considered in terms of climate change potential (referred to as CCP and in terms of CO_2 equivalents).

Table 6. Strengths and weaknesses identified during life-cycle assessment of building design trade-offs for environmental impact and hazard performance.

Building	Strengths	Weaknesses
Control	• Lower upfront environmental impact from structural design than shallow and deep green roof buildings • Lower collapse risk in rare earthquake events than green roof buildings	• Worse performance under frequent earthquake events than shallow and deep green roof buildings
Shallow green roof	• Lower annual operating energy than control building • Storm water management/heat island mitigation • Lower environmental impact at high-frequency earthquake events than control and retrofit buildings	• Larger upfront environmental impact from structural materials than control building • Worse performance in rare earthquake events (potential for significant economic losses and environmental impact) than control
Deep green roof	• Lowest annual operating energy of all four buildings • Added occupant comfort from access to green roof • Lower environmental impact at high-frequency earthquake events than control and retrofit buildings	• Worse performance in rare earthquakes (potential for significant economic losses and environmental impact) than control
Retrofit green roof	• Lower upfront environmental impact than shallow and deep green roof buildings • Midlife reduction in operating energy	• Larger potential economic and material losses and environmental impact at all levels of ground shaking intensity than control

Second, the environmental impact from temporary and permanent irrigation of the green roof systems, as well as changes in storm water management, was excluded from the scope. Inclusion of this component in the analysis would better quantify the benefits of the green building for comparison with the other buildings. Third, additional experimental studies are needed to better represent the response of green roof systems under seismic excitation. More broadly, as many communities seek to 'build green' while facing a simultaneous risk of major hazard events, greater experimental testing and computational modelling are needed to study the behaviour of green building features under extreme hazard events.

Finally, the specific building-to-building findings presented in this paper may not necessarily be generalised across all building configurations and uses. However, the results provide valuable relative metrics of performance for evaluating the trade-offs associated with green roof buildings in high seismic areas. Refinements in the procedures may alter some of the specific quantitative values, but not the relative positions of the comparison between buildings. In addition, the case study serves to illustrate the green-resilience framework, providing a novel tool that can be employed for various building types, locations and types of hazards.

6. Conclusions

This paper presents a new framework for the joint assessment of building environmental impact and hazard performance, in the context of green building design choices and earthquake hazard risk. The framework is illustrated through an examination of the seismic performance of office buildings with green roofs with different characteristics, examined through non-linear dynamic analysis and assembly-based loss analysis and environmental impact assessment.

The findings of this study demonstrate key trade-offs between design choices for green building objectives and performance goals for post-hazard functionality, which are summarised in Table 6 and Figure 13. The results of each building's performance in six different arenas, collapse capacity, construction cost, post-hazard cost, construction CCP, post-hazard CCP and operating energy CCP, are normalised against the values for the control building. The results presented here are for a ground shaking intensity of $Sa(T_1 = 1.0 \text{ s}) = .86 \text{ g}$, but similar plots can be made for each hazard level.

The results from these buildings illustrate the challenging trade-offs presented for building design decision-makers, as each building offers varying economic, environmental and hazard performance strengths and weaknesses. Although the control building performs moderately well under low probability, high-intensity earthquake events, in terms of post-earthquake economic losses and environmental impact from repair activities, these strengths are counterbalanced by the building's relatively worse performance under high probability, low-intensity earthquake events and its large annual operating energy consumption.

Conversely, all three of the green roof buildings offer significant environmental benefits during the life-cycle of a building, provided an earthquake does not occur. This is a scenario under which green buildings, and especially those with green roof systems, are normally considered. The results of

our case study, however, suggest that significant environmental impact may result from repair actions due to low probability, high-intensity earthquakes, making the green roof buildings, in effect, less 'green' than their control counterpart when considering post-hazard repairs. In addition, the green roof buildings come with additional material expenditures due to the larger beam and column sizes required to support the more massive roofs under gravity and seismic loads.

The framework presented in this paper advances the science of building design and assessment, through an integrated, multi-criteria life-cycle assessment procedure for buildings that includes green building design decision-making. The twenty-first-century building stock faces the challenge and opportunity of a growing number of green buildings, coupled with greater vulnerability to extreme earthquake events from expanding populations and dense urban environments. In order to support urban development objectives for both sustainability and resilience, the proposed framework encourages decision-makers to assess the environmental impacts and hazard resistance of their buildings in tandem, rather than as isolated considerations. Future work will explore other design decisions that contribute to the trade-offs associated with green-resilient building design.

Acknowledgements

Any opinions, findings, and conclusions or recommendations expressed in this material are those of the authors and do not necessarily reflect the views of the National Science Foundation. Discussions with Dr. Paul Chinowsky, Matthew Comber, Dr. Sherri Cook, Dirk Kestner, Dr. Wil Srubar, Dr. David Veshosky and Dr. John Zhai were very helpful in developing the ideas herein.

Funding

This material is based upon work supported by the National Science Foundation [grant number 1234503].

References

Almufti, I., & Willford, M. (2013). *REDi rating system: Resilience-based earthquake design initiative for the next generation of buildings.* London: Arup.

American Society of Civil Engineers. (2010). *Minimum design loads for buildings and other structures: ASCE/SEI 7-10.* Reston, VA: Author.

Applied Technology Council. (2012a, September). *Seismic performance of assessment of buildings.* Applied Technology Council and Federal Emergency Management Agency (FEMA P-58) 1&2. Redwood City, CA: Author.

Applied Technology Council. (2012b, December). *Seismic performance of assessment of buildings.* Applied Technology Council and Federal Emergency Management Agency (FEMA P-58) 4. Redwood City, CA: Author.

ASHRAE. (2014). *Advanced energy design guide for small to medium office buildings.* Atlanta, GA: Author.

Baker, J. W. (2011). Conditional mean spectrum: Tool for ground-motion selection. *Journal of Structural Engineering, 137,* 322–331.

Baker, J. W. (2015). Efficient analytical fragility function fitting using dynamic structural analysis. *Earthquake Spectra, 31,* 579–599.

Beck, J. L., Porter, K. A., Shaikhutdinov, R., Au, S. K., Moroi, T., Tsukada, Y., & Masuda, M. (2002). *Impact of Seismic risk on lifetime property values,* Final Report. Richmond, CA: Consortium of Universities for Research in Earthquake Engineering.

Berke, P. R. (1995). Natural-hazard reduction and sustainable development: A global assessment. *Journal of Planning Literature, 9,* 370–382.

Bilec, M. M., Ries, R. J., & Matthews, H. S. (2010). Life-cycle assessment modeling of construction processes for buildings. *Journal of Infrastructure Systems, 16*, 199–205.

Bokalders, V., & Block, M. (2010). *The whole building handbook*. London: Earthscan.

BREEAM. (2015). *What is BREEAM?* Retrieved from www.breeam.org

Bruneau, M., Chang, S. E., Eguchi, R. T., Lee, G. C., O'Rourke, T. D., Reinhorn, A. M., … von Winterfeldt, D. (2003). A framework to quantitatively assess and enhance the seismic resilience of communities. *Earthquake Spectra, 19*, 733–752.

California Energy Commission. (2006). *California commercial end-use survey (CEUS)*. San Diego, CA: Itron.

Carbon Leadership Forum. (2012). *North American product category rules for ISO 14025 type III environmental product declarations and/or GHG protocol conformant product "carbon footprint" of concrete*. Seattle, WA: University of Washington.

Carmody, M. O., Jasarevic, M., Omenzetter, P., Clifton, G. C., & Fassman, E. A. (2009). *Seismic response of green roofs*. 2009 New Zealand Society for Earthquake Engineering Conference, Christchurch, NZ.

Ciroth, A., & Srocka, M. (2008). How to obtain a precise and representative estimate for parameters in LCA. A case study for the functional unit. *International Journal of LCA, 13*, 265–277. doi:http://dx.doi.org/10.1065/lca2007.06.345

Comber, M., Poland, C., & Sinclair, M. (2013). Environmental impact seismic assessment: Application of performance-based earthquake engineering methodologies to optimize environmental performance. *Proceedings of Structures Congress, 2012*, 910–921. doi:http://dx.doi.org/10.1061/9780784412367.081

Ching, F. (2014). *Building construction illustrated*. Hoboken, NJ: Wiley.

Cook, S. M. (2014). *Sustainable wastewater management: Modeling and decision strategies for unused medications and wastewater solids* (PhD thesis). University of Michigan, Ann Arbor, MI, USA.

Court, A., Simonen, K., Webster, M., Trusty, W., & Morris, P. (2012). Linking next-generation performance-based seismic design criteria to environmental performance (ATC-86 and ATC-58). *Structures Congress, 2012*, 922–928. doi:http://dx.doi.org/10.1061/9780784412367.082

Department of Energy. (2012a). *2011 Buildings energy data book*. Silver Spring, MD: D&R International.

Environmental Protection Agency. (2008). *Life cycle assessment: Principles and practice*. Cincinnati, OH: Scientific Applications International Corporation.

Feese, C., Li, Y., & Bulleit, W. M. (2014). Assessment of seismic damage of buildings and related environmental impacts. *Journal of Performance of Constructed Facilities, 29*, 04014106 1–10 (2015). doi:http://dx.doi.org/10.1061/(ASCE)CF.1943-5509.0000584

General Services Administration. (2011). *A report of the United States general services administration the benefits and challenges of green roofs on public and commercial buildings*. Washington, DC: Author.

Goedkoop, M., Oele, M., Leijting, J., Ponsioen, T., & Meijer, E. (2013). *Introduction to LCA with SimaPro*. PréSustainability. Retrieved from http://www.pre-sustainability.com/

Goulet, C., Haselton, C. B., Mitrani-Reiser, J., Beck J. L., Deierlein, G., Porter, K. A., & Stewart, J. P. (2007). Evaluation of the seismic performance of a code-conforming reinforced-concrete frame building – From seismic hazard to collapse safety and economic losses. *Earthquake Engineering & Structural Dynamics, 36*, 1973–1997.

Gromala, D. S., Kapur, O., Kochkin, V., Line, P., Passman, S., Reeder, A., & Trusty, W. (2010). *Natural hazards and sustainability for residential buildings*. FEMA P-798. Washington, DC: FEMA.

Haselton, C. B., & Deierlein, G. G. (2007). *Assessing seismic collapse safety of modern reinforced concrete moment frame buildings* (TR No. 156). Palo Alto, CA: John A. Blume Earthquake Engineering Center, Stanford University.

Haselton, C. B., Liel, A. B., Deierlein, G. G., Dean, B. S., & Chou, J. H. (2011). Seismic collapse safety of reinforced concrete buildings. I: Assessment of ductile moment frames. *Journal of Structural Engineering, 137*, 481–491.

Haselton, C. B., Mitrani-Reiser, J., Goulet, C., Deierlein, G. G., Beck, J., Porter, K. A., … Taciroglu, E. (2008). *An assessment to benchmark the seismic performance of a code-conforming reinforced concrete moment-frame building* (Report 2007/12). Berkeley: Pacific Engineering Research Center, University of California.

Hendrickson, C., Horvath, A., Joshi, S., & Lave, L. (1998). Economic input–output models for environmental life-cycle assessment. *Environmental Science & Technology, 32*, 184–191. doi:http://dx.doi.org/10.1021/es983471i

Hossain, K. A., & Gencturk, B. (2014). Life-cycle environmental impact assessment of reinforced concrete buildings subjected to natural hazards. *Journal of Architectural Engineering*. doi:http://dx.doi.org/10.1061/(ASCE)AE.1943-5568.0000153, A4014001.

Ibarra, L. F., Medina, R. A., & Krawinkler, H. (2005). Hysteretic models that incorporate strength and stiffness deterioration. *Earthquake Engineering Structural Dynamics, 34*, 1489–1511.

International Code Council. (2009). *2009 International building code*. Country Club Hills, IL: Author.

ISO, IEC. (2006). 14044:2006: Environmental management – life cycle assessment – requirements and guidelines. *International Organization for Standardization*, Geneva, Switzerland.

Jalayer, F., & Cornell, C. A. (2009). Alternative non-linear demand estimation methods for probability-based seismic assessments. *Earthquake Engineering & Structural Dynamics, 38*, 951–972.

Junnila, S., & Horvath, A. (2003). Life-cycle environmental effects of an office building. *Journal of Infrastructure Systems, 9*, 157–166.

Lawson, A. M., Bersani, K. S., Fahim-nader, M., & Guo, J. (2002, December). Benchmark input–output accounts of the United States 1997. *Survey of Current Business*, 19–109.

Lin, T., Harmsen, S. C., Baker, J. W., & Luco, N. (2013). Conditional spectrum computation incorporating multiple causal earthquake and ground-motion prediction models. *Bulletin of the Seismological Society of America, 103*, 1103–1116.

Menna, C., Asprone, D., Jalayer, F., Protoa, A., & Manfredi, G. (2012). Assessment of ecological sustainability of a building subjected to potential seismic events during its lifetime. *International Journal of Life Cycle Assessment, 18*, 504–515. doi:http://dx.doi.org/10.1007/s11367-012-0477-9

Mitrani-Reiser, J., Haselton, C., Goulet, C., Porter, K., Beck, J., & Deierlein, G. (2006). Evaluation of the seismic performance of a code-conforming reinforced-concrete frame building – part II: Loss estimation. *Earthquake Engineering & Structural Dynamics, 36*, 1973–1997. doi:http://dx.doi.org/10.1002/eqe.694

Moehle,J., & Deierlein, G. G. (2004). *A framework methodology for performance-based earthquake engineering*. Proceedings of 13th World Conference on Earthquake Engineering, Paper No. 679, Vancouver, BC, Canada.

National Roofing Contractors Association. (2007). *The NRCA green roof systems manual*. Rosemont, IL: Author.

Padgett, J. E., & Tapia, C. (2013, December). Sustainability of natural hazard risk mitigation: Life cycle analysis of environmental indicators for bridge infrastructure. *Journal of Infrastructure Systems, 19*, 395–408.

Peck, S., & Kuhn, M. (2003). *Design guidelines for green roofs*. Ottawa: Ontario Association of Architects.

PEER. (2014). *The open system for earthquake engineering simulation*. Pacific Earthquake Engineering Research Center, CA. Retrieved from www.opensees.berkeley.org

Porter, K., Kennedy, R., & Bachman, R. (2007). Creating fragility functions for performance-based earthquake engineering. *Earthquake Spectra, 23*, 471–489.

Porter, K., Kiremidjian, A. S., & LeGrue, J. S. (2001). Assembly-based vulnerability of buildings and its use in performance evaluation. *Earthquake Spectra, 17*, 291–312.

Ramirez, C. M., Liel, A. B., Mitrani-Reiser, J., Haselton, C. B., Spear, A. D., Striner, J., … Miranda, E. (2012). Expected earthquake damage and repair costs in reinforced concrete frame buildings. *Earthquake Engineering & Structural Dynamics, 41*, 1455–1475.

Sailor, D. J. (2008). A green roof model for building energy simulation programs. *Energy and Buildings, 40*, 1466–1478.

Saiz, S., Kennedy, C., Bass, B., & Pressnail, K. (2006). Comparative life cycle assessment of standard and green roofs. *Environmental Science & Technology, 40*, 4312–4316.

Schwab, A. K., & Brower, D. J. (1999). *Sustainable development and natural hazards mitigation*. Raleigh: North Carolina Division of Emergency Management, USA.

Sharrard, A. L., Matthews, H. S., & Ries, R. J. (2008). Estimating construction project environmental effects using an input–output-based hybrid life-cycle assessment model. *Journal of Infrastructure Systems, 14*, 327–336.

Shook, D. Personal correspondence, 2013.

Thormark, C. (2007). *Energy and resources, material choice and recycling potential in low energy buildings.* Paper presented at the International CIB Conference SB07 Sustainable Construction, Materials and Practices. Lisbon, Portugal.

United States Green Building Council. (2009). *Green building and LEED core concepts.* Washington, DC: Author.

United States Green Building Council. (2015). *LEED credit library.* LEED BD+C: v4 – LEEDv4. Washington, DC: Author. Retrieved from www.usgbc.org/credits

U.S. Energy Information Administration. (2012). *Annual energy outlook 2012.* Washington, DC: Department of Energy.

Vieira, P. S., & Horvath, A. (2008). Assessing the end-of-life impacts of buildings. *Environmental Science and Technology, 42*, 4663–4669.

Weidema, B. P., Bauer, C., Hischier, R., Mutel, C., Nemecek, T., Reinhard, J., … Wernet, G. (2013). *Overview and methodology: Data quality guideline for the ecoinvent database version 3.* Ecoinvent Report 1(3). St. Gallen: The Ecoinvent Centre.

Weiler, S. K., & Scholz-Barth, K. (2009). *Green roof systems.* Hoboken, NJ: Wiley.

Welsh-Huggins, S. J., & Liel, A. B. (2014a). *Integrating green and resilient building design for enhanced disaster recovery.* Proceedings of the 3rd International Conference on Urban Disaster Reduction, Earthquake Engineering Research Institute, Boulder, CO, USA.

Welsh-Huggins, S. J., & Liel, A. B. (2014b). *Integrating hazard-induced damage and environmental impacts in building life-cycle assessments.* Proceedings of the 2014 International Symposium of Life-Cycle Civil Engineering, Tokyo, Japan.

Werner, W., & Burns, J. (2012). Quantification and optimization of structural embodied energy and carbon. *Structures Congress, 2012*, 929–940. doi: http://dx.doi.org/10.1061/9780784412367.083

Practical application of life-cycle management system for shore protection facilities

Hiroshi Yokota

ABSTRACT

Shore protection facilities have a long lifetime and must be expected to meet demands for providing people living coastlines with safety and security. The principal demands for those facilities seem to be simple but have not been practically so easy to be kept over the requirements. The reasons for those are that facilities are rather big and are exposed to severe environments for materials. Those facts may cause difficulties in even visual inspection. In addition, as it would be very important, rise of the seawater level, etc. due to global warming may affect the function and performance, which has to be taken into account rehabilitation planning. To overcome such difficulties, it is important to pursue coordination between design and maintenance based on the procedure of the life-cycle management through which sustainability indicators would be maximum/minimum. This paper presents the concept and the framework of the life-cycle management system for those facilities and introduces methodologies of the management system particularly feasible to practical maintenance.

Introduction

Social and economic activities concentrate in coastal areas in many countries. A shore protection facility is expected to prevent those areas from disasters caused by storm surge, tsunami, high waves, coastal erosion, etc. as well as to facilitate coast utilisation and environmental preservation. Typical structural types are embankment, revetment and parapet, detached breakwater, etc., an example of which is shown in Figure 1 (Japan Society of Civil Engineers [JSCE], 2000). While many facilities or structures have been newly built in coastal areas, some of them require remedial actions including repair, strengthening, upgrading or renovation because they have been exposed to very severe physical actions as well as harsh environmental actions. Physical actions such as waves and storm surges may cause damages to structures. In addition, materials tend to deteriorate rapidly in marine environments and degradation of structural performance or even structural collapse may be consequences.

At the initial design of a structure, designers make several assumptions, in which the most severe conditions are likely to be assumed for structural design in order to ensure structural performance over the requirement. Serious deterioration may be caused by insufficient durability design with optimistic assumptions against materials deterioration and by lack of proper maintenance after construction of the structure. To meet these facts, it is extremely important to pursue collaboration between the durability design work and subsequent maintenance work. Based on the performance-based design concept, structural performance should be ensured during the design service life.

For this purpose, it is necessary to provide the methodologies to keep the structural performance over the minimum limits during the design and maintenance stages. In addition, even if no damages or deterioration occur, structures in coastal areas may have insufficient structural performance because of seawater level rise due to global warming, etc. The life-cycle management is a process which seeks to ensure that structural functions and performance would meet their requirements and that the service life of a structure would equal or exceed its design life, while taking into account the sustainability indicators such as life-cycle cost and environmental impact.

This paper firstly presents the concept and the framework of the life-cycle management system for shore protection facilities and structures. Then, simplified inspection strategy is proposed focusing on crack width induced in concrete members as one of the important management tools. Structural performance evaluation based on crack inspection is also described in this paper. Finally, remedial actions to meet upgrading structural performance requirements are discussed. These issues would be particularly feasible to practical maintenance for shore protection facilities and structures.

Life-cycle management

Concept of life-cycle management

The service life of a structure is made up of all the activities including planning, basic and detailed designs, execution including material selection, production and construction, maintenance

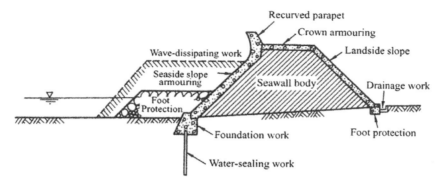

Figure 1. Typical cross section of embankment as an example of shore protection facilities and structures (JSCE, 2000).

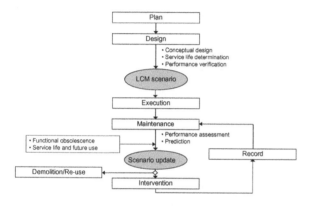

Figure 2. Life-cycle management system.

Figure 3. Function, performance requirements and performance indices.

including assessment and remedial action, and decommissioning. The life-cycle management (Yokota & Hashimoto, 2013) is an integrated concept to assist in activities managing the total life cycle of structures to realise sustainability as shown in Figure 2. The International Federation for Structural Concrete (its name is well known FIB) defines the life-cycle management in its latest model code (FIB, 2010) as

> the overall strategy to be used in managing a structure through its development and service life, with the aim of improving its efficiency from a business/engineering point of view, ensuring that it meets the associated performance requirements defined at the time of design or as may be subsequently modified during the service life of the structure.

During the initial design stage, the durability design will be applied to predict the performance degradation. While lots of alternatives can exist, the fundamental concept on how the structural performance should be ensured must be well considered based on conditions, design service life, structural characteristics, material properties, difficulties in assessment and remedial action, social and economic importance, etc.

Maintenance is the major strategy to counter the degradation, which is carried out to assess the present conditions of structure and to quantify the level of structural performance. In addition, by predicting the future progress of structural performance degradation, one should choose the most appropriate remedial action in which the decision-making indicators are optimised under budget capping (Yokota, Iwanami, & Kato, 2012). During the maintenance stage, maintenance engineers will initially follow the scenario ('LCM scenario' in Figure 2)

that had been assumed at the design stage. In other words, the output of the design has to be verified with the maintenance work because progress of deterioration would not follow the design assumptions. This is related to 'Scenario update' in the figure. The LCM scenario should be updated with reflecting the actual situation of the structure and changes in conditions.

Function and performance requirements

Functions of a shore protection facility are principally to conserve coastal areas and protect the hinterland from high waves, storm surges, tsunami, etc. In addition, some facilities are required to preserve marine environments and ecosystem. To ensure those functions, it is important to ensure performance such as stability (safety), serviceability, environmental preservation, durability and restorability. The crest level of the facility is one of the performance indices that should be higher than the requirement level for ensuring stability or safety. Accordingly, the crest level of the facility is the easiest point to be designed and inspected. The hierarchy of the function, the performance requirements and the performance indices is shown in Figure 3.

Sustainability indicators

To evaluate the LCM scenario, costs or life-cycle costs have been generally used as an indicator. The scenario having the lowest cost/life-cycle cost should be selected as the most appropriate scenario. However, from the viewpoint of sustainability, not only indicators of economic aspects but also indicators of social and environmental aspects should be considered in future. These indicators, as examples (ISO, 2015), should be determined in consideration of the following items:

- use of energy and material resources,
- emissions to air, water and soil,
- production and management of wastes,
- species and ecosystem,
- landscape,
- community and territorial system, etc.

Efficient method of assessment

Characteristics of cracks in un-reinforced concrete members

While various kinds of structures have been constructed in coastal areas, the most popular structural type is reinforced concrete. Accordingly, the methodologies on structural performance evaluation on a deteriorated reinforced concrete structure have been intensively developed to date. In addition to reinforced concrete, un-reinforced concrete has been widely applied for coastal structures as parapet of seawalls, costal embankments, etc. However, it is not exaggerated to say that structural performance evaluation on un-reinforced concrete has been almost forgotten. Accurate, reliable evaluation on performance degradation of an un-reinforced concrete structure should be established soon because it has played important roles for shore protection.

Cracks in an un-reinforced concrete structure are the main symptom of deterioration; thus, crack widths, lengths, locations and numbers of cracks are generally concentrated as one of the inspection items for regular maintenance of the structure. Moreover, the presence of a penetrating crack through the cross section of a structure is one of the most concerns. A shore protection facility is rather long and large, and therefore unfortunately requires a lot of inspection work at the maintenance stage even only focusing on cracks.

Crack numbers, positions, widths and depths in many un-reinforced concrete parapets have been investigated (Furuya, Yokota, Komatsu, Hashimoto, & Kato, 2014). Figure 4 shows the probability of vertical crack occurrence in parapets, which was analysed with the investigated data. The number of vertical cracks was distributed mostly between 0 (no cracks) and 2 regardless of the location and the dimensions of parapets. The data distribution seemed to follow the Poisson distribution as shown in Equation (1). Accordingly, the conformance degree of probability distributions of the numbers of vertical cracks to the Poisson distribution was estimated using the Kolmogorov–Smirnov statistic (Eadie, Drijard, James, Roos, & Sadoulet, 1971):

$$f_p(x) = \exp(-\mu)\frac{\mu^x}{x!} \tag{1}$$

where μ is the mean number of cracks occurring a specified region (the Poisson parameter) and $x = 0, 1, 2, \dots$.

It was found that the Poisson distribution is a suitable distribution to describe the vertical crack numbers probability

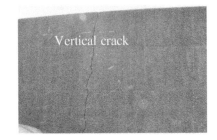

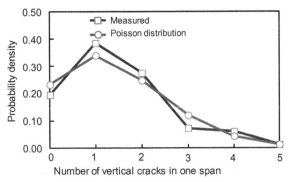

Figure 4. Probability of vertical crack occurrence in parapets.

distributions. The mean value of vertical crack numbers in one span (between longitudinal construction joints) of the parapet was calculated 1.46. The probabilities of occurring at least one crack and two cracks are approximately .78 and .37, respectively, as shown in Figure 4.

Herein, the maximum width of vertical cracks throughout the facility is focused because it seems that structural performance degradation of an un-reinforced concrete structure depends on the width of crack. The frequency distribution and the probability density functions of crack widths are shown in Figure 5, which were measured in two parapets of embankments: Facility C-1 and Facility S-1. The substructure type, overall length and span length of C-1 are a sheet pile type, 845 and 9.5 m; and those of S-1 are a gravity type, 491 and 7.5 m, respectively. The crack widths measured in coastal un-reinforced concrete structures were roughly distributed in the region of 2 mm or less as shown in Figure 5.

The distribution of crack width measured is discussed if it could be expressed with extreme statistics including the Gumbel, the Frechet and the Weibull distributions. The relationship between the cumulative distribution function of each extreme distribution and the return period was quantified as shown in Equations (2–5), and eventually, the maximum crack widths were calculated:

$$x_{g,\,\max} = \alpha_g - \beta_g \cdot \ln\left[-\ln\left(\frac{T-1}{T}\right)\right] \tag{2}$$

$$x_{f,\,\max} = \exp\left[\frac{1}{\alpha_f}\left[\ln\beta_f - \ln\left\{-\ln\left(\frac{T-1}{T}\right)\right\}\right]\right] \tag{3}$$

$$x_{w,\,\max} = \exp\left[\frac{1}{\alpha_w}\ln\left\{-\ln\left(\frac{1}{T}\right)\right\} + \ln\beta_w\right] \tag{4}$$

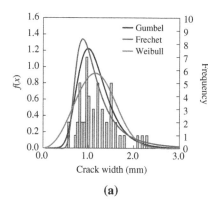

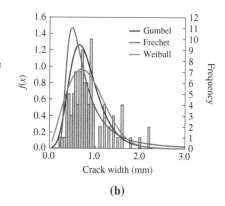

(a) **(b)**

Figure 5. Frequency and probability density function of crack widths occurring in the spans of parapets. (a) Facility C-1. (b) Facility S-1.

Table 1. Parameters for extreme statics analyses.

	Gumbel		Frechet		Weibull	
	a_g	β_g	a_f	β_f	a_w	β_w
S-1	.648	.291	2.113	.321	2.148	.935
C-1	.987	.300	3.290	.823	3.065	1.303

$$T = \frac{m \cdot L}{l} \quad (5)$$

where x_{max} is the estimated maximum crack width, α and β are distribution parameters, subscripts of these symbols g, f and w denote for the Gumbel, the Frechet and the Weibull distributions, respectively, T is the return period, L is the total length of a facility, and l and m are the length and numbers of vertical cracks in the inspected area, respectively. The distribution parameters that showed most conformable to the distributed data (as shown in Figure 5) are summarised in Table 1.

The conformance degree of probability distribution on the extreme value analysis data to the Gumbel distribution, the Weibull distribution and the Frechet distribution was evaluated with the Kolmogorov–Smirnov test as mentioned earlier. In conclusion (Furuya et al., 2014), it was found that the Gumbel distribution is the most suitable distribution to describe the extreme value analysis data.

Estimation of maximum crack width with limited numbers of data

Estimation of the maximum crack width of long coastal structures with partial measurements may contribute to energy-saving maintenance work. Certainly, in case that damage or deterioration may concentrate a certain part of facility due to longitudinal shapes in relation to the planar profile of coast lines, such the part may be selected for an inspection point. However, it is necessary to choose limited numbers of inspection points in case that almost uniform deterioration may occur.

The maximum crack width was estimated by the Gumbel distribution in Equation (2). Figure 6 illustrates the relationship between the sampling rate (the span numbers of inspected area divided by the total span numbers) and the maximum crack width,

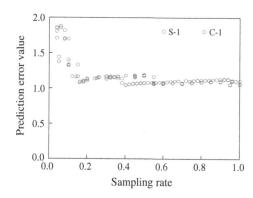

Figure 6. Relationship between sampling rate and prediction error value estimated by the Gumbel distribution.

where the maximum crack width obtained by the inspection of each facility is assumed as a true value. The estimated maximum crack widths approached the true value with an increase in the sampling rate. When the sampling rates become about .2, the prediction error values converge with approximately 1.05–1.2. When the maximum crack width is estimated using extreme statistics, the extreme value analysis data can be fit to Gumbel distribution. It can be concluded that inspection on at least about 20% of the entire facility (the sampling rate is about .2) would give the most probable maximum crack width. In other words, the true maximum crack width would be obtained by the measured maximum crack width (the measured maximum crack width multiplied by 1.05–1.2) with the minimum error.

Effect of cracks on structural safety

When a wide, deep crack occurs, safety that is one of the important structural performance requirements of shore protection facility may be more or less degraded. Based on the investigations by Furuya, Yokota, and Hashimoto (2012), the almost linear relationship has been found between crack width and crack depth with the correlation coefficient (R) of .679. If the crack depth can be estimated with easily measured physical parameters like crack width, it is very useful to evaluate structural performance. However, there is a concern how much the error of estimation might be.

The crack depth is increased as increase in not only crack width but also in the sizes of parapet. The influence of parapet

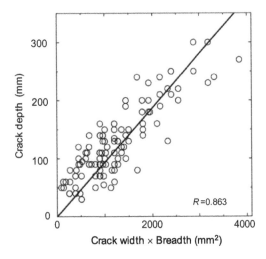

Figure 7. Relationship between parapet breadth × crack width and crack depth.

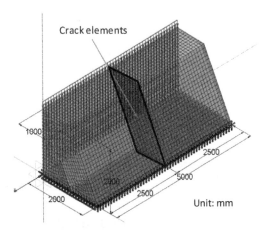

Figure 8. Structural analysis model of a parapet with a vertical crack in its midspan.

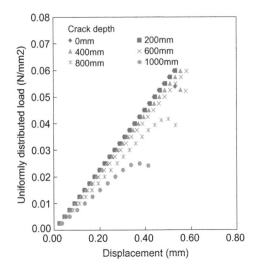

Figure 9. Load-displacement curves of un-reinforced concrete parapet with various cracks.

The estimated peak load is defined as the load-carrying capacity as structural performance. Since the crack depth can be calculated with the crack width using the relationship as shown in Figure 7, the relationship between the crack width and structural performance degradation is obtained. Being normalised with the structural performance of the parapet without initial cracks, structural performance degradation ratio is obtained. When horizontal forces such as wave forces were taken into account, the normalised structural performance of the parapet with an initial crack width of 7 mm was .96. Under vertical forces such as settlement and uplift pressure application, the normalised structural performance decreases as the crack width is more than 4 mm. It is considered that the influence of crack width on the structural performance of un-reinforced concrete structures under vertical forces is more significant than in the case of horizontal forces. Therefore, when the condition grade of un-reinforced concrete structure is evaluated, it is necessary to consider the width and the number of cracks simultaneously.

It is concluded, from the viewpoint energy-saving inspection and performance evaluation with easily obtained inspection data, investigating crack widths in just about 20% of all the spans of an entire facility would give the maximum crack width with reasonable accuracy as described earlier. Since almost the linear relationship exists between crack width and crack depth in consideration of structure dimensions, performance evaluation of un-reinforced concrete structure can be made based on the width and numbers of existing cracks. Such the trial enables us to execute more efficient maintenance work for shore protection facilities.

sizes on the crack depth estimation is examined. The relationship between the product of the breadth of parapet and the crack width and crack depth is shown in Figure 7. The correlation coefficient was .863. Accordingly, the crack depth can be reasonably estimated with crack width measured during inspection.

The structure analysis model of the parapet with one vertical initial crack, as shown in Figure 8, was prepared for nonlinear finite element analysis to calculate the load-carrying capacity relating to the safety of structure. The analysis models depend on the depth, width and the numbers of cracks. Vertical forces such as differential settlement and downward waves were applied as examples of external actions vertically applied. The smeared crack model was applied for the analysis.

The load-displacement curves are shown in Figure 9. The displacement in the figure is that parallel to the direction of force at the top centre of the model. The load-displacement curves of the parapet with initial cracks were almost linear before reaching the peak load regardless of the crack depth (width). After the peak load, failure progressed rapidly as smeared cracks occurred in the concrete. The peak load decrease began when the crack depth reached to 600 mm. The peak load had 50% decreased in the 1000-mm-deep case. Moreover, the stiffness, which is shown in the initial slope of load-displacement curve, had 7% decreased.

Structural details for ensuring stability during earthquakes

Failure mechanism due to tsunami

Lots of shore protection facilities in the Tohoku Region, Japan, were partially or totally failed due to the earthquake-related disasters caused by March 2011 Tohoku Region Pacific Coast Earthquake. The earthquake motion did not cause serious

Figure 10. Typical failure of coastal embankments. (a) Washout of covering concrete blocks. (b) Washout of concrete parapets (MLIT, 2011a).

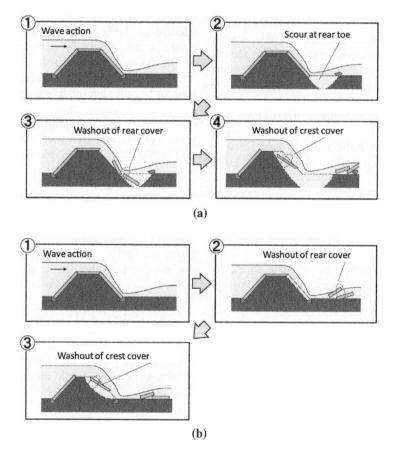

Figure 11. Typical failure progress of coastal embankment due to upcoming tsunami. (a) Triggered by scouring at the toe of rear slope. (b) Triggered by breakage of rear slope cover (MLIT, 2011b).

damages to the facility, but subsequent huge tsunami inflicted various catastrophic damages on them.

The failure modes of structures induced by subsequent tsunami can be categorised into the following five patters: (i) washout of entire structure, (ii) washout of covering concrete blocks, (iii) failure and/or washout of parapet, (iv) scouring and (v) movement due to seismic motions. Figure 10(a) and (b) illustrates patterns (ii) and (iii), respectively (Ministry Land, Infrastructure, Transport and Tourism [MLIT], 2011a). MLIT (2011b) reported that the percentage of entirely collapsed coastal embankments would be reduced from 29 to

6% if rear toe had been well protected. The embankments in which toe of rear slope had been carefully strengthened was hard to reach entire washout. Moreover, the deeper the overflow water, the more failure the percentage. When the overflow water depth was more than 4 m, most of the structures entirely collapsed.

Figure 11(a) depicts the typical failure progress of coastal embankments triggered by scouring at the toe of rear slope: (1) water stream of tsunami overtops the crest of the structure, (2) water travels down the rear slope to reach the ground at the toe of the slope with high speed, which causes scouring, (3) failure of rear slope cover and subsequent washout of backfill occur and then (4) washout of crest cover and further washout of backfill occur. The figure shows the upcoming wave case but the similar progress is expected in the backrush case.

Figure 11(b) displays the typical failure progress of coastal embankments triggered by breakage of rear slope covering blocks: (1) water stream of tsunami overtops the crest of the structure, (2) high-speed water stream washes out rear slope cover blocks and then (3) washout of crest cover and washout of backfill through the cover block joints. The other failure progress includes washout of concrete parapet due to tsunami (wave) forces.

Countermeasures against tsunami causing progressive failures

Lessons from the 2011 Tohoku Region Pacific Coast Earthquake suggest that already existing shore protection facilities may be required upgrades to meet such catastrophic tsunami actions. Almost all existing structures have been designed in consideration of likely tsunami having about 100-year return period. The structure should have enough ability to avoid progressive failure even when higher tsunami, say its return period of longer than several hundred years or more, than the design one attacks it. Structural robustness is such the ability to be realised by upgrading the existing structure. If the structure is designed to ensure structural robustness, it can keep its original cross section, which enables to reduce disaster in the hinterland. Also more time may be given for evacuation. Furthermore, remaining of even a part of structure enables to realise rapid restoration work and reduce the risk of subsequent failure leading to save restoration cost.

The following countermeasures are effective to ensure structural robustness for catastrophic tsunami: (1) to strengthen the toe of rear slope to avoid scouring, (2) to make front and rear slopes more gentle to reduce the speed of water down flow, (3) to thicken and strengthen covering concrete blocks, (4) to connect adjacent concrete blocks, (5) to remove concrete parapet and to raise earth fill to the crest level required, (6) to insert anchor bar between the parapet and the embankment, and so on.

Sea level rise

A shore protection facility has a long lifetime and must be expected to meet demands during its lifetime that cannot be foreseen. One of the concerns for shore protection facilities is likely sea level rise due to global warming (Intergovernmental Panel on Climate Change [IPCC], 2013). In addition, the IPCC indicates that the typhoon or storm intensity likely becomes high due to global warming as well. Accordingly, design tides and waves have to be determined in consideration of the above effects at the target year of performance verification. In other words, the levels of performance requirements tend to be high year by year. Kawai, Moriya, Mizutani, and Yokota (2012) proposed criteria for the determination of design values.

Another problem is that the effect of global warming should spread to all the existing facilities. It means that taking action on countermeasures to all the facilities would be impossible because of budget restrictions. Under such difficulties, proper levels should be determined from the viewpoint of the balance between soft and hard countermeasures. The hard countermeasure includes to heighten the crest level of facilities and/or to enlarge breadth of the facilities, while the soft countermeasure is to consider efficient ways of evacuation and/or warning to people living there. The process of determination of countermeasures for sea level rise and/or high intensity of typhoons/storms will be discussed later.

Proposed life-cycle management for shore protection facilities

The asset management is defined in PAS 55 (British Standards Institution [BSI], 2004) as 'systematic and coordinated activities and practices through which an organization optimally and sustainably manages its asset systems, their associated performance, risks and expenditure over their life cycles for the purpose of achieving its organizational strategic plan.' General maintenance work for a shore protection facility has focused on each structural component or even one facility. However, even if only limited parts of the facility fails, disaster may occur in wide areas as schematically shown in Figure 12. Accordingly,

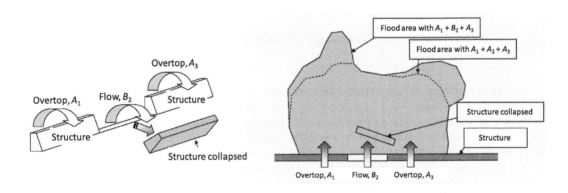

Figure 12. Partial collapse of a structure may bring flood in wider areas.

Table 2. Grading criteria based on the appearance of a shore protection facility (MAFF et al., 2014: originally in Japanese; translated by the author).

Grade	a	b	c	d
Crest level	Lower than the requirement	–	–	Higher than the requirement
Crack	Penetrating cracks (5 mm wide or more)	Multi-directional, non-penetrating cracks (several millimetres wide)	One-directional, non-penetrating cracks (several millimetres wide)	Minute or class (1 mm wide or less) No visible cracks
Damage/peeling	Widely, deeply spreading damage	Deeply spreading damage	Surface damage	Small, local damages No damages
Joint movement	Overturn defect	Widely opening joint Percolation of water	Gap No water percolation	Tiny unevenness No opening, no unevenness

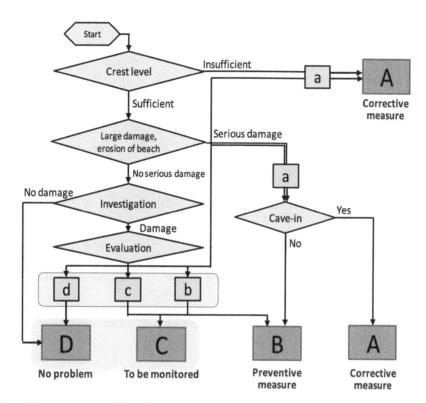

Figure 13. Decision-making flow with the grade evaluated through visual inspection (MAFF et al., 2014: originally in Japanese; translated by the author).

the life-cycle management for a shore protection facility should be extended to a group of facilities based on the concept of the asset management.

Shore protection facilities are built in harsh environments for materials and difficult to realise easy inspection. The regular inspection has been generally done for only limited parts of the facility and even some damages occurring in such limited area may not lead to entire collapse of the facility. The damage in the facility is generally initiated by excessive loads due to storms and earthquakes and propagated during the event of subsequent wave actions, etc. According to the latest inspection manual for shore protection facilities (MAFF, Fisheries Agency, & MLIT, 2014), structural performance evaluation is carried out mainly with visual inspection according to Table 2.

The crest level, crack formation, damage/peeling of concrete and joint movement between the adjacent spans are focused for the deterioration grade. If these grades can be associated with structural performance or capacity, it may be easy to evaluate and predict the current and future function and performance degradation. Though this association is difficult at the moment because of lack of data, in the meantime, the decision-making flow is given, as presented in Figure 13, for counter action against

likely performance degradation depending on the visually judged grade. By inputting the results of crack evaluation on un-reinforced concrete member described in this paper, Table 2 can be updated more objectively. It is because crack widths specified in the table such as 5 mm wide and 1 mm wide do not have objective bases. When the analytical results, for example as shown in Figure 9, can be taken into account, the grading criteria can be updated with more accuracy and reliability. This makes to upgrade the evaluation process of a damaged structure, that is, to provide with definite bases during 'Evaluation' in Figure 13.

The life-cycle management should be done with understanding the characteristics of shore protection facilities on the basis of likely performance degradation caused by deterioration, upgrade in seismic performance due to higher tsunami and seismic motion, and sea level rise due to global warming. To meet such demands, management should be effectively realised for seeking to achieve objectives shown in Figure 14. Against deterioration of a facility, the assessment of it can be done with some criteria as well as structural analyses. Future progress of performance degradation should be predicted as well. Based on these results, remedial action should be planned to ensure functions and

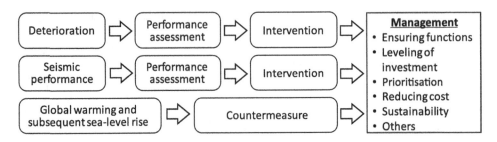

Figure 14. Life-cycle management for shore protection facilities from the viewpoints of deterioration, seismic performance and environmental changes.

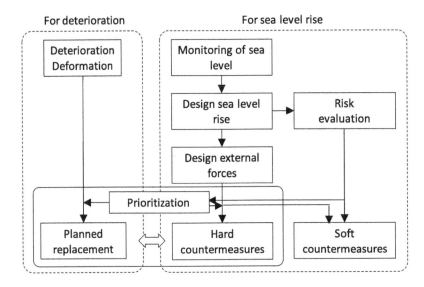

Figure 15. Procedure for determination of countermeasures for deterioration and for sea level rise (CDIT, 2011; originally in Japanese; translated by the author).

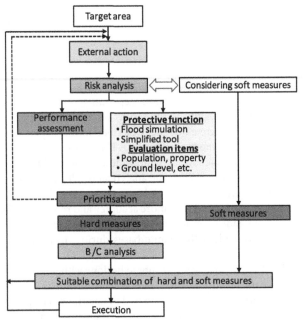

Figure 16. Procedure for making adaptation strategies for environmental changes (CDIT, 2011; originally in Japanese; translated by the author).

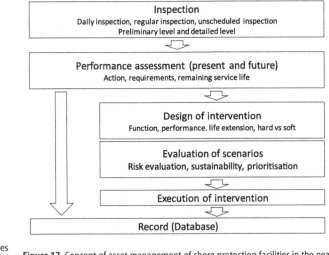

Figure 17. Concept of asset management of shore protection facilities in the near future.

performance over the requirements. On the other hand, against the sea level rise, the trend of the sea level is determined with doing monitoring, etc. and remedial actions should be planned.

Accordingly, the coordination of these remedial actions should be discussed. Figure 15 depicts the procedure for determinations of remedial actions for deterioration and for sea level rise. The best timings of each countermeasure are generally not the same; thus, prioritisation between the two is well considered. Also, hard countermeasures and soft countermeasures should be well balanced as mentioned earlier. In particular, Figure 16 presents the procedure for adaptation strategy to meet future environmental changes such as sea level rise and increase in external actions.

In conclusion, Figure 17 illustrates the proposed concept of asset management of shore protection facilities. Structural performance assessment and risk evaluation are particularly important both at present and in future. Using proper sets of sustainability indicators, the most effective scenario should be selected and remedial action should be carried out.

Conclusions

Compared with other civil infrastructure such as bridges, methodologies and specific techniques for effective maintenance for shore protection facility are delayed in developing. We have a limited amount of time to find a reasonable way to upgrade the facility to meet likely sea level rise and catastrophic tsunami. More research and data collection of inspection are strongly expected. Once field data are accumulated with enough quantity and quality, the life-cycle management procedure can be improved more effectively and reasonably.

Acknowledgments

Dr Katsufumi Hashimoto, Assistant Professor of Hokkaido University, Dr Ema Kato, Research Group Head of Port and Airport Research Institute, and Messrs. Koichi Furuya and Shuhei Komatsu, Ex-student of Hokkaido University, are fully acknowledged for their great contributions on this research. The author thanks members of Working Group on Harbor and Coastal Structures in Special Committee on Great East Japan Earthquake, Japan Concrete Institute and JSCE Committee on Asset Management for Shore Protection Facility for beneficial discussion.

Disclosure statement

No potential conflict of interest was reported by the author.

Funding

The part of this work was supported by JSPS KAKENHI [grant number 24360174].

References

BSI. (2004). *PAS 55: Asset management–Part 1: Specification for the optimised management of physical infrastructure assets*. London: Author.

CDIT. (2011). *Manual for making adaptation strategies for sea-level rise considering the deterioration of coastal structures*. Tokyo: Author.

Eadie, W. T., Drijard, D., James, F. E., Roos, M., & Sadoulet, B. (1971). *Statistical methods in experimental physics* (pp. 269–271). Amsterdam: North-Holland.

FIB. (2010). *fib Model code for concrete structures 2010*. Berlin: Wilhelm Ernst & Zohn.

Furuya, K., Yokota, H., & Hashimoto, K. (2012). *Structural performance assessment of un-reinforced concrete coastal structures*. Proceedings of the 1st International Conference on Performance-based and Life-cycle Structural Engineering, PLSE 2012, Hong Kong, 5–7 December 2012.

Furuya, K., Yokota, H., Komatsu, S., Hashimoto, K., & Kato, E. (2014). *Prediction of extreme crack width of concrete coastal facilities based on extreme statistics*. Proceedings of 4th International Symposium on Life-Cycle Civil Engineering, IALCCE 2014, Tokyo, 16–19 November 2014.

IPCC. (2013). *Summary for policymakers of the IPCC Report "Climate Change 2013 – The physical science basis"*. Cambridge: Cambridge University Press.

ISO. (2015). *ISO/TS 21929-2: 2015 Sustainability in building construction – Sustainability indicators – Part 2: Framework for the development of indicators for civil engineering works*. Geneva: Author.

JSCE. (2000). *Design manual for coastal facilities 2000*. Tokyo: Maruzen.

Kawai, H., Moriya, Y., Mizutani, N., & Yokota, H. (2012). Discussion on future marine force criteria for life extension plan of coastal defence facility. *Proceedings of Japan Society of Civil Engineers, B3, 68*: 995–1000.

MLIT. (2011a). *Committee material on countermeasures against tsunami in coast*. Tokyo: Author.

MLIT. (2011b). *Basic view on restoration of damaged coastal embankments etc. caused by 2011 Tohoku Region Pacific Coast Earthquake and tsunami*. Tokyo: Author.

MAFF, Fisheries Agency, & MLIT. (2014). *Maintenance manual for coastal facilities*. Tokyo: Ministry of Land, Infrastructure, Transport and Tourism.

Yokota, H., Iwanami, M., & Kato, E. (2012). *Strategic maintenance of infrastructure in coastal areas*. Proceedings of International Symposium on Recent and Future Technologies in Coastal Development, Yokohama, 14–16 December 2012.

Yokota, H., & Hashimoto, K. (2013). Life-cycle management of concrete structures. *International Journal of Structural Engineering, 4*, 138–145.

Framework for maintenance management of shield tunnel using structural performance and life cycle cost as indicators

Jianhong Wang, Atsushi Koizumi and Hiroshi Tanaka

ABSTRACT

In the past decades, many shield tunnels have been constructed for use in important lifeline infrastructure such as roads, railways, water supply systems and sewers. A good performance of the tunnel requires appropriate maintenance associated with a substantial cost; therefore, a rational maintenance approach is required. Preventive maintenance has been previously applied in an efficient and effective manner. However, the evaluation of structural performance, as well as the choosing an appropriate maintenance management strategy, is still a challenge. This paper proposes a maintenance framework for the shield tunnel. Structural performance and life cycle cost are evaluated as the major indicators accounting for the major deterioration (e.g. steel corrosion, cracking, etc.). The structural performance-based strength cost for an expected service time is used as the criterion to select a rational maintenance plan. Furthermore, structural reliability analysis is performed to validate the proposed framework and investigate the influence of deterioration model. The proposed framework was successfully applied to the maintenance of a trunk sewerage pipe for choosing a rational maintenance plan.

1. Introduction

In the past decades, many tunnels have been built for servicing urban functions such as transportation, water supply and sewage. However, some of these tunnels suffer from degradation in serviceability or strength performance. In many service tunnels, structural problems have been found to be due to deterioration, and repair and/or rehabilitation work have been implemented (Asakura & Kojima, 2003; Environmental Planning Bureau of Yokohama [EPBY], 2013; Mutou & Kawabata, 2011). Therefore, rational and effective maintenance is required for such utilities, especially for those operating beyond their designed life (Frangopol & Liu, 2007; Richards, 1998).

Generally, deterioration depends on the structural type, material and operational conditions, and the corresponding maintenance strategy varies accordingly (Davies, Clarke, Whiter, & Cunningham, 2001; Federal Highway Administration [FHWA], 2004). In addition, the purpose of a tunnel usually decides what aspect of maintenance is the most important. For example, in tunnels used for transporting people, concrete cracking is always the vital issue, whereas in tunnels used for water supply and sewerage utilities, strength degradation is the most important aspect. Moreover, for a shield tunnel, the deterioration condition of the primary lining is greatly affected by whether to install the secondary lining; therefore, the strength degradation of a shield tunnel with a secondary lining is not clearly understood although leaks and rust are often observed *in situ*.

Infrastructure management discussion should account for deterioration model and maintenance philosophies (Das, 1999; Frangopol, Kallen, & Van Noortwijk, 2004; Yuan, Jiang, & Liu, 2013). Maintenance philosophies are classified under three categories based on maintenance timing: corrective, preventive and predictive (Figure 1). Corrective maintenance involves activity performed to repair a fault or restore a failed facility to operational or safe conditions; i.e. the maintenance work is carried out after failure. Both preventive and predictive maintenance philosophies are intended to ensure the safety of facilities, but the former prevents incipient failure by employing periodic inspection and repair, whereas the latter is based on condition-based prediction using probability theory. More details on these maintenance philosophies can be obtained from the literature published by the Department of Defense (DoD, 2008).

In Japan, preventive maintenance has been widely implemented in the management of civil infrastructure such as bridges and tunnels (Asakura & Kojima, 2003; Kimura, 2013; Mashimo, 2002; Ministry of Land, Infrastructure, Transport & Tourism [MLIT], 2015a). However, so far, the evaluation and prediction of structural performance over the service life of infrastructure, as well as the selection of a rational maintenance plan, are still a challenge. These aspects are extremely important for the maintenance of trunk sewer tunnels. Trunk sewer tunnels, which act as basic municipal lifelines to collect and transport sanitary waste

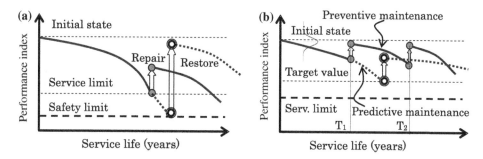

Figure 1. Schematic images of maintenance philosophies. (a) Corrective maintenance and (b) preventive and predictive maintenance.

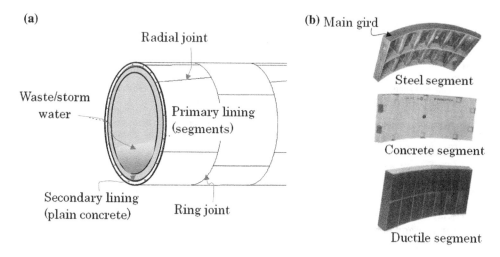

Figure 2. Trunk sewer built by shield tunnelling method. (a) Structural profile of tunnel and (b) segment types for primary lining.

and rainwater, always require high drainage performance and structural integrity.

However, hydrogen sulphide commonly causes severe corrosion across tunnel cross sections and reduces structural performance. In practice, to provide a smooth internal surface and protect the primary lining, a shield tunnel-based trunk sewer (Figure 2) is usually fitted with a plain concrete lining after the installation of the segmental lining. Consequently, the deterioration condition of the primary lining is unknown because it is covered internally by the secondary lining and externally by ground, which increases the difficulty of performing maintenance works. On the other hand, the maintenance of ageing sewerage pipes is becoming an urgent issue (Chughtai & Zayed, 2008; Fenner, 2000; Ministry of Land, Infrastructure, Transport & Tourism [MLIT], 2015b). In particular, local government agencies are gravely concerned about the maintenance of shield tunnel-based trunk sewers because many of them will pass the design life of 50 years (Bureau of Sewerage of Tokyo [BST], 2012). Therefore, condition-based predictive maintenance must be applied to shield tunnels with secondary linings.

The present paper focuses on sewerage shield tunnels and discusses their maintenance based on the predictive maintenance philosophy. A maintenance framework using structural performance and life cycle cost (LCC) as indicators is proposed and successfully applied to a 50-year-old shield tunnel-based trunk sewer. Furthermore, to investigate the uncertainty in the

deterioration model, reliability analysis is performed by the probabilistic description of the related random variables. The statistical method is also expected to improve the assumed deterioration model by the further inspection data.

2. Deteriorations

2.1. Deterioration condition

The deterioration condition of a shield tunnel can be defined based on the degradation intensity and amount of deterioration factors, which are listed in Table 1 (Davies et al., 2001; Yuan et al., 2013). The deterioration factors for shield tunnels are classified into environmental and structural factors and vary with different members. In general, the primary lining of a shield tunnel with a secondary lining is less vulnerable to degradation and damage. Four deterioration stages are defined based on the inspected condition of the secondary lining, and the corresponding deterioration of the primary lining is assumed based on related empirical and/or technical knowledge. The deterioration conditions are defined in Table 2. Herein, leakage, crack and rust stain as the main factors affecting the secondary lining, are used to define the deterioration condition. Because the deterioration condition is basically estimated from the field observation and survey data, greater accuracy may be expected by collecting more field data.

Table 1. Deterioration factors of shield tunnel.

		Tunnels without secondary linings	Tunnels with secondary linings
Environmental factor		Leakage, wetness, concrete seepage, chemical corrosion, concrete corrosion, steel corrosion, ground voids, groundwater	
Structural factor	Lining	Pattern cracking, segment deflection, ovality, settlement, joint stagger, lining uplift, overload	Secondary lining: Concrete corrosion, spalling, cracking, seepage, wetness, leakage, etc.
	Concrete segments	Spalling, cracking, reinforcement corrosion, delamination, crashing, cavitation	
	Steel segments	Steel corrosion, stains, deformation	Primary lining (structural lining): Leakage, steel/reinforcement corrosion
	Joints	Opening, bolt corrosion, missing sealant, wrapping, loose bolts	

Table 2. Definition of deterioration conditions of shield tunnel with secondary lining.

Deterioration	Secondary lining (inspected)	Primary lining (assumed)
State 1	• Minor fine cracking due to construction • No leakage	Good condition
State 2	• Fine cracks and minor concrete spalls • No rust • No leakage	• Good condition, but likely to deteriorate • Corrosion on outside surface of steel segment with low corrosion loss
State 3	• Leakage in cracks or joints; water drips down slowly and intermittently • Light rust stain	• Corrosion on both sides of steel segment with medium corrosion loss • No corrosion of reinforcement
State 4	• Leakage; water drips continually • Heavy rust stain • Flexural cracking up to 1.5 mm, number of concrete spalls	• Corrosion on both sides of steel segment with high corrosion loss • Reinforcement corrosion based on results of field surveys

2.2. Deterioration models

A deterioration model plays an important role in the estimation of structural performance and prediction of residual life. Successful modelling of deterioration depends on the selection of the major deterioration factor and the clarification of the influence of each factor on structural performance. Once the major deterioration factor is determined, related data should be collected, in addition to aspects such as the design condition, used material, environmental condition and maintenance history, to obtain a reliable deterioration model. Particularly, maintenance countermeasures must be reflected in the deterioration model, considering that they can directly affect the degradation of tunnel performance and the service life. The deterioration factors associated with structural performance are as follows:

(1) Concrete cracking and corrosion

Cracking is a common phenomenon associated with concrete and may be induced by material shrinkage, external forces or other environmental effects. Although cracks have the little impact on the flexural strength of reinforced concrete, they can accelerate corrosion of the primary lining. In addition, the degradation of shear resistance capacity should be noted because cracks reduce the cross-sectional area of a concrete member (Wang, Shi, & Nakano, 2013). On the other hand, hydrogen sulphide can corrode concrete by reducing the cross-section of a member and its material strength and eventually lead to the degradation of structural integrity (e.g. Davies et al., 2001). However, concrete cracking and corrosion in secondary lining can be disregarded for a shield tunnel with a secondary lining because the post-installed secondary lining is considered as a nonstructural member in the structural design of a shield tunnel (Japan Society of Civil Engineers [JSCE], 2007).

(2) Steel or reinforcement corrosion

Steel corrosion is affected by many aspects including moisture, availability of oxygen, temperature, bacteria and nutrients (water quality), and alkalinity (high alkalinity reduces corrosion). Generally, corrosion losses occur more easily in highly wet conditions and have a nonlinear trend; therefore, the corrosion model should be defined in different phases (Ahmad, 2003). For a tunnel lining, the corrosion loss in the steel segment or reinforcement can be estimated by an empirical expression for corrosion (Alamilla, Espinosa-Medina, & Sosa, 2009; Ricker, 2007):

$$t_c = k_c T^n \qquad (1)$$

where t_c is the corrosion loss in different dimensions (thickness, height and diameter), k_c is the annual corrosion loss ratio, T is the service years and n is the exponential corrosion constant. The corrosion constants k_c and n should be determined based on the empirical corrosion data; however, they may be defined according to related design specifications in practice.

In the current study, the exponential corrosion constant is considered to be $n = 1.0$, which is the maximum value and the steel corrosion loss ratio (k_c) is set based on design specifications (Japan Society of Civil Engineers [JSCE], 2010); according to these specifications, $k_c = .02$ and .03 mm/year for land tunnels below and above the groundwater level, and $k_c = .1$ mm/year for exposed tunnels. If leakage is observed in the secondary lining, corrosion at both sides should be taken into account; else, only external corrosion through the soil should be considered. Furthermore, although the corrosion may be relatively non-uniform or localised in practice, it is reasonable to consider the most critical case for safe design, in which the corroded section is just subjected to the maximum internal forces. In addition, the corrosion model should consider the initial, speeding and steady stages of corrosion, individually.

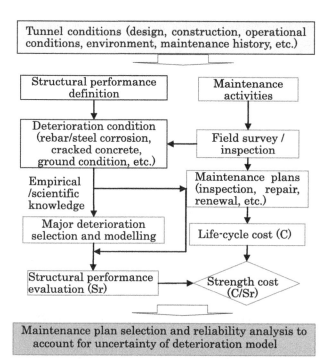

Note: maintenance plan selection criterion is the minimum strength cost for design service life.

Figure 3. Framework for maintenance management of shield tunnels.

For reinforced concrete segments, the steel bar's corrosion due to carbonation can be disregarded because the secondary lining has enough thickness (e.g. larger than 150 mm) to protect the steel bar. However, the corrosion of steel reinforcement subjected to chloride attacks must be considered based on field observation.

(3) Ground condition (loads and supports)

Groundwater level changes, ground loosening and subsidence should be considered in the case of longitudinal cracks observed in the secondary lining. Accurate conditions of support and load must be applied for the structural analysis. Any cavities in the surrounding ground should be carefully considered (Meguid & Dang, 2009). It should be noted that leakage might loosen the surrounding ground and even lead to a void around sewerage pipes.

Generally, leakage and seepage need to be considered as vital factors that substantially affect the aforementioned deterioration models and the tunnel performance. Leakage can speed up steel corrosion and loosen the ground by the water influx into or out of the tunnel through cracks and cold joints, whereas seepage only accelerates steel corrosion by an increase in the moisture passing through the concrete. Deterioration modelling should use several cycles of condition data to identify trends and predict the future by extrapolating the trends. A minimum of three cycles of inspection data is required for a deterioration model.

3. Maintenance framework

3.1. Maintenance framework tasks

Although tunnel degradation varies depending on various factors, structural performance is always required to ensure

that a tunnel is in safe condition. In addition, the maintenance of a shield tunnel with a secondary lining is limited by budget constraints. Therefore, a maintenance framework was considered using structural performance and LCC as indicators, as shown in Figure 3. Tasks within the framework include: (1) collecting information regarding tunnel conditions, (2) defining the required structural performance, (3) identifying the deterioration conditions by field inspection and selecting the major deterioration factor, (4) assuming the maintenance plans, (5) modelling the deterioration and evaluating the structural performance, (6) estimating the LCC and strength cost and (7) selecting a rational maintenance plan and validating the plan by reliability analysis.

The tunnel conditions involve all of the design, construction, operation and repair. In addition, the major deterioration should be selected, and its model should be assumed according to the condition definition (Table 2). Tunnel strength uses the safety ratio of load-carrying capacity to load effects as the indicator, with respect to a static load or an earthquake event. However, considering that an earthquake event is just a temporary load, the maintenance plan selection only considers the strength cost under static load. The structural performance under seismic load is used to discuss the additional countermeasure of seismic retrofitting. For example, if a tunnel is estimated to fail due to an earthquake event during the service life, additional seismic retrofitting must be implemented.

3.2. Strength evaluation and related structural analysis approach

Structural performance is evaluated using a design method recommended by the current design code for shield tunnels (JSCE, 2007; Japan Sewage Works Association [JSWA], 2009). The safety of a segment and joint is checked using the allowable stress design method for a static load and Level-1 earthquake event and using the ultimate limit-state design method for a Level-2 earthquake event. Level-1 is a lower design level earthquake for the basic function of the facility and Level-2 is the higher design level earthquake for life safety (Youssef, Jeffrey, Birger, & John, 2001). The application of the traditional design method recommended by the design code for strength evaluation makes it easy to compare results with the initial design results and to be accepted by engineers and clients.

The internal forces are estimated by the numerical analysis of the ground spring-supported beam-spring model as shown in Figure 4. The evaluation of load effects due to an earthquake event uses the seismic deformation method (Hashash, Hook, Schmidt, John, & Yao, 2001; Kawashima, 1999) considering the scientific evidence that a buried structure always moves along with the surrounding ground during an earthquake. The seismic deformation method includes (1) the seismic analysis of the ground performed using dynamic analysis software such as 'Shake' (Schnabel, Lysmer, & Seed, 1972), which gives the ground response results of the maximum relative displacement between the crown and invert of the tunnel, and (2) structural analysis performed by imposing a ground response on a beam-spring model to evaluate the internal forces.

The strength of a shield tunnel is defined by the safety factor $S_r = R_d/S_d$ and the tunnel safety is judged when $S_r \geq 1$. Here, R_d is the nominal resistance capacity, $R_d = R(f_k, \gamma_m, \gamma_b)$, evaluated

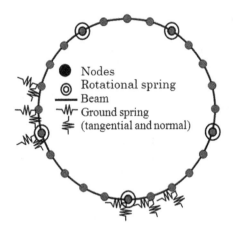

Figure 4. Structural model of numerical analysis.

from the mean value of the material strength f_k, material factor γ_m and member factor γ_b, and S_d is the load effects, $S_d = S(\gamma_a, \gamma_f, F_k)$, calculated from the design (mean) value of the load F_k, analysis factor γ_a and load factor γ_f. More details can be obtained from the design code of shield tunnels (JSCE, 2007). For a shield tunnel, the load effects of the bending moment (M), axial force (N) and shear force (S_f) are targeted with respect to segments and joints. Moreover, the deterioration is considered by reducing the dimensions of cross-section of a member.

3.3. Cost evaluation

Life cycle costs are normally defined as the sum of the costs incurred by an agency over a design service life, including construction and maintenance aspects such as periodic inspections, repairs and rehabilitation (Frangopol, Lin, & Estes, 1997). Disregarding other costs such as failure cost, the simplified life cycle cost (C_T) can be calculated as

$$C_T = C_{in}/(1+I)^T + C_m T + \sum_{i=1}^{N} C_r \tag{2}$$

where C_T is also the expected total cost over a service time (T), C_{in} is the initial construction cost, I is the discount rate, C_m is the periodic inspection cost (average cost per year), T is the design service period, C_r is the cost of repairs or renewal and N is the number of times repairs or rehabilitation are undertaken.

3.4. Reliability analysis

Because there is considerable uncertainty associated with corrosion model, material strength, load effects and LCC, reliability analysis should be carried out by considering the deviation of each variable (Akiyama, Frangopol, & Suzuki, 2012; Frangopol et al., 2004). The approach based on the second-moment description may be applied using the mean value and variance or the standard deviation of each variable. The available methods of reliability analysis include the first-order reliability method (FORM), second-order reliability method (SORM), point estimate method

and Monte Carlo simulation method (e.g. Melchers, 1987; Tee, Khan, & Chen, 2013). In the current study, the simpler FORM is used. For a limit function $Z(X_1, X_2, \dots X_n)$ expressed by a normal distribution, its mean value and standard deviation can be approximated as follows:

$$\bar{Z} \approx Z(\bar{X}_1, \bar{X}_2, \bar{X}_3, \cdots, \bar{X}_n) \tag{3}$$

$$\sigma^2(Z) \approx \sum_{i=1}^{n} \left[\sigma_{X_i}^2 \left(\frac{\partial Z}{\partial X_i} \right)^2_{X_i = \bar{X}_i} \right] \tag{4}$$

Based on standard structural reliability theory, the reliability index (β) and the probability of failure (P_f) can be given as:

$$\beta = \frac{\bar{Z}}{\sigma(Z)} \tag{5}$$

$$P_f = P[Z \leq 0] = 1 - \Phi(\beta) \tag{6}$$

where Φ is the cumulative distribution function for the standard normal distribution. In addition, the relative contribution of each variable X_i to the uncertainty of the failure function $Z(X)$ can be simply evaluated by:

$$\alpha_i^2 = \frac{\frac{\partial Z}{\partial X_i} \sigma(X_i)}{\sigma^2(Z)} \tag{7}$$

where α_i^2 is also known as the sensitivity coefficient; i.e. it expresses the importance of X_i to the structural reliability. In the current study, the limit state functions $Z = R-S$ are defined with $Z = 0$ as the failure surface; the functions are defined with respect to a segment subjected to a combination force of bending and axial forces and with respect to a radial joint subjected to shear forces.

The geometric coefficients are calculated by the following equations in terms of the moment of inertia and cross-sectional area of the segment and radial joint bolts (W_s, A_s and W_b, A_b):

$$W_s = (H_0 - t_c)^2 (B_0 - t_c)/6 \tag{8a}$$

$$A_s = (H_0 - t_c)(B_0 - t_c) \tag{8b}$$

$$A_b = \pi(\varphi_b/2 - t_c)^2 \tag{9a}$$

$$W_b = A_b h_b/2 \tag{9b}$$

where H_0 and B_0 are the height and width of the main gird of the segment, respectively, and φ_b and h_b are the bolt diameter and the inertia distance to the extreme end of the radial joint, respectively. Moreover, t_c is the total corrosion loss during the expected service time, and its mean value and standard deviation have been evaluated previously.

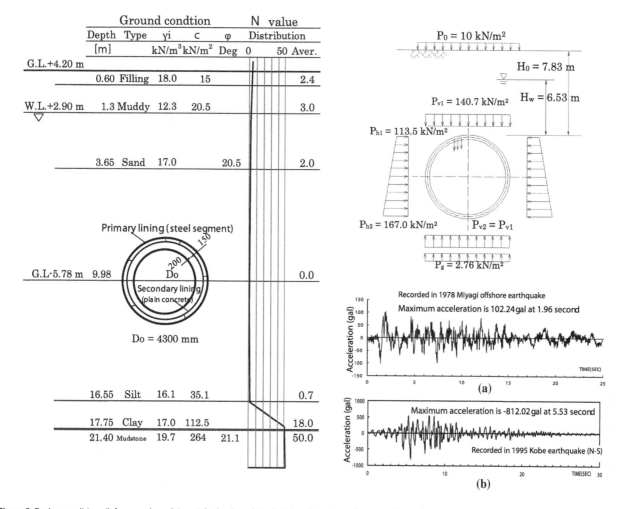

Figure 5. Design conditions (left: ground condition; right: load condition). (a) Level 1 earthquake event. (b) Level 2 earthquake event.

Table 3. Material properties (design values).

Load condition	Material strength	Steel plate (SS400)	Bolt (4.6)	Others
Static load and Level-1 seismic event	Tensile stress σ_a (N/mm²)	160	120	Young's modulus $E = 2.1 \times 10^5$ N/mm²
	Shear stress τ_a (N/mm²)	90	90	
Level-2 seismic event	Yield strength f_y (N/mm²)	235	240	
	Shear strength (N/mm²)	135	230	
	Tensile strength (N/mm²)	–	400	

4. Example of application to sewer tunnel

4.1. Target trunk sewer

For application of the proposed framework, an approximately 50-year-old shield tunnel-based trunk sewer with an outer diameter of 4300 mm was selected. The design conditions of the ground and structure were collected from the design document, as shown in Figure 5. The tunnel was assembled using 150-mm-thick, 900-mm-wide steel segments, which serviced as the primary lining, and using plain concrete, which acted as the secondary lining. More dimensional details can be obtained from standard steel segment for shield tunnelling (Japan Sewage Works Association [JSWA], 1973). The material properties of the steel segment and joint are listed in Table 3; the load conditions are given in Figure 5 in terms of static load and earthquake acceleration waves. According

to the related design specifications (e.g. JSCE, 2007; JSWA, 2009), the acceleration waves of the Level 1 and 2 seismic events use the one recorded in the 1978 Miyagi offshore earthquake and the Kobe wave (NS) recorded in the 1995 Hyogoken-Nanbu earthquake, respectively.

Field inspection was carried out previously, and the health situation was identified as shown in Figure 6. Leakage and heavy rust were observed at each cold joint in the secondary lining, and localised cracking and spalling were also found in the area adjacent to the cold joints. The material of the secondary lining (concrete) was verified to be in a good state by a field boring test. In addition, partial loss of invert concrete was found in several sites. Overall, the trunk sewer was adjudged to be in a good serviceable condition. However, factors that concern the client are the deterioration condition of the primary lining and its structural performance.

Figure 6. Current condition of secondary lining (left: leakage and rust in cold joint; right: boring test).

Table 4. Maintenance plan and corresponding corrosion ratios.

		Steel corrosion ratio k_c (mm/year)			
Cases	Time (years)	Case 0	Case 1	Case 2	Case 3
Plan		None	Cold joint repair in 30th year	Cold joint repair in 60th year	Sewer renewal in 60th year
Condition					
State 1	0–10	.00	0	0	0
State 2	11–30	.02	.02	.02	.02
State 3	31–60	.04	.02	.04	.04
State 4	61–200	.10	.04	.06	.06

Note: Cold joints in the secondary lining begin leaking in the 30th year.

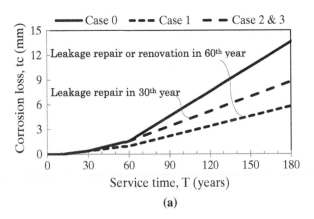

(a)

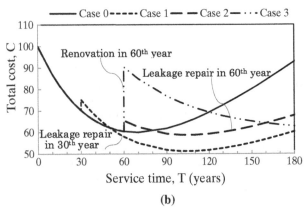

(b)

Figure 7. Modelling of corrosion and total cost over service life. (a) Corrosion model and (b) cost model.

4.2. Deterioration modelling and LCC

Based on the field inspection results, the deterioration model for steel corrosion was selected. Four maintenance plans (countermeasures) are assumed based on the recorded leakage time and preventive maintenance strategy: case 0 is the basic case in which only periodical inspection is carried out without any countermeasure; case 1 takes the countermeasure of repairing the cold joint immediately after leakage occurred in the 30th year; in case 2, the cold joint is repaired in the 60th year; and in case 3, the tunnel is renovated in the 60th year. The variables of corrosion are set based on the aforementioned corrosion model with respect to the maintenance plans as shown in Table 4, where corrosion was considered for both sides after leakage. The corrosion losses are evaluated by Equation (1), as shown in Figure 7(a).

Meanwhile, the relative costs for each item are assumed. The initial construction cost (C_{in}) is set as 100, and leakage repair and renewal costs (C_r) are 5 and 30, respectively. The annual inspection cost (C_m) is assumed to be an average value over the design period: .5 for the period before repair or renewal, .25 for that after leakage repairs and .0 for that after renovation. Repair or renewal is only carried out once ($N = 1$), and a discount rate (I) of 2% is adopted. The LCCs are calculated by Equation (2), as shown in Figure 7(b).

4.3. Analysis results and discussion

Numerical analysis was performed to evaluate the load effects of the trunk sewer under static and seismic loads. The structural performance of the tunnel for case 0 was evaluated in terms of

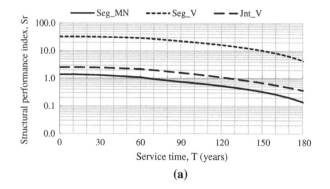

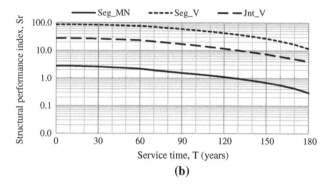

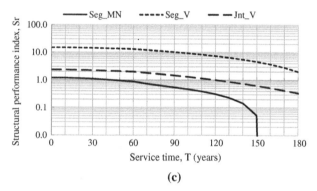

Figure 8. Structural performance of segment and radial joint (Case 0). (a) Under static loads, (b) under Level-1 earthquake load and (c) under Level-2 earthquake load.

the safety index of the segment and joint under the bending moment, axial force and shear forces. The results are shown in Figure 8, where the notations Seg_MN, Seg.V and Jnt_V denote the segment subjected to the combination of bending moment and axial force, segment under shear force and radial joint under shear force, respectively. The radial joint subjected to the combination of bending moment and axial force is disregarded because the joint bolt is under compression. From Figure 8, it is evident that the segment under bending moment and axial force is the most critical for both static and earthquake loads, and the safety index is the smallest when the tunnel is subjected to the Level-2 earthquake event.

Because failure in any member is considered as structural failure according to the design criterion, the structural performance of a tunnel considers the most critical one of all the indicators representing the members. Figure 9 shows the structural performance of the tunnel under the static load

and Level-1 and Level-2 earthquake events with respect to assumed maintenance plans. The corresponding strength costs were also estimated for each maintenance plan as shown in Figure 10.

Figure 9 indicates that the sewerage pipe fails after 45 years for cases 0, 2 and 3, and after approximately 60 years for case 1 if subjected to a Level-2 seismic event. Similarly, if the tunnel is subjected to a Level-1 seismic event, its failure occurs after approximately 68 years for cases 0, 2 and 3, and after approximately 100 years for case 1. However, when the tunnel is subjected to only static loads, the trunk sewer provides safe service for over 120 years even in the absence of any countermeasure (case 0), and for approximately 170 years for the other maintenance plans. Furthermore, it should be noted that failure ($S_r < 1$) does not mean a structural collapse of the complete lining but rather indicates stress on the critical member beyond the designed material strength. Additionally, since an earthquake is a potential event having an occurrence probability, the structural performance of the sewerage tunnel under earthquake loads may be only referred to as a possible risk. Consequently, the target trunk sewer was judged to be safe for 100 years based on the safety criteria of structural performance under static loads, but an earthquake retrofit countermeasure should be considered urgently to address the high failure risk due to an earthquake event.

Figure 10 illustrates that the strength cost is the lowest for case 1 wherein the cold joint in the secondary lining is repaired soon after the occurrence of leakage. The strength costs decrease in the initial 50 years and increase in the following service years for cases 0, 2 and 3. However, the strength cost for case 1 decreases until the 70th year after a small increase in the 30th year because of the leakage repair. Therefore, the sooner the leakage of the cold joint is stopped, the less the strength cost. Unfortunately, for case 3, the renovation of the existing pipe provides benefits in terms of the strength cost only after a service of 160 years. In addition, the strength cost increases sharply after the 60th year of operation if no countermeasure is taken.

4.4. Reliability analysis results

The uncertainty associated with the corrosion model was investigated by the probabilistic approach, which is a useful measure to improve the corrosion model using available empirical/inspection data. In addition, the variation in the load effects was considered. The reliability of the tunnel was evaluated by the FORM for the input data listed in Table 5. In this table, the critical cases of segments subjected to the maximum bending moment and axial force due to static loads are presented for different corrosion losses. The limit state function is computed for the main girder with height $H_0 = 150$ mm and width $B_0 = 32$ mm subjected to bending moment and axial force:

$$Z = f_k - \left[\frac{M}{W_s} + \frac{N}{A_s} \right] \qquad (10)$$

The evaluated failure probability over the service time and the relative contribution of corrosion loss to the variance of the failure function are shown in Figure 11(a) and (b), respectively. Figure 11(a) indicates that case 0 (without any countermeasure) has the highest probability of failure during the entire service

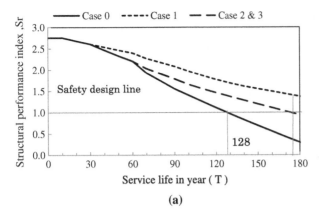

(a)

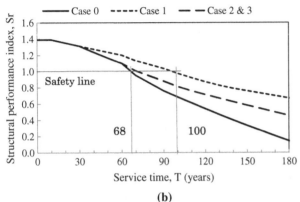

(b)

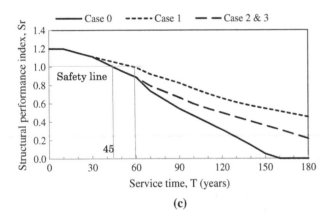

(c)

Figure 9. Structural performance of tunnel with respect to service life. (a) Under static loads, (b) under Level-1 earthquake load and (c) under Level-2 earthquake load.

time, compared to the other maintenance plans (cases 1, 2 and 3). This also means that an increase in the corrosion rate decreases the structural reliability, assuming that the corrosion model in

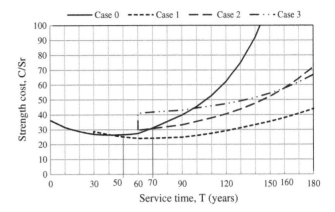

Figure 10. Strength cost with respect to service life.

each case varies only in terms of the corrosion rate. Moreover, the exponential constant affects the failure probability of the tunnel significantly; e.g. the failure probability increases from 9.0×10^{-6} to .45 as n increases from .5 to 1.0, for case 0 in the 120th year. The failure probability of the target sewer tunnel, which was initially designed as 1.0×10^{-6}, will increase by .05 after a service of 100 years if no maintenance is taken.

However, if a failure probability of .1 (10%) is considered as the threshold level for secure service (Babu & Srivastava, 2010), the service life for the worst case (case 0) is approximately 104 years, which is less than that evaluated by the design method (128 years). The reduction in the service life is reasonable because the reliability analysis accounted for the uncertainty associated with the material strength, the internal forces and the corrosion loss caused by the coefficients of variation. On the other hand, from Figure 11(b), it is observed evidently that the contribution of the corrosion loss to the probability of failure increases with the service life and that the contribution is significant after a service of 100 years for case 0. The contribution of corrosion loss (t_c) increases as k_c becomes larger, and a similar tendency is found for different exponential constants (n). However, the exponential constant (n) has a more significant contribution; e.g. when $n = .5$, the sensitivity of corrosion loss is less than .002 even in the 180th year, with respect to a failure probability of less than 2.0×10^{-5}.

As expected, the results for the contribution of corrosion loss and the failure probability show that the corrosion loss increases and the resistance capacity decreases with an increase in the corrosion rate, exponential constant and service time. In addition, because the design method, which uses one characteristic value to describe an uncertain variable, is usually considered as the Level-I method of structural reliability, and because the FORM is a higher-level reliability analysis method that uses two values of mean and variance to describe the uncertain variable, the critical

Table 5. Input data for reliability analysis of target tunnel.

Random variable			Distribution	
Description		Type	Mean	C.O.V*
f_y	Material strength	Normal	240 MPa	.1
M	Bending moment	Normal	5.47 kN·m	.2
N	Axial force	Normal	286.0 kN	.2
k_c	Corrosion loss rate	Normal	Case 0, case 1, and cases 2, 3 (Ref. Table 4)	.1
n	Exponential constant	Normal	1.0 and .5	.1

*C.O.V = Coefficient of variation.

52

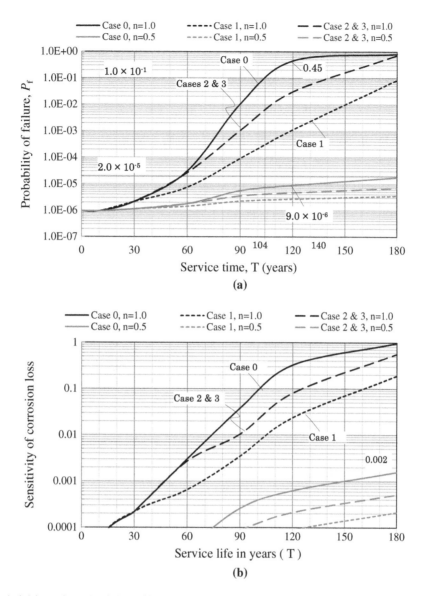

Figure 11. Results of structural reliability analysis. (a) Probability of failure with respect to service time and (b) sensitivity of corrosion loss with respect to service time.

result obtained by the FORM may validate the structural performance evaluated by the two methods. By applying the proposed framework, the structural performance can be estimated to select the best maintenance plan for the expected service life and a seismic retrofit strategy can be applied to avoid unexpected collapse or failure in the pipe network. Moreover, the deterioration model can be improved using probability theory if inspection data are available.

5. Conclusions

In this study, the maintenance of shield tunnels with secondary linings was investigated. A maintenance framework using the structural performance and LCC as indicators was proposed and successfully applied to select a maintenance plan for a 50-year-old trunk sewer tunnel. The maintenance framework was validated by reliability analysis, in terms of the evaluation method of the structural performance and the deterioration model.

From the results of this study, the following major conclusions can be drawn:

(1) As the most concerning parameters of maintenance, the effects of the structural performance and LCC on the deterioration of the tunnel are evaluated by the proposed framework. Therefore, the maintenance framework may be applied as an effective and efficient means for decision-making while selecting a maintenance plan for urban tunnels.

(2) A structural design method and a reliability analysis method (e.g. FORM) are valid and useful in structural performance evaluation. However, it should be noted that the latter is more likely to provide a critical result than the design method because the uncertainty associated with the deterioration variables is easily taken into account by the approximate method.

(3) An application example of the proposed framework indicated that early leakage repair is desirable in

tunnel maintenance because leakage significantly reduces performance by inducing and accelerating the corrosion loss of the primary lining.

In future, the deterioration models should be improved by developing more expertise in material science and condition data. Particularly, technologies for surveying the primary lining are expected to be developed soon. In addition, Level-III and Level-IV methods for reliability analysis should be considered to improve the proposed framework in future studies.

Acknowledgements

The authors are thankful to Professor S. Kimura (Kanazawa Institute of Technology) for his valuable comments on the LCC analysis as well as to Bureau of Sewerage of Tokyo and Environmental Planning Bureau of Yokohama for providing the related materials.

Disclosure statement

No potential conflict of interest was reported by the authors.

References

Ahmad, S. (2003). Reinforcement corrosion in concrete structures, its monitoring and service life prediction–a review. *Cement and Concrete Composites, 25*, 459–471.

Akiyama, M., Frangopol, D. M., & Suzuki, M. (2012). Integration of the effects of airborne chlorides into reliability based durability design of reinforced concrete structures in a marine environment. *Structure and Infrastructure Engineering, 8*, 125–134.

Alamilla, J. L., Espinosa-Medina, M. A., & Sosa, E. (2009). Modelling steel corrosion damage in soil environment. *Corrosion Science, 51*, 2628–2638.

Asakura, T., & Kojima, Y. (2003). Tunnel maintenance in Japan. *Tunnelling and Underground Space Technology, 18*, 161–169.

Babu, S. G. L., & Srivastava, A. (2010). Reliability Analysis of Buried Flexible Pipe-Soil Systems. *Journal of Pipeline Systems Engineering and Practice, 1*, 33–41.

Bureau of Sewerage of Tokyo (BST). (2012). *Report on examination of performance of trunk sewers built by shield tunneling method*. Tokyo: Tokyo Metropolitan Government.

Chughtai, F., & Zayed, T. (2008). Infrastructure condition prediction models for sustainable sewer pipelines. *Journal of Performance of Constructed Facilities, 22*, 333–341.

Das, P. C. (1999). Prioritization of bridge maintenance needs. In D. M. Frangopol (Ed.), *Case studies in optimal design and maintenance planning of civil infrastructure systems* (pp. 26–44). Reston, VA: ASCE.

Davies, J. P., Clarke, B. A., Whiter, J. T., & Cunningham, R. J. (2001). Factors influencing the structural deterioration and collapse of rigid sewer pipes. *Urban Water, 3*, 73–89.

Department of Defense (DoD). (2008). Condition based maintenance plus DoD guidance. Deputy Under Secretary of Defense for Logistics and Material Readiness. Washington, DC: Deputy Under Secretary of Defense for Logistics and Material Readiness.

Environmental Planning Bureau of Yokohama (EPBY). (2013). *Report on evaluation of seismic performance of existing sewerage shield*. Yohama: Environmental Planning Bureau of Yokohama, Japan.

Frangopol, D. M., Kallen, M. J., & Van Noortwijk, J. M. (2004). Probabilistic models for life-cycle performance of deteriorating structures: Review and future directions. *Progress in Structural Engineering and Materials, 6*, 197–212.

Frangopol, D. M., Lin, K. Y., & Estes, A. C. (1997). Life-cycle cost design of deteriorating structures. *Journal of Structural Engineering, 123*, 1390–1401.

Frangopol, D. M., & Liu, M. (2007). Maintenance and management of civil infrastructure based on condition, safety, optimization, and life–cycle cost. *Structure and Infrastructure Engineering, 3*, 29–41.

Federal Highway Administration (FHWA). (2004). *Highway and rail transit tunnel maintenance and rehabilitation manual*. Washington, DC: U.S. Department of Transportation Federal Highway Administration.

Fenner, R. A. (2000). Approaches to sewer maintenance: a review. *Urban Water, 2*, 343–356.

Hashash, Y. M., Hook, J. J., Schmidt, B., John, I., & Yao, C. (2001). Seismic design and analysis of underground structures. *Tunnelling and Underground Space Technology, 16*, 247–293.

Japan Society of Civil Engineers (JSCE). (2007). *Standard Specifications for Tunneling: Shield Tunnel* (2006th ed.). Tokyo: Japan Society of Civil Engineers.

Japan Society of Civil Engineers (JSCE). (2010). Tunnel library No.23–Segment design (Revision), Chapter 8. Tokyo: Japan Society of Civil Engineers.

Japan Sewage Works Association (JSWA). (1973). *Standard segment for shield tunnelling*. Tokyo: Japan Sewage Works Association.

Japan Sewage Works Association (JSWA). (2009). *Specification and instruction on seismic design of sewage utilities*. Tokyo: Japan Sewage Works Association.

Kawashima, K. (1999). *Seismic design of underground structures in soft ground, a review*. Proceedings of the International Symposium on Tunneling in Difficult Ground Conditions, Tokyo, Japan.

Kimura, S. (2013). *Managements of service level and operational risk depending on the service of the tunnel*. Proceedings of the Seventh China–Japan Conference on Shield Tunnelling, Xian.

Mashimo, H. (2002). State of the road tunnel safety technology in Japan. *Tunnelling and Underground Space Technology, 17*, 145–152.

Meguid, M. A., & Dang, H. K. (2009). The effect of erosion voids on existing tunnel linings. *Tunnelling and Underground Space Technology, 24*, 278–286.

Melchers, R. E. (1987). *Structural reliability analysis and prediction*. Ellis Horwood, Chichester: Wiley.

Ministry of Land, Infrastructure, Transport and Tourism (MLIT). (2015a). The maintenance of national road network in Japan. Retrieved from http://www.mlit.go.jp/road/road_e/03key_challenges/1-2-2.pdf

Ministry of Land, Infrastructure, Transport and Tourism (MLIT). (2015b). Retrieved from http://www.mlit.go.jp/crd/sewerage/policy/03.html

Mutou, Y., & Kawabata, M. (2011). Maintenance of railroad shield tunnel. Proceeding of the 6th Japan–China Conference on Shield–driven Tunneling, Nagaoka.

Richards, J. A. (1998). Inspection, maintenance and repair of tunnels: International lessons and practice. *Tunnelling and Underground Space Technology, 13*, 369–375.

Ricker, R. E. (2007). *Analysis of pipeline steel corrosion data from NBS (NIST) studies conducted between 1922–1940 and relevance to pipeline management*. Gaithersburg, MD: US Department of Commerce, Technology Administration, National Institute of Standards and Technology.

Schnabel, P. B., Lysmer, J. and Seed, H. B. (1972). *SHAKE: A computer program for earthquake response analysis of horizontally layered sites* (Report No. UCB/EERC-72/12) (p. 102). Berkeley: Earthquake Engineering Research Centre, University of California.

Tee, K. F., Khan, L. R., & Chen, H. (2013). Probabilistic failure analysis of underground flexible pipes. *Structural Engineering Mechanism, 47*, 167–183.

Wang, J., Shi, Z., & Nakano, M. (2013). Strength degradation analysis of an aging RC girder bridge using FE crack analysis and simple capacity–evaluation equations. *Engineering Fracture Mechanics, 108*, 209–221.

Yuan, Y., Jiang, X., & Liu, X. (2013). Predictive maintenance of shield tunnels. *Tunnelling and Underground Space Technology, 38*, 69–86.

Youssef, M. A. H., Jeffrey, J. H., Birger, S., & John, I. Y. (2001). Seismic design and analysis of underground structures. *Journal of Tunnelling and Underground Space Technology, 16*, 247–293.

Simplified structural deterioration model for reinforced concrete bridge piers under cyclic loading[1]

Anirudh S. Rao, Michael D. Lepech, Anne S. Kiremidjian and Xiao-Yan Sun

ABSTRACT

Deterioration due to reinforcement corrosion represents a significant cause of damage to reinforced concrete bridges all over the world. Although numerous studies have identified the substantial influence that corrosion can have on the failure of concrete structures under seismic loads, there has been comparatively less work done on the modelling of structural vulnerability due to corrosion. Accurate and computationally efficient structural modelling of corrosion deterioration is an essential prerequisite for structural reliability or fragility prediction, or life cycle cost and impact estimation. In this paper, a simplified approach for modelling reinforced concrete bridge columns that have undergone deterioration due to chloride-induced reinforcement corrosion is presented. While several models of increasing complexity are evaluated, a simplified non-linear analytical model that accounts primarily for the reduction in the steel cross-section due to corrosion is the central focus of the paper. The accuracy of the simple model is validated by comparing analytical results with experimental tests on reinforced concrete columns subjected to accelerated corrosion. The model results show excellent agreement with the experimental tests, making it particularly suitable for applications in vulnerability or reliability analysis where numerous non-linear dynamic analyses are necessary, or for further downscale or upscale integration with materials-level or system-level models.

Introduction

Corrosion of steel in reinforced concrete members combined with poor concrete durability is a primary source of structural deterioration for civil engineering infrastructure worldwide including bridges, piers, seawalls, dams and other structures (Vu & Stewart, 2000). The U.S. Federal Highway Administration (FHWA) estimates the annual direct cost of corrosion of bridges in the United States to be around $8.3 billion (Koch, Brongers, Thompson, Virmani, & Payer, 2002). In recent years, advanced levels of deterioration have been observed in reinforced concrete highway bridges exposed to seawater or spray in coastal locations and due to the application of deicing salts in northern locations (Hartt, Powers, Leroux, & Lysogorski, 2004).

Deterioration due to corrosion has typically been modelled using phenomenological processes proposed by Tuutti (1982), wherein steel corrosion in reinforced concrete is segmented into a series of states beginning with (1) initial transport of chlorides (or other corrosives) through the concrete cover, (2) de-passivation of the reinforcing steel, (3) formation of oxidation products that cause radial cracking around the steel rebar and (4) accelerated oxidation ultimately causing concrete cover spalling. This conceptual multi-stage deterioration process is described

in Figure 1. The stages of this phenomenon have been quantified, modelled, and reviewed in part or whole by others (Liu & Weyers, 1998; Middleton & Hogg, 1998; Neville, 1995). Thus, structural vulnerability is likely to change with time if a structure is undergoing deterioration due to corrosion.

Recognising the importance of the issue, many researchers have looked at the effect of reinforcement corrosion on the reliability of reinforced concrete structures subjected to primarily dead and live gravity loads (e.g. Akiyama, Frangopol, & Suzuki, 2012; Alipour, Shafei, & Shinozuka, 2013; Biondini & Vergani, 2015; Chiu & Chi, 2013; Frangopol, Lin, & Estes, 1997; Papakonstantinou & Shinozuka, 2013; Stewart & Rosowsky, 1998; Val, Stewart, & Melchers, 1998; Vu & Stewart, 2000). These studies have found that the corrosion of reinforcing steel can lead to significant reductions in reliability over time in reinforced concrete members and structural systems. A number of researchers have also looked at the seismic performance of RC bridges and other structures that have undergone corrosion of reinforcement (e.g. Akiyama & Frangopol, 2013; Akiyama, Frangopol, & Matsuzaki, 2011; Biondini, Camnasio, & Palermo, 2014; Biondini, Palermo, & Toniolo, 2011; Celarec, Vamvatsikos, & Dolšek, 2011; Choe, Gardoni, Rosowsky, & Haukaas, 2008, 2009; Choe, Gardoni, & Rosowsky, 2010; Ghosh & Padgett, 2010;

[1]This paper is part of the keynote lecture presented by Professor Anne Kiremidjian on 'Time-dependent earthquake risk assessment modeling considering sustainability metrics' at the IALCCE 2014, Tokyo, Japan.

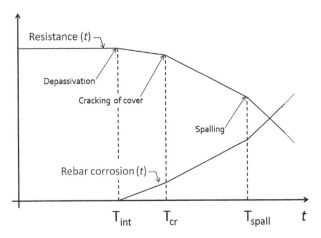

Figure 1. Conceptual multi-stage deterioration process showing time, *t*, vs. reduction in structural resistance over time.
Notes: Also shown conceptually is the inverse relationship between structural resistance and amount of rebar corrosion over time.

Simon, Bracci, & Gardoni, 2010; Titi & Biondini, 2014; Yalciner, Sensoy, & Eren, 2012).

While previous researchers have shown the significant impact that corrosion can have on the failure of structures under seismic loads, structural deterioration due to corrosion has been typically modelled in ways that are computationally difficult to incorporate into vulnerability or reliability analysis where numerous non-linear dynamic analyses are necessary, or are difficult to integrate across scales. For example, Kashani, Lowes, Crewe, and Alexander (2014) have implemented a relatively complex corrosion damage model for a non-linear fibre beam-column element in OpenSees. However, the Kashani et al. model is limited in its use for computationally intensive reliability analysis and simple integration with models of adjacent scales. The novelty of this study is the creation of a simple structural deterioration model that can be incorporated with computationally intensive reliability analysis (Rao, 2014). Additionally, this model is novel such that it can be integrated with smaller scale models (i.e. materials-level), as demonstrated by Michel, Geiker, Stang, and Lepech (2015) and Lepech et al. (2015), and larger scale models (i.e. building-level or system-level), as demonstrated by Rao, Lepech, and Kiremidjian (2013). Thus, enabling high fidelity, predictive modelling of structural reliability or vulnerability, along with life cycle cost and impact estimation.

In this study the effect of deterioration due to corrosion is explicitly incorporated in the model for the dynamic behaviour of the structure. The intent is to use this model in subsequent bridge column and bridge structure vulnerability assessments through fragility analysis. One of the most significant improvements of the proposed model over existing approaches is the consideration of corrosion effects on reinforced concrete behaviour that have been validated with experimental tests, leading to a more realistic representation of structural deterioration and bringing the structural model closer to actual field conditions over time. To the best of the authors' knowledge, the validation of the structural modelling of corrosion in reinforced concrete bridge columns presented in this paper with large-scale accelerated corrosion and cyclic loading of experimental specimens is the first of its kind.

Structural modelling of corroded RC columns

The ultimate goal of the model presented in this paper is for it to be used in the development of vulnerability functions for an RC column subjected to seismic loads that has undergone deterioration due to corrosion. For this purpose, it is first necessary to develop a model for dynamic analysis of uncorroded reinforced concrete columns that captures their properties and dynamic behaviour. Thus, the next subsection presents the assumptions and formulation of the dynamic model for an uncorroded column and the subsequent subsection incorporates the assumptions and model for the corroded column.

Modelling the uncorroded column

The analytical modelling of the uncorroded bridge column is conducted using the Open System for Earthquake Engineering Simulation (OpenSees) software (Mazzoni, McKenna, Scott, & Fenves, 2009), a finite element method based object-oriented software framework for simulating the seismic response of structural systems. OpenSees was chosen because it provides an extensive library of material models, non-linear elements and solvers that can be easily interchanged, and it has been widely adopted by the research community, making it highly conducive for the modelling and simulation of complex structural systems subject to seismic loading.

Other researchers (e.g. Biondini et al., 2011) have successfully adopted the use of lumped plasticity models for fragility analysis of corroded reinforced concrete structures, which can be more computationally efficient than fibre beam element models. Such models are not considered for this work even though they are highly computationally efficient since they do not facilitate easy integration with adjacent scale models, such as those demonstrated by Michel et al. (2015) and Lepech et al. (2015). Thus, the column for this study is modelled as a non-linear distributed-plasticity, fibre beam-column element, which is a flexibility-based element that considers the spread of plasticity along the length of the element. Five integration points are used along the length of the column element.

The column cross-section is discretised into a number of inelastic fibre segments as shown in Figure 2, based on the recommended uniformly distributed radial discretisation configuration recommended by Berry and Eberhard (2008). The reinforcement steel is modelled using the uniaxial Giuffre-Menegotto-Pinto 'Steel02' material, which provides isotropic strain hardening. Typical hysteresis curves for Steel02 are shown in Figure 3.

The confined core concrete fibres and unconfined cover concrete fibres are modelled using the uniaxial 'Concrete01' material in OpenSees, which provides a parabolic stress–strain response for the concrete in compression up to the maximum compressive strength, beyond which the strength deteriorates linearly to a residual strength. Concrete01 provides no tensile strength. OpenSees also provides the materials 'Concrete02' which allows for tensile strength with linear tension softening and 'Concrete03' which provides tensile strength with non-linear tension softening. Concrete01 was preferred as it led to fewer convergence issues during dynamic analysis.

Implementation of the Concrete01 material in OpenSees, including the determination of appropriate element length,

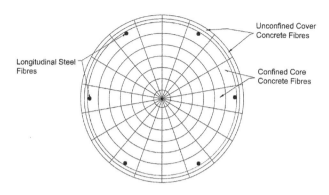

Figure 2. Discretisation of cross-section into inelastic fibres.

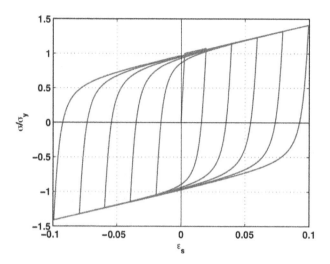

Figure 3. Typical hysteresis curves for 'Steel02' (Berry & Eberhard, 2008).

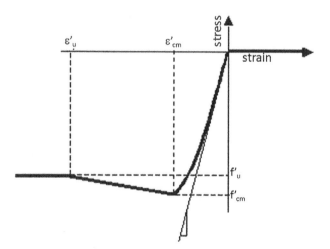

Figure 4. 'Concrete01' material properties (Mazzoni et al., 2009).

followed Berry and Eberhard (2008). The material properties and behaviour of Concrete01 are shown in Figure 4. The maximum compressive strength of the confined concrete, f'_{cc}, and the corresponding strain, ε_{cc}, are calculated using the theoretical stress–strain model proposed by Mander, Priestley, and Park (1988) as follows:

$$f'_{cc} = f'_{co}\left(-1.254 + 2.254\sqrt{1 + \frac{7.94 f'_l}{f'_{co}}} - 2\frac{f'_l}{f'_{co}}\right) \quad (1)$$

$$\varepsilon_{cc} = 0.002\left(1 + 5\left(\frac{f'_{cc}}{f'_{co}} - 1\right)\right) \quad (2)$$

where:

$$f'_l = \frac{1}{2}k_e \rho_s f_{yh} \quad (3)$$

$$k_e = \frac{1 - \frac{s'}{2d_s}}{1 - \rho_{cc}} \quad (4)$$

where f'_{co} is the unconfined concrete strength, f'_l is the confining lateral pressure provided by the transverse reinforcement, f_{yh} is the yield strength of transverse reinforcement, k_e is a coefficient measuring confinement effectiveness, ρ_{cc} is the ratio of longitudinal rebar cross-sectional area to the area of the concrete core, s' is the clear spacing of transverse reinforcement and d_s is the diameter of the core. An ultimate strain of .005 is specified for the concrete model.

An elastic shear section is combined with the section of the beam-column element to capture the shear component of the total deflection of the column, Δ_{shear}, that is calculated using elastic theory as:

$$\Delta_{shear} = \frac{4}{3}\left(\frac{V \cdot H_{col}}{G \cdot A_G}\right) \quad (5)$$

where V is the shear force in the column, H_{col} is the height of the column, A_G is the cross-sectional area of the column and G is the shear modulus of concrete, defined as:

$$G = \frac{E_c}{2(1 + v)} \quad (6)$$

where v is the Poisson's ratio for concrete, assumed as .2 an, E_c is the Young's modulus of concrete, defined per ACI 318 (2000) as:

$$E_c = 57\sqrt{f'_c} \quad (7)$$

Additionally, a zero-length section is introduced at the base of the column to account for the component of the deformation due to bond-slip. The zero-length element is defined by a master-node and a slave-node at the same location. The master-node has all degrees of freedom restrained and the slave-node has its translational degrees of freedom constrained to be equal to those of the master node. The tensile stress–strain model used for the reinforcement is based on the three-parameter bond-stress model presented by Ranf (2007) and illustrated in Figure 5.

The model describes the dependence of the bond stress (τ) on the axial stress in the longitudinal reinforcement (σ) as follows:

$$\tau(\sigma) = \begin{cases} \lambda_e \frac{\sigma}{\sigma_d}\sqrt{f'_c} & \sigma \leq \sigma_d \\ \lambda_e \sqrt{f'_c} & \sigma_d \leq \sigma \leq \sigma_y \\ \lambda_i \sqrt{f'_c} & \sigma_y \leq \sigma \end{cases} \quad (8)$$

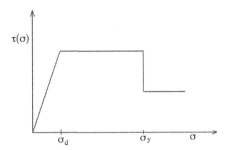

Figure 5. Bond stress model (after Ranf, 2007).

where σ_d is the bar stress needed to fully develop the bond, σ_y is the yield stress of the reinforcement and λ_e and λ_i are the elastic and inelastic bond-stress parameters. Berry and Eberhard (2008) found the values $\sigma_d = .25\sigma_y$, $\lambda_e = .90$ and $\lambda_i = .45$ to be optimal following a calibration exercise for the model. These values are therefore used for modelling the bond-slip component of the deformation in the column.

In addition to the above model for bond-slip, the bond–slip models implemented in OpenSees (i.e. BARSLIP and Bond_SP01) were also considered initially. The BARSLIP model was deemed unsatisfactory for the purposes of this study because it includes predefined bond strengths, which cannot be changed to reflect corrosion deterioration. The model by Ranf (2007) was preferred over Bond_SP01 because the three parameters that are used in that model have been calibrated by Berry and Eberhard (2008) using a database of 300 column tests. On the other hand, no calibration studies were found for the three parameters used in the Bond_SP01 model.

Modelling the deterioration due to corrosion

Corrosion of reinforcement damages concrete structures in multiple ways (Lounis, 2005). The area of both longitudinal and transverse steel is reduced due to the formation of rust. The expansive rust products generate tensile stresses in the surrounding concrete, which may lead to cracking of the concrete cover. The formation of cracks leading up to the surface may lead to an acceleration of the local rate of corrosion, causing pitting corrosion. The highly localised reduction in steel cross-section at these corrosion pits may cause stress concentration in the steel resulting in a loss of ductility in the structure. Eventually, the cover cracking and spalling may lead to a reduction in the concrete cross-sectional area.

The effects of corrosion are assumed to be uniform about the perimeter of the column and all rebars are assumed to undergo the same reduction in diameter at any given point of time during the service life of the bridge. The diameter of rebars in the corroded columns is computed as:

$$d_{b,\mathrm{corr}} = \sqrt{\frac{4A_{b,\mathrm{corr}}}{\pi}} \tag{9}$$

where

$$A_{b,\mathrm{corr}} = A_{b0} \times \left(1 - \frac{X}{100}\right) \tag{10}$$

where $A_{b,\mathrm{corr}}$ is the cross-sectional area of the corroded rebars, A_{b0} is the cross-sectional area of the pristine rebars, before the start of corrosion and X is a measure of the amount of corrosion undergone by the column and expressed in terms of per cent mass loss of reinforcing steel.

In practice, the deterioration of bridge columns is usually concentrated at the ends of the column. The top end of a column can be susceptible to higher corrosion if it is located beneath a faulty deck joint from which water containing chloride ions can leak through. The bottom end of the column witnesses a relatively higher degree of chloride ion accumulation, caused by the splashing of water laden with deicing salts by passing vehicles or by splashing seawater if the bridge is located in a marine environment.

However, to the best knowledge of the authors, there have been no experimental studies looking at the behaviour of bridge columns that have undergone such localised corrosion with which a model could be validated. The majority of tests on deteriorating bridge columns that have been conducted to date involve a uniform corrosion of the reinforcing steel bars along the height of the column under accelerated corrosion conditions. Thus, for modelling purposes in this paper, the reduction in rebar diameter is assumed to be uniform along the height of the column, allowing for the validation of the model with existing experimental results.

The cracking and spalling of the cover concrete caused by the expansive rust products is captured by reducing the strength of the cover concrete elements using the model described by Coronelli and Gambarova (2004) as follows:

$$f_{\mathrm{co}}^* = \frac{f_{\mathrm{co}}'}{1 + K\frac{\varepsilon_1}{\varepsilon_{\mathrm{co}}}} \tag{11}$$

where K is a parameter related to bar diameter and roughness ($K \approx .1$ for medium-diameter ribbed bars), $\varepsilon_{\mathrm{co}}$ is the strain at the peak compressive stress f_{co}', and the strain ε_1 is calculated as:

$$\varepsilon_1 = \frac{D_{\mathrm{col}}^* - D_{\mathrm{col}}}{D_{\mathrm{col}}} \tag{12}$$

where D_{col} is the width of the column before corrosion cracking and D_{col}^* is the column width increased by corrosion cracking and rust expansion. The increase in width of the column is approximated by:

$$D_{\mathrm{col}}^* - D_{\mathrm{col}} = n_{\mathrm{bars}} w_{\mathrm{cr}} \tag{13}$$

where n_{bars} is the number of bars undergoing compression, and w_{cr} is the sum of the widths of all crack openings for a given corrosion level, which can be computed using the equation proposed by Molina et al. (1993) as follows:

$$w_{\mathrm{cr}} = 2\pi t \tag{14}$$

where

$$t = \left(v_{\mathrm{rs}} - 1\right)Y \tag{15}$$

where Y is the depth of the corrosion attack (equal to the reduction in bar radius) and v_{rs} is the ratio of the volume of the corroded steel bar including the expansive rust products to the volume of the virgin steel bar. In this paper, the value of v_{rs} is assumed to be equal to 2, as in Molina et al. (1993).

The reduction in reinforcement steel diameter, the loss of confinement due to cracking and spalling of the concrete cover and formation of porous rust products at the steel-concrete interface may also lead to changes in the bond strength between steel and concrete. Several experimental tests have been conducted on the influence of corrosion on bond strength (e.g. Almusallam, Al-Gahtani, & Aziz, 1996; Al-Sulaimani, Kaleemullah, Basunbul, & Rasheeduzzafar, 1990; Amleh, 2000; Berra, Castellani, & Coronelli, 1997; Cabrera, 1996; Cabrera & Ghodussi, 1992; Cairns, Plizzari, Du, Law, & Franzoni, 2005; Clark & Saifullah, 1993, 1994; Fang, Lundgren, Chen, & Zhu, 2004; Fang, Lundgren, Plos, & Gylltoft, 2006; Rodriguez, Ortega, & Casal, 1994; Stanish, 1997; Wang & Liu, 2004). However, there are large variations in the change in bond strength as a function of degree of corrosion as reported by these various experimental studies. For instance, as noted by Fang et al. (2004) and Wang and Liu (2004), the effects of corrosion on bonding in well-confined concrete are negligible. On the other hand, Amleh (2000) and Fang et al. (2006) report that the bond strength decreases substantially at higher levels of corrosion, whereas Berra et al. (1997) noted a moderate increase in bond strength with increasing corrosion when efficient skin reinforcement was provided in order to confine the bars.

Thus, based on the existing literature, no clear relationship between the bond strength and corrosion level can be ascertained. There is also a lack of theoretical models that can correctly predict the loss in bond strength due to corrosion. Due to these reasons, certain assumptions had to be made in order to study the effect of corrosion on the bond between the rebar steel and concrete. A rough comparison of residual bond strength at different levels of corrosion as reported by some of the above studies can be obtained in FIB (2000), and a scatterplot of these results is shown in Figure 6. Based on these results, the increase in bond strength at lower levels of corrosion ($X = 2\%$ and below) is assumed to be negligible, and with further increase in the level of corrosion, the bond strength is assumed to decrease according to the following empirical equation:

$$f_b^* = \begin{cases} 1.0, & X \leq 2\% \\ 1.5 \times X^{-0.6}, & X > 2\% \end{cases} \quad (16)$$

where f_b^* is the residual bond strength after corrosion expressed as a fraction of the bond stress before corrosion. The variation in bond strength with corrosion as approximated by

Equation (16) is shown in Figure 6 alongside the experimental results it is based on.

Validation with experimental tests

In order to validate the assumptions made in the corrosion modelling approach described above, analytical and experimental test results of several reinforced concrete columns were compared. Two separate sets of experimental tests are used for validation of the structural modelling of corroded reinforced concrete columns. The first set of tests comprised eight columns having the same specifications, which were subjected to different axial loads and varying degrees of accelerated corrosion in an experimental study conducted at the State Key Laboratory of Coastal and Offshore Engineering at Dalian University of Technology in China (Li & Gong, 2008). The second set of tests comprised one uncorroded column and one corroded column and are taken from a larger set of tests conducted at the Newmark Structural Engineering Research Laboratory of the Department of Civil and Environmental Engineering at University of Illinois at Urbana-Champaign (Aquino & Hawkins, 2007).

Description of tests used for validation

The Dalian tests

The experimental reinforced concrete column test specimens in the Dalian study (Li & Gong, 2008) were 1 m tall and had a diameter of .26 m. The longitudinal reinforcement consisted of 6×16 mm bars; the distribution of longitudinal bars is shown in the column cross-section in Figure 7. The transverse reinforcement consisted of 8 mm stirrups at a spacing of .1 m. The concrete cover was .03 m. The geometric and structural properties of the test specimens are summarised in Table 1.

Corrosion was induced in the test specimens by the application of an external current. The specimens to be corroded were placed in a 3.5% NaCl solution and a current of 1.0 mA/cm^2 was applied with a corrosion-resistant plate immersed in the solution serving as the cathode and the reinforcement steel cage serving as the anode. At the end of the accelerated corrosion process, longitudinal cracks were observed running parallel to the steel bars in the corroded column specimens, as shown in Figure 8, and the authors noted that the corrosion damage was uniform along the length of the bars. A constant axial load was applied with the help of hydraulic jacks. Table 2 displays the different axial load ratios and corrosion mass loss ratios for the series of tested column specimens.

The corroded and uncorroded test columns were subjected to reversed cyclic loading using the triangular waveform shown in Figure 9 and hysteretic and skeleton curves for the columns

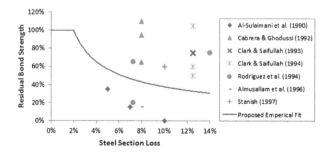

Figure 6. Variation in bond strength with corrosion.

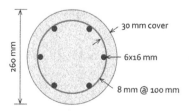

Figure 7. Cross section of the Dalian test specimens (after Li & Gong, 2008).

Table 1. Geometric and structural properties of the Dalian test specimens (Li & Gong, 2008).

Height, H	1 m
Diameter, D	.26 m
Longitudinal reinforcement ratio, ρ	2.3%
Longitudinal rebar diameter, d_0	16 mm
Transverse rebar diameter, d_s	8 mm
Nominal yield strength of steel, f_y	373.2 MPa
Concrete compressive strength, f_{ck}	32.4 MPa

Figure 8. Corrosion-damaged specimen (from Li & Gong, 2008).

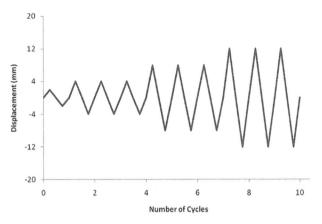

Figure 9. Lateral displacement history used in the Dalian tests (after Li & Gong, 2008).

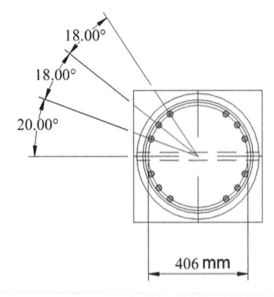

Figure 10. Cross-section of the Illinois test specimens (from Aquino 2002).

were obtained. The initial displacement of 2 mm was followed by three full cycles at each subsequent displacement level. Each displacement level was higher than the previous level by 4 mm.

The Illinois tests

The test columns in the Illinois study (Aquino 2002) were 2.45 m tall and had a diameter of .50 m. The longitudinal reinforcement consisted of 12 #8 bars spliced to 8 #8 steel bars; the distribution of longitudinal bars is shown in the column cross-section in Figure 10. The transverse reinforcement consisted of #3 stirrups at a spacing of .2 m. The concrete cover was .025 m. The geometric and structural properties of the test specimens are summarised in Table 3. One of the columns was left as built (no environmental exposure) to serve as the control specimen. The column to be corroded was placed in a 3% NaCl solution and copper mesh positioned around the column served as the cathode during the accelerated corrosion process, whereas the reinforcement cage served as the anode. The corroded column had a 7.8% mass loss. Vertical cracks running parallel to the longitudinal bars formed on the surface of the concrete during the corrosion process, and a visual inspection of the exposed reinforcement revealed uniformity in the corrosion damage. Figure 11 shows the corrosion-damaged column specimen where

the longitudinal cracking and accumulation of rust products are visible.

Subsequently, the triangular waveform shown in Figure 12 was used for the lateral cyclic load testing of both the uncorroded control specimen and the corroded specimen. The observed load–deformation hysteretic response of the control column was the same in both directions. The yield load was 141 kN and the yield displacement was 20 mm and the column failed at a load of 165 kN and a corresponding displacement of 87.5 mm. The response of the corroded column stayed symmetric until a load of 133 kN and a corresponding displacement of 21 mm. Beyond this point, a drop in load in the positive direction was observed,

Table 2. Axial load ratios and mass loss ratios for the Dalian test specimens (after Li & Gong, 2008).

Specimen	A		B			C		D
	A_0	A_{12}	B_0	B_{21}	B_{22}	C_0	C_{32}	D_{43}
Mass loss due to corrosion (%)	.0%	9.4%	.0%	5.1%	9.4%	.0%	9.4%	14.7%
Axial load ratio (ALR)		.15		.25			.40	.32

Table 3. Geometric and structural properties of the Illinois test specimens (Aquino 2002).

Height, H	2.45 m
Diameter, D	.50 m
Longitudinal rebar diameter, d_0	25.4 mm
Transverse rebar diameter, d_s	9.53 mm
Nominal yield strength of steel, f_y	413.7 MPa
Nominal strength of concrete, f_c'	31.7 MPa

Table 4. Modelling cases.

Modelling case	Phenomena considered		
	Reduction in rebar cross-section	Change in concrete material behaviour	Change in bond-strength
I	Yes	No	No
II	Yes	Yes	No
III	Yes	No	Yes
IV	Yes	Yes	Yes

Figure 11. Corrosion damaged specimen (from Aquino 2002).

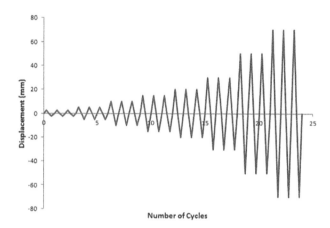

Figure 12. Lateral displacement history used in the Illinois tests (after Aquino 2002).

whereas the load in the negative direction was observed to be still increasing. Thus, beyond this point, the loading was applied in the negative direction only, using three half cycles instead of the planned three full cycles at each displacement increment. The longitudinal cracks caused due to the corrosion were observed to open significantly on the tension side during the positive half cycle of loading.

Model verification

OpenSees models were created for both the uncorroded and the corroded specimens for both the Dalian tests and the Illinois tests using the modelling strategies proposed in the previous section. Four different modelling cases were considered. The simplest of these was Case I, where we considered only the reduction in steel bar cross-section. Case II and Case III also included the changes in concrete material behaviour and changes in bond strength,

respectively, whereas Case IV considered all of the above phenomena. These modelling cases are documented in Table 4. Additional modelling strategies were investigated, but were not sufficiently significant to incorporate into the model. These additional strategies included considerations of a loss of ductility in corroded reinforcing steel and a more complex accounting of concrete damage surrounding the corroding reinforcement.

Specifically, effects of concrete splitting, cover loss and change in effective moment of inertia were considered. While each of these phenomenon are important deterioration mechanisms, the inclusion of most of them did not substantially increase the accuracy of the model presented using the measures of model accuracy proposed by Berry and Eberhard (2008) discussed below. More detail on all of the modelling strategies investigated can be found in Rao (2014).

The OpenSees models for each of the four cases were then subjected to the same cyclic load patterns that were used in the original tests. Displacement-driven analyses were carried out until the failure displacement identified from the tests was reached in the OpenSees analysis. The predicted cyclic response of the columns obtained from the OpenSees models was then compared with the cyclic response obtained from the tests to gauge the accuracy of the models.

Measures of accuracy

In order to evaluate the cyclic response of the OpenSees models, the following two measures of model accuracy proposed by Berry and Eberhard (2008) were adopted:

- *Normalised Hysteretic Force Error, E_{force}:* This quantity estimates the root mean square error in the OpenSees prediction of the lateral force at corresponding displacements, normalised by the maximum force measured:

$$E_{force} = \sqrt{\frac{\sum_{i=1}^{n}\left(F_{test}^{i} - F_{OpenSees}^{i}\right)^2}{\left(\max\left(F_{test}\right)\right)^2 n}} \quad (17)$$

where F_{test} and $F_{OpenSees}$ are the lateral forces measured during the test and predicted by the OpenSees model at corresponding displacements, and n is the total number of data points in the loading history.

- *Normalised Hysteretic Energy Error, E_{energy}:* This quantity estimates the error between the OpenSees estimation of the hysteretic energy $\Omega_{OpenSees}$ and the measured hysteretic energy Ω_{test}, normalised by the measured hysteretic energy:

$$E_{energy} = \frac{\Omega_{test} - \Omega_{OpenSees}}{\Omega_{test}} \quad (18)$$

Table 5. Benchmark error statistics evaluated for a set of 37 bridge columns (from Berry & Eberhard, 2008).

	E_{force} (%)	abs(E_{energy}) (%)	E_{energy} (%)
Mean	16.13	26.02	−23.69
Min	6.63	1.34	−109.97
Max	44.71	109.97	13.36

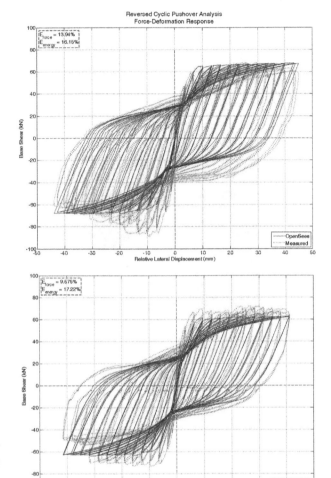

Figure 13. Hysteresis curves for columns A_0 (top) and A_{12} (bottom) from the Dalian tests (Li & Gong, 2008).

The hysteretic energy can be calculated using the trapezoidal rule for estimating the area enclosed within the hysteresis loops as in the following equation.

$$\Omega = \sum_{i=1}^{n-1} \frac{F_{i+1} - F_i}{2} \left(\Delta_{i+1} - \Delta_i \right) \tag{19}$$

where F_i and Δ_i are the lateral force and displacement associated with the ith step, and n is the total number of data points in the loading history.

A positive value of E_{energy} indicates that the amount of energy dissipated during the test is underestimated by the OpenSees model and a negative value of E_{energy} indicates that the amount of energy dissipated is overestimated.

Comparison of analytical and experimental results

The error statistics reported by Berry and Eberhard (2008) in their evaluation of their proposed distributed-plasticity modelling strategy for reinforced concrete bridge columns were used as a benchmark to assess the accuracy of the results obtained in this study. Berry and Eberhard (2008) compared the calculated and measured cyclic response for 37 bridge columns and the key error statistics of that evaluation are provided in Table 5. The mean error in hysteretic force for the bridge column database in that study was 16.13% and the mean absolute error in hysteretic energy was 26.02%.

The hysteresis curves obtained from the test and the corresponding OpenSees analyses using Modeling Case I for the eight specimens from the Dalian tests (Li & Gong, 2008) of varying axial load ratio and degree of corrosion are plotted in Figures 13–16 and the key error statistics for the cyclic response of the proposed model are provided in Table 6.

Figure 17 depicts the results from the four modelling cases compared with the measured hysteresis curves for column specimen A_{12} from the Dalian test. Similarly, the hysteresis curves obtained from the test and the corresponding OpenSees analyses using Modeling Case I for the two columns from the Illinois study (Aquino, 2002) are plotted in Figure 18 and the error statistics are provided in Table 7.

The mean error in hysteretic force calculated for the Dalian specimens using Modeling Case I was 12.9% and the mean absolute error in hysteretic energy was 25.3%. These error values are lower than the corresponding values reported by Berry and Eberhard (2008), indicating that the simplified analytical model taking into consideration only the reduction in rebar cross-sectional area is a very good fit. Column specimen 'D$_{43}$', which was subjected to a high axial load ratio (.32) coupled with a high degree of corrosion (14.7% steel mass loss) was the source of the highest error in both hysteretic force (15.97%) and hysteretic energy (53.21%). However, this specimen was tested as an extreme case and the combination of such a high axial load ratio and degree of corrosion is unlikely to be seen in an actual bridge column. The hysteresis curves from the Dalian tests are not symmetrical about the load axis. This is probably due to higher lateral deformations and earlier yielding of one side of the column as compared to the other. This asymmetric behaviour cannot be predicted by the material models used in the analytical study.

The error in hysteretic force for the uncorroded column in the Illinois test using Modeling Case I was 8.44% and the error in hysteretic energy was only 4.77%, affirming that the model is a good fit for the uncorroded column. The error in hysteretic force for the corroded column in that test was 10.44% and the error in hysteretic energy was 38.18%. The higher error in hysteretic energy is probably due to the large opening of the corrosion cracks during the loading process, indicating that the effect of corrosion cracking in the cover concrete on structural behaviour merits further investigation. However, the error in hysteretic force was lower than the mean error in hysteretic force reported by Berry and Eberhard (2008), indicating that the simplified analytical model is able to predict the hysteretic force response of the corroded column reasonably well.

Key error statistics for the cyclic response of the Dalian test specimens using Modeling Cases II, III and IV are also provided in Table 6, and the mean errors in hysteretic force and energy

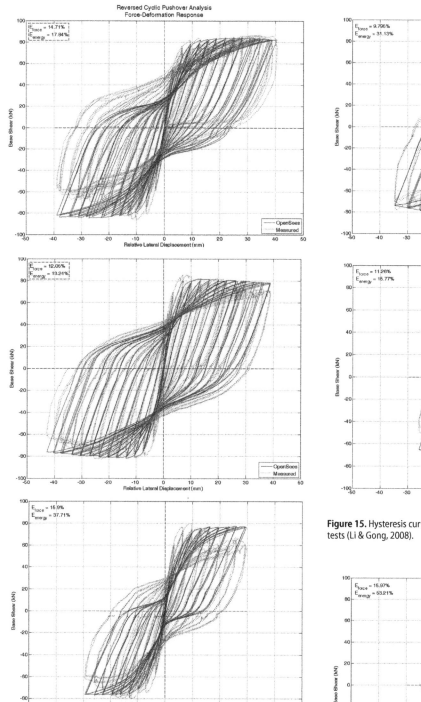

Figure 14. Hysteresis curves for columns B_0 (top), B_{21} (centre) and B_{22} (bottom) from the Dalian tests (Li & Gong, 2008).

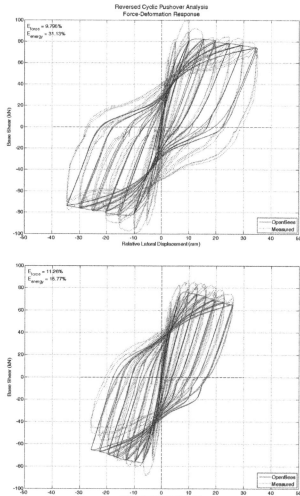

Figure 15. Hysteresis curves for columns C_0 (top) and C_{32} (bottom) from the Dalian tests (Li & Gong, 2008).

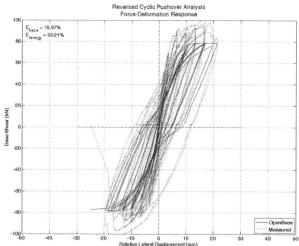

Figure 16. Hysteresis curves for column D_{43} from the Dalian tests (Li & Gong, 2008).

are similar to the values obtained using Modeling Case I. Thus, the consideration of changes in concrete material behaviour and bond strength deterioration does not appear to lead to any significant improvements in the accuracy of the model. Similarly, error statistics for the cyclic response of the Illinois test specimens using Modeling Cases II, III and IV are listed in Table 7. A slight

reduction in the error in hysteretic force is observed in all three cases compared to the simplified modelling approach of Case I;

Table 6. Error statistics for the Dalian tests (Li & Gong, 2008).

	Modelling case							
	I		II		III		IV	
Column specimen	E_{force} (%)	E_{energy} (%)	E_{force} (%)	E_{energy} (%)	E_{force} (%)	E_{energy} (%)	E_{force} (%)	E_{energy} (%)
A_0	13.94	16.15	13.94	16.15	13.94	16.15	13.94	16.15
A_{12}	9.58	17.22	10.02	16.94	11.21	21.72	15.69	25.49
B_0	14.71	17.84	14.71	17.84	14.71	17.84	14.71	17.84
B_{21}	12.05	13.24	11.39	16.85	11.47	14.89	11.52	18.34
B_{22}	15.9	37.71	13.57	41.04	13.13	39.1	16.93	43.57
C_0	9.80	31.13	9.80	31.13	9.80	31.13	9.80	31.13
C_{32}	11.26	15.77	14.78	19.52	11.53	17.29	15	20.48
D_{43}	15.97	53.21	20.56	58.6	16.59	51.75	21.65	56.65
Mean	12.90	25.28	13.60	27.26	12.80	26.23	14.91	28.71

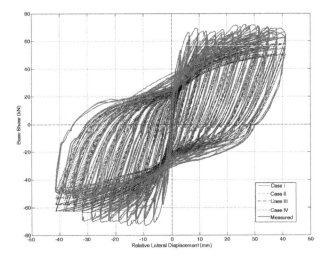

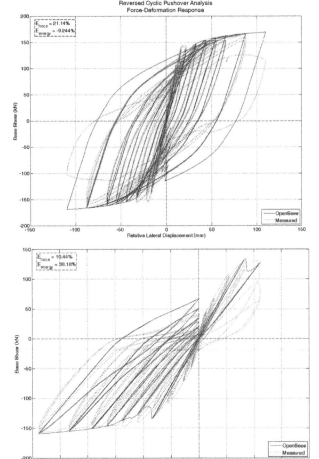

Figure 17. Hysteresis curves for column A_{12} from the Dalian tests (Li & Gong, 2008): Four modelling cases.

however, this is also accompanied by an increase in the error in hysteretic energy in all three cases. Overall, the simplified modelling approach used in Case I that accounts only for the reduction in steel cross-section provides a reasonable degree of accuracy and can be used with confidence for modelling the behaviour of corroded concrete columns subject to cyclic lateral loads.

Conclusions

This paper presents a simplified approach for modelling corroded reinforced concrete bridge columns subjected to cyclic loads and validates the approach by comparison with experimental tests. The columns are modelled as a non-linear distributed-plasticity, fibre beam-column element in OpenSees using the uniaxial Giuffre-Menegotto-Pinto material Steel02 for the reinforcing steel fibres and Concrete01 for the core concrete and cover concrete fibres. Deformations due to shear and bond-slip between

Figure 18. Hysteresis curves from the test and the OpenSees analysis compared for the two column specimens from the Illinois test (Aquino, 2002).

the steel and concrete are taken into account in the model. The modelling of the corroded columns accounts primarily for the

Table 7. Error statistics for the Illinois tests (Aquino, 2002).

	Modelling case							
	I		II		III		IV	
Column specimen	E_{force} (%)	E_{energy} (%)	E_{force} (%)	E_{energy} (%)	E_{force} (%)	E_{energy} (%)	E_{force} (%)	E_{energy} (%)
Uncorroded specimen	8.44	4.77	8.44	4.77	8.44	4.77	8.44	4.77
Corroded specimen	10.44	38.18	10.12	47.65	9.91	44.76	9.84	54.5

reduction in steel cross-section due to corrosion, but changes in the concrete material behaviour due to cracking and spalling of the cover and changes in the bond strength due to corrosion were also investigated. The model was then validated by comparison with two separate sets of experimental tests. Error statistics of normalised hysteretic force and normalised hysteretic energy, used in previous evaluations of distributed-plasticity modelling strategies, were used as a benchmark to assess the accuracy of the proposed model.

The simplified analytical model taking into consideration only the reduction in rebar cross-sectional area shows very good agreement with test results. The consideration of changes in concrete material behaviour or bond strength deterioration does not lead to any significant improvements in the accuracy of the model. However, there is clearly a need for thorough experimental studies in order to better understand bond strength degradation and its controlling factors and for theoretical models that can predict the loss of bond more accurately. One of the main advantages of the non-linear analytical model presented in this paper is that it is simple, yet it captures the effect of corrosion on reinforced concrete bridge columns reasonably well. Moreover, to the best of the authors' knowledge, it is the first time that a proposed structural model of a corroded reinforced concrete column has been validated with experimental data and shown to have accuracy comparable to similar models developed for uncorroded columns. These properties of the proposed modelling approach for corroded concrete columns make it particularly suitable for applications in vulnerability analysis where repeated non-linear dynamic analyses are needed.

Future work will involve further integrating this model both downscale (i.e. materials-level) and upscale (i.e. building-scale and system-scale), accounting for brittle failure modes of reinforced concrete columns and non-circular column geometries which are not yet considered, and using the proposed modelling approach in conjunction with corrosion models to develop realistic time-dependent seismic vulnerability and reliability functions for deteriorating bridges; with the broader aim of integrating this research with life cycle assessment models in order to create a comprehensive tool for determining the influence of alternative design, repair and maintenance options on the sustainability of deteriorating infrastructure systems exposed to extreme events.

Acknowledgements

The authors would like to thank the financial support of National Natural Science Foundation of China (50808158), the National High Technology Research and Development Program of China (2007AA04Z437), the Blume Earthquake Engineering Research Center, the Leavell Fellowship, and the Terman Faculty Fellowship at Stanford University. The authors would also like to thank Professor Wilkins Aquino from the Department of Civil and Environmental Engineering, Duke University, Durham, NC, and Jinbo Li and Professor Jinxin Gong from The State Key Laboratory of Coastal and Offshore Engineering, Dalian University of Technology, Dalian City, Liaoning Province, People's Republic of China for the use of their data from the cyclic load tests. The research presented in this paper was partially supported through U.S. NSF Grants CMMI-106756 and NEESR-105651. Any opinions, findings, conclusions or recommendations expressed in this material are those of the authors and do not necessarily reflect the views of the U.S. National Science Foundation.

Disclosure statement

No potential conflict of interest was reported by the authors.

Funding

This work was supported by the National Science Foundation Division of Civil, Mechanical and Manufacturing Innovation [grant numbers 105651, 106756]; National Natural Science Foundation of China [grant number 50808158]; National High Technology Research and Development Program of China [grant number 2007AA04Z437]; Blume Earthquake Engineering Research Center, Leavell Fellowship and Terman Faculty Fellowship at Stanford University.

References

Akiyama, M., & Frangopol, D. M. (2013). Long-term seismic performance of RC structures in an aggressive environment: Emphasis on bridge piers. *Structure and Infrastructure Engineering, 10,* 865–879.

Akiyama, M., Frangopol, D. M., & Matsuzaki, H. (2011). Life-cycle reliability of RC bridge piers under seismic and airborne chloride hazards. *Earthquake Engineering and Structural Dynamics, 40,* 1671–1687.

Akiyama, M., Frangopol, D. M., & Suzuki, M. (2012). Integration of the effects of airborne chlorides into reliability-based durability design of reinforced concrete structures in a marine environment. *Structure and Infrastructure Engineering, 8,* 125–134.

Alipour, A., Shafei, B., & Shinozuka, M. (2013). Capacity loss evaluation of reinforced concrete bridges located in extreme chloride-laden environments. *Structure and Infrastructure Engineering, 9,* 8–27.

Almusallam, A. A., Al-Gahtani, A. S., & Aziz, A. R. (1996). Effect of reinforcement corrosion on bond strength. *Construction and Building Materials, 10,* 123–129.

Al-Sulaimani, G. J., Kaleemullah, M., Basunbul, I. A., & Rasheeduzzafar. (1990). Influence of corrosion and cracking on bond behavior and strength of reinforced concrete members. *ACI Structural Journal, 87,* 220–231.

Amleh, L. (2000). *Bond deterioration of reinforcing steel in concrete due to corrosion.* Ph.D. Dissertation, McGill University, Montreal, Canada.

Aquino, W. (2002). *Long-term performance of seismically rehabilitated corrosion-damaged columns.* Ph.D. dissertation, University of Illinois at Urbana-Champaign, IL.

Aquino, W., & Hawkins, N. M. (2007). Seismic retrofitting of corroded reinforced concrete columns using carbon composites. *ACI Structural Journal, 104,* 348–356.

Berra, M., Castellani, A., & Coronelli, D. (1997). *Bond in reinforced concrete and corrosion of bars.* Proceedings of the 7th International Conference on Structural Faults and Repair, Engineering Technics Press, Edinburgh, UK, pp. 349–356.

Berry, M. P., & Eberhard, M. O. (2008). *Performance modeling strategies for modern reinforced concrete bridge columns* (Pacific Earthquake Engineering Research Center Report 2007/07). Berkeley, CA: University of California.

Biondini, F., Camnasio, E., & Palermo, A. (2014). Lifetime seismic performance of concrete bridges exposed to corrosion. *Structure and Infrastructure Engineering, 10,* 880–900.

Biondini, F., Palermo, A., & Toniolo, G. (2011). Seismic performance of concrete structures exposed to corrosion: Case studies of low-rise precast buildings. *Structure and Infrastructure Engineering, 7,* 109–119.

Biondini, F., & Vergani, M. (2015). Deteriorating beam finite element for nonlinear analysis of concrete structures under corrosion. *Structure and Infrastructure Engineering, 11,* 519–532.

Cabrera, J. G. (1996). Deterioration of concrete due to reinforcement steel corrosion. *Cement and Concrete Composites, 18,* 47–59.

Cabrera, J. G., & Ghodussi, P. (1992). *Effect of reinforcement corrosion on the strength of steel concrete bond.* Proceedings of the International Conference on Bond in Concrete – from Research to Practice, Riga, Latvia, pp. 10.11–10.24.

Cairns, J., Plizzari, G. A., Du, Y., Law, D. W., & Franzoni, C. (2005). Mechanical properties of corrosion-damaged reinforcement. *ACI Materials Journal, 102*, 256–264.

Celarec, D., Vamvatsikos, D., & Dolšek, M. (2011). Simplified estimation of seismic risk for reinforced concrete buildings with consideration of corrosion over time. *Bulletin of Earthquake Engineering, 9*, 1137–1155.

Chiu, C. K., & Chi, K. N. (2013). Analysis of lifetime losses of low-rise reinforced concrete buildings attacked by corrosion and earthquakes using a novel method. *Structure and Infrastructure Engineering, 9*, 1225–1239.

Choe, D.-E., Gardoni, P., Rosowsky, D. V., & Haukaas, T. (2008). Probabilistic capacity models and seismic fragility estimates for RC columns subject to corrosion. *Reliability Engineering & System Safety, 93*, 383–393.

Choe, D.-E., Gardoni, P., Rosowsky, D. V., & Haukaas, T. (2009). Seismic fragility estimates for reinforced concrete bridges subject to corrosion. *Structural Safety, 31*, 275–283.

Choe, D.-E., Gardoni, P., & Rosowsky, D. V. (2010). Fragility increment functions for deteriorating reinforced concrete bridge columns. *Journal of Engineering Mechanics, 136*, 969–978.

Clark, L. A., & Saifullah, M. (1993). Effect of corrosion on reinforcement bond strength. In: M. Forde (Ed.), *Proceedings of 5th International Conference on Structural Faults and Repair* (Vol. 3, pp. 113–119). Edinburgh: Engineering Technics Press.

Clark, L. A., & Saifullah, M. (1994). Effect of corrosion rate on the bond strength of corroded reinforcement. *Corrosion and corrosion protection of steel in concrete* (pp. 591–602). Sheffield: Sheffield Academic Press.

Coronelli, D. & Gambarova, P. (2004). Structural assessment of corroded reinforced concrete beams: modeling guidelines. *ASCE Journal of Structural Engineering, 130*, 1214–1224. doi:10.1061/(ASCE)0733-9445(2004)130:8(1214)

Fang, C., Lundgren, K., Chen, L., & Zhu, C. (2004). Corrosion influence on bond in reinforced concrete. *Cement and Concrete Research, 34*, 2159–2167.

Fang, C., Lundgren, K., Plos, M., & Gylltoft, K. (2006). Bond behaviour of corroded reinforcing steel bars in concrete. *Cement and Concrete Research, 36*, 1931–1938.

Federation Internationale du Beton (FIB). (2000). *Bond of reinforcement in concrete: state-of-art report. Bulletin, 10*, 160–167.

Frangopol, D. M., Lin, K.-Y., & Estes, A. C. (1997). Reliability of reinforced concrete girders under corrosion attack. *Journal of Structural Engineering, 123*, 286–297.

Ghosh, J., & Padgett, J. E. (2010). Aging considerations in the development of time-dependent seismic fragility curves. *Journal of Structural Engineering, 136*, 1497–1511.

Hartt, W. H., Powers, R. G., Leroux, V., & Lysogorski, D. K. (2004). *Critical literature review of high-performance corrosion reinforcements in concrete bridge applications, FHWA-HRT-04-093.* McLean, VA: Federal Highway Adminisrator.

Kashani, M. M., Lowes, L. N., Crewe, A. J., & Alexander, N. A. (2014). *Implementation of corrosion damage models in nonlinear fibre beam-column element.* 10th U.S. National Conference on Earthquake Engineering: Fontiers of Earthquake Engineering. July 21–25, 2014. Anchorage, AK.

Koch, G. H., Brongers, P. H., Thompson, N. G., Virmani, Y. P., & Payer, J. H. (2002). *Corrosion costs and preventive strategies in the United States, FHWA-RD-01-156.* Washington, DC: FHWA.

Lepech, M., Rao, A., Kiremidjian, A., Michel, A., Stang, H., & Geiker, M. (2015). Multi-physics modeling and multi-scale deterioration modeling of reinforced concrete part II: Coupling corrosion and damage at the structural scale. In K. Dahl, H. Stang, & M. Bræstrup (Eds.), *Proceedings of 2015 fib Symposium: Innovation and Design* (pp. 20–22). Lausanne: Federation International du Beton.

Li, J., & Gong, J. (2008). Influences of rebar corrosion on seismic behavior of circular RC columns. *China Journal of Highway and Transport, 21*, 55–60.

Liu, Y., & Weyers, R. E. (1998). Modeling the time-to-corrosion cracking in chloride contaminated reinforced concrete structures. *ACI Materials Journal, 95*, 675–680.

Lounis, Z. (2005). Uncertainty modeling of chloride contamination and corrosion of concrete bridges. In N. O. Attoh-Okine & B. M. Ayyub, (Eds.), *Applied research in uncertainty modeling and analysis* (pp. 491–511). USA: Springer-Verlag.

Mander, J. B., Priestley, M. J. N., & Park, R. (1988). Theoretical stress–strain model for confined concrete. *Journal of Structural Engineering, 114*, 1804–1826.

Mazzoni, S., McKenna, F., Scott, M. H., & Fenves, G. L. (2009). *Open system for earthquake engineering simulation user command-language manual.* Berkeley, CA: Pacific Earthquake Engineering Research Center.

Michel, A., Geiker, M., Stang, H., & Lepech, M. (2015). Multi-physics modeling and multi-scale deterioration modeling of reinforced concrete part I: Coupling transport and corrosion at the material scale. In K. Dahl, H. Stang, & M. Bræstrup (Eds.), *Proceedings of 2015 fib Symposium: Innovation and Design* (pp. 18–20). Lausanne: Federation International du Beton.

Middleton, C. R., & Hogg, V. (1998). *Review of deterioration models used to predict corrosion in reinforced concrete structures.* Technical report, Cambridge: Cambridge University Department of Engineering.

Molina, F., Alonso, C., & Andrade, C. (1993). Cover cracking as a function of rebar corrosion: Part 2 - Numerical model. *Materials and Structures, 26*, 532–548.

Neville, A. (1995). Chloride attack of reinforced concrete: An overview. *Materials and Structures, 28*, 63–70.

Papakonstantinou, K. G., & Shinozuka, M. (2013). Probabilistic model for steel corrosion in reinforced concrete structures of large dimensions considering crack effects. *Engineering Structures, 57*, 306–326.

Ranf, R. T. (2007). *Model selection for performance-based earthquake engineering of bridges* (PhD thesis). University of Washington, USA.

Rao, A. (2014). *Structural deterioration and time-dependent seismic risk analysis* (PhD thesis). Department of Civil and Environmental Engineering. Stanford University, Stanford, CA, USA.

Rao, A. S., Lepech, M. D., & Kiremidjian, A. S. (2013). *Time-dependent risk assessment of deteriorating reinforced-concrete bridges for sustainable infrastructure design.* Proceedings of the 11th International Conference on Structural Safety & Reliability, June 16–20, 2013, New York, NY, USA.

Rodriguez, J., Ortega, L., & Casal, J. (1994). Corrosion of reinforcing bars and service life of reinforced concrete structures: Corrosion and bond deterioration. *International Conference on Concrete Across Borders, Odense, Denmark, 2*, 315–326.

Simon, J., Bracci, J. M., & Gardoni, P. (2010). Seismic response and fragility of deteriorated reinforced concrete bridges. *Journal of Structural Engineering., 136*, 1273–1281.

Stanish, K. (1997). *Corrosion effects on bond strength in reinforced concrete* (MSc thesis). Department of Civil Engineering, University of Toronto, Canada.

Stewart, M. G., & Rosowsky, D. V. (1998). Time-dependent reliability of deteriorating reinforced concrete bridge decks. *Structural Safety, 20*, 91–109.

Titi, A., & Biondini, F. (2014). Probabilistic seismic assessment of multistory precast concrete frames exposed to corrosion. *Bulletin of Earthquake Engineering, 12*, 2665–2681.

Tuutti, K. (1982). *Corrosion of steel in concrete.* Stockholm: Swedish Cement and Concrete Research Institute.

Val, D. V., Stewart, M. G., & Melchers, R. E. (1998). Effect of reinforcement corrosion on reliability of highway bridges. *Engineering Structures, 20*, 1010–1019.

Vu, K. A. T., & Stewart, M. G. (2000). Structural reliability of concrete bridges including improved chloride-induced corrosion models. *Structural Safety, 22*, 313–333.

Wang, X., & Liu, X. (2004). Modeling bond strength of corroded reinforcement without stirrups. *Cement and Concrete Research, 34*, 1331–1339.

Yalciner, H., Sensoy, S., & Eren, O. (2012). Time-dependent seismic performance assessment of a single-degree-of-freedom frame subject to corrosion. *Engineering Failure Analysis, 19*, 109–122.

Development of time-dependent fragility functions for deteriorating reinforced concrete bridge piers[1]

Anirudh S. Rao, Michael D. Lepech and Anne Kiremidjian

ABSTRACT

This is the second article of an effort to develop a time-dependent framework for earthquake risk assessment. In this paper, a model is developed for estimating fragility functions that depend on the degree of structural deterioration. The deterioration model presented in the first paper is used to model the change in the structure over time and due to exposure to different environmental conditions. The effect of deterioration on structural demand and capacities is considered in the evaluation of the fragility functions. Three reinforced concrete columns are designed to correspond to seismic design criteria in place in 1960, 1980 and 2000. Fragility functions are developed for non-deteriorating and deteriorating columns based on their age and environmental exposure. It is shown that the time of construction, which reflects the seismic design criteria, and the environmental conditions dictate the amount of structural deterioration. Older columns in highly corrosive environments show the highest vulnerability. Thus, ignoring the degree of deterioration can lead to a significant underestimation of damage and loss.

Introduction

Seismic fragility analysis involves understanding and probabilistically quantifying the propensity of a structure or structural component to undergo damage during an earthquake. Seismic fragility analysis forms an integral part of the Pacific Earthquake Engineering Research (PEER) Center framework for seismic risk analysis, along with seismic hazard analysis and consequence analysis. As a result, much work has been done in the past two decades to develop methods to generate fragility functions for both structural and non-structural components. However, the onset of corrosion deterioration can lead to a shifting of the fragility curves for a component over time, contributing to a time-dependent seismic risk for the component.

In recent years, the researchers have begun addressing the effects of deterioration on the seismic fragility of structures. Billah and Alam (2015) provide a review of the various methods of fragility development including the heuristic, empirical and analytical approaches. Biondini, Palermo, and Toniolo (2011) study the performance of concrete decks exposed to corrosion and subjected to seismic excitations. Structural deterioration has been considered by Akiyama, Frangopol, and Matsuzaki (2011), Akiyama, Frangopol, and Suzuki (2012) in the development of an overall risk model for RC concrete piers and for reliability-based durability design of such structures. In both of these studies, a hazard model is considered to capture the degree of chloride concentration at the location of the bridge relative to the coastline.

Rao, Lepech, Kiremidjian, and Sun (in press) provide an extensive discussion on the fundamental differences in the corrosion modelling presented here and those used in previous studies.

In this paper, an approach for seismic fragility development for bridge columns subjected to corrosion deterioration is presented first. A case study of three California highway bridge columns designed according to different seismic design criteria reflecting three construction time periods is used to investigate the changes that occur in both the damage state capacities and the seismic demand for corroding bridge columns. Prescriptive equations using physics or empirical relationships based on experimental evidence are used to model the damage state capacities for the columns, whereas multiple non-linear dynamic stripe analysis is used to obtain a probabilistic seismic demand model for the deteriorating columns. Seismic fragility functions for the deteriorating columns are then generated for different levels of reinforcement corrosion.

The effect of corrosion on the probability of collapse at different intensity levels is examined for the three case study columns. An increase in the probability of collapse at a given intensity level is observed for all three columns, with the oldest column exhibiting the highest probability of collapse amongst the three columns for a given intensity of ground motion and a given level of corrosion deterioration. The oldest of the three columns, which is designed according to pre-1971 design standards, is also shown to have the highest mean annual frequency of collapse.

---

[1]This paper is part of the keynote lecture presented by Prof. Anne Kiremidjian on 'Time-dependent earthquake risk assessment modeling considering sustainability metrics' at the IALCCE 2014, Tokyo, Japan.

Deterioration modelling of concrete columns

Reinforced concrete, RC, is the most popular construction material, however the carbon steel used for reinforcement makes it highly susceptible to corrosion deterioration. Amongst the many models for RC deterioration modelling, the approach proposed by Frangopol and Das (1999), Thoft-Christensen (1999) and Frangopol, Kong, and Gharaibeh (2000, 2001) is most appropriate for use within the PEER framework. Their structural reliability-based formulation uses time-dependent reliability index, $\beta(t)$, that considers the time to initiation of deterioration, the rate of deterioration, the time to the next maintenance and the effect of maintenance on the reliability index as random variables.

Most recently, Akiyama et al. (2011, 2012) developed formulation for risk assessment of RC columns subjected to both chloride exposure and earthquake excitation. While Akiyama et al. (2011, 2012) evaluate the seismic performance for different levels of corrosion under the premise that such corrosion can be established by X-ray investigations, the current paper uses a phenomenological model that includes the various stages of the corrosion process – depassification leading to corrosion initiation, concrete cover cracking and concrete spalling as described in Rao et al. (in press). In addition, the current paper considers the degree of corrosion from airborne chlorides and corrosion due to seawater splashing on the columns. An advantage of the Akiyama et al. approach, however, is the probabilistic hazard treatment of the airborne chloride concentration that leads to different levels of corrosion. Ideally, a more robust model can be developed that incorporates the mechanistic model used by Rao et al. and the corrosion hazard model by Akiyama et al.

As stated previously, the deterioration due to corrosion is modelled using phenomenological processes originally proposed by Tuutti (1982) and later extended by others. The deterioration model is described in Rao et al. (in press) and is used to develop an approach for modelling corroded reinforced columns

subjected to cyclic loads. The model was validated with experimental data and an accuracy of the non-linear behaviour of corroded columns is shown to be of the same order of typical non-linear analysis of conventional non-corroded columns (see Rao et al., in press). In this paper, statistical analysis is performed for three example columns to develop probability distributions for corrosion initiation and degree of corrosion. In order to capture exposure to varying environmental conditions, the three columns are placed within: (a) the coastal splash zone and (b) the coastal and marine atmosphere zone (distance to the coastline $d_{coast} = 200$ m). In the splash zone, the columns are subjected to direct salt spray from the tidal water and have high concentration of chlorides on the surface with a mean of 7.35 kg/m^3 and coefficient of variation (COV) of .7 of surface chloride concentration (after Val & Stewart, 2003).

Detailed structural models for reinforced concrete columns designed under three different design codes are considered to illustrate the methodology developed in this paper. The three columns correspond to RC columns designed in 1960, referred to as C_{60}, in 1980 referred to as C_{80} and in 2000 referred to as C_{00}. The dates correspond to major changes in the seismic design codes in California. Figure 1 shows the cross-sectional properties of the three columns reflecting the increase in number and cross-sectional area of longitudinal rebar, and decrease in the tie spacing along the height of the columns. Subsequent applications will be for these three columns located in Oakland, California, thus the selection of these designs.

Corrosion initiation

Corrosion is initiated when cracks in the concrete cover allow chlorides to reach the reinforcing bard of an RC column. A Monte Carlo analysis was carried out in order to obtain the distribution for the time to initiation of corrosion, T_{init}, for the three columns C_{60}, C_{80} and C_{00} using the mechanistic model described

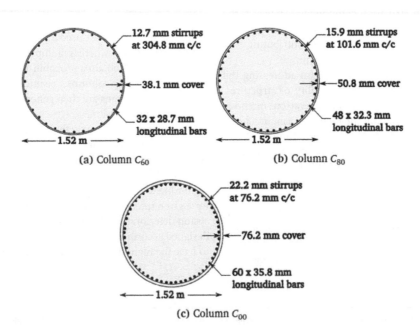

(a) Column C_{60}

(b) Column C_{80}

(c) Column C_{00}

Figure 1. Cross sections of three example columns. (a) Column C_{60}, designed according to a 1960s code; (b) Column C_{80}, designed according to a 1980s code, (c) Column C_{00}, designed according to a 2000's code.

in Rao et al. (in press). For the purposes of this analysis, the only difference in the three columns is the thickness of concrete cover provided to the columns. Column C_{60} has a mean cover thickness of 38.1 mm, column C_{80} has a mean cover thickness of 50.8 mm and column C_{00} has a mean cover thickness of 76.2 mm. The effect of distance to the coast, d_{coast}, is examined by running the analysis for the three columns for two different locations relative to the coast: (1) the splash zone and (2) the coastal or marine atmosphere zone (d_{coast} = 200 m). Columns located in the splash zone are subjected to direct salt spray from the tidal water, and have a high concentration of chlorides on the surface.

A mean value of 7.35 kg/m3 and a COV of .70 were used for the surface chloride concentration for the splash zone (after Val & Stewart, 2003). Columns located in the coastal or marine atmosphere zone, though not subject to direct salt spray, are still exposed to a high level of atmospheric chlorides due to proximity to the ocean. The mean and COV of the surface chloride concentration, C_0, for this zone were estimated. A lognormal probability density was used to model the surface chloride concentration (Val & Stewart, 2003). A uniform probability density from .9 kg/m³ to 1.8 kg/ m³ was used for critical chloride concentration at the steel reinforcement surface, $Ccrit$, based on the Caltrans Corrosion Guidelines (California Department of Transportation, 2012). A mean of 5.11×10^{-12} m²/s and a COV of 1.42 were used for the diffusion coefficient, Dc, based on the findings from Weyers et al. (1994), which were obtained from analysis of 252 concrete samples from 49 California bridges. A normal distribution was truncated for values smaller than zero and used to model Dc, and the diffusion coefficient was assumed to be time-independent. The concrete cover thickness was modelled using a normal distribution, with a COV equal to .17 (Mirza & MacGregor, 1982). A detailed description of the corrosion initiation model is given in Rao (2014).

A sample size of 10,000 was generated using the Latin hypercube sampling scheme to estimate the distribution of time to corrosion initiation. Lognormal, Weibull and Gamma distributions were fitted to the output of the simulations. The lognormal distribution was found to consistently fit the data the best based on Kolmogorov–Smirnov, Cramer–von Mises and Anderson Darling goodness of fit tests. Figure 2 shows the three distributions fitted to the simulated times to corrosion initiation for the C_{00} column. Similar analyses were performed for the C_{60} and C_{80}

columns. Table 1 summarises the results of the three statistical tests for each of the fitted distributions. From the figure and the statistical tests it is shown that the lognormal distribution has the best fit to the data.

Figures 3(a) and (b) present the distributions for the C_{60} columns in the splash zone and the coastal zone. From these figures, the effect of the higher exposure to chlorides in the splash zone is clearly seen as the time to corrosion initiation occurs considerably earlier than that at the coastal/marine zone. Table 2 provides the mean value of the time to corrosion initiation

Table 1. Goodness of fit test results for the Lognormal, Gamma and Weibull distributions fitted to the simulation results for the time to corrosion initiation for the C_{00} column located in the splash zone.

Statistic	Lognormal	Gamma	Weibull
Kolmogorov–Smirnov	.013	.053	.070
Cramer–von Mises	.128	4.202	10.447
Anderson–Darling	.687	24.284	70.169

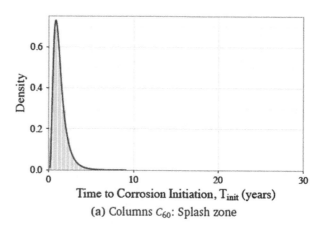

(a) Columns C_{60}: Splash zone

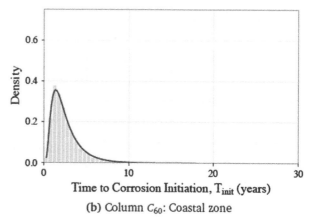

(b) Column C_{60}: Coastal zone

Figure 3. Distribution of the time to corrosion initiation for the C_{60} column located in (a) the splash zone and (b) in the coastal/marine zone.

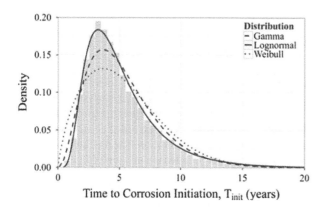

Figure 2. Comparison of distribution fits for time to corrosion initiation for C_{00} column in the splash zone.

Table 2. Mean and standard deviation of the fitted lognormal distribution for T_{init}.

Column	Splash zone		Marine atmospheric zone	
	$\mu_{\tau_{init}}$ (years)	$\sigma_{\tau_{init}}$ (years)	$\mu_{\tau_{init}}$ (years)	$\sigma_{\tau_{init}}$ (years)
C_{60}	1.35	.66	2.86	1.84
C_{80}	2.40	1.17	5.08	3.27
C_{00}	5.39	2.62	11.44	7.39

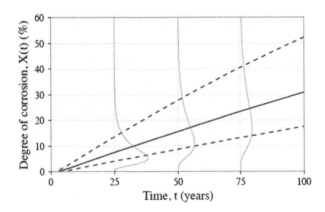

Figure 4. Distributions for the degree of corrosion, $W(t)$ for the C_{60} column as a function of time.

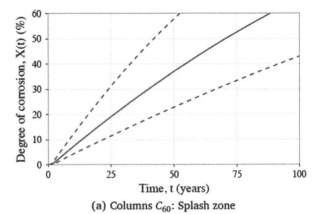

(a) Columns C_{60}: Splash zone

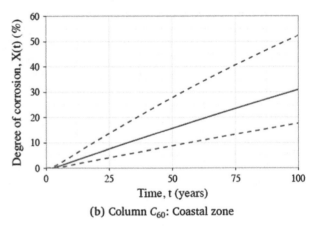

(b) Column C_{60}: Coastal zone

Figure 5. Degree of column deterioration (%) steel mass loss for the C_{60} column in (a) the splash zone and (b) the coastal/marine atmospheric zone. (Note that $X(t) = W(t)$).

and the corresponding standard deviation for the three types of columns studied located in the coastal and splash zones. As expected, the mean time to corrosion initiation increases with the newer columns, which have greater concrete cover.

Degree of corrosion

Similar to the approach for time to corrosion initiation estimation, a Monte Carlo simulation approach was used for the development of the time-dependent distributions for the degree of corrosion, denoted as $W(t)$, represented in this case by the amount of steel mass loss. The same three columns exposed to the two environmental regions were used for that purpose. The longitudinal rebar diameters are 28.7, 32.3 and 35.8 mm, respectively, for the C_{60}, C_{80} and C_{00} columns. The degree of corrosion is a function of the rate of corrosion typically measured in terms of the corrosion current density and corrosion initiation.

Figure 4 presents the results of the Monte Carlo simulation for the degree of corrosion, $W(t)$, for the C_{60} column as a function of time located in the coastal zone. Lognormal, Gamma and Weibull distributions were again tested against the simulated data. Both the Kolmogorov–Smirnov and Cramer–von Mises statistical goodness of fit test showed that the lognormal distribution provides the best fit. Thus, lognormal distributions were fitted to the data for each year as shown on Figure 4. The differences in the corrosion rates between columns in the coastal and the plash zones are shown in Figure 5. Table 3 lists the parameters of the lognormal distribution for each column in the two environmental zones for times $t = 25, 50, 75$ and 100 years after construction.

In the development of the degree of corrosion it is assumed that there are no repairs over the forecast time and corrosion is allowed to continue uninhibited. Since the objective of the analysis presented herein is to find the effect of deterioration on the risk over time, this assumption is appropriate.

Seismic fragility functions for deteriorating structures

Seismic fragility is most often expressed as a function of the engineering demand parameter (EDP), or as a function of the intensity measure, *IM*. The approach yields EDP-based seismic

Table 3. Mean and standard deviation of the fitted lognormal distribution for $X(t)$.

Column	Time, t (years)	Splash zone $\mu_{X(t)}$ (%)	Splash zone $\sigma_{X(t)}$ (%)	Marine atmospheric zone $\mu_{X(t)}$ (%)	Marine atmospheric zone $\sigma_{X(t)}$ (%)
C_{60}	25	21.36	11.35	8.97	5.89
	50	40.18	19.38	18.37	11.52
	75	55.63	25.56	27.03	16.27
	100	71.36	43.35	34.91	20.10
C_{80}	25	18.36	9.93	7.57	6.15
	50	35.76	17.73	15.74	10.08
	75	50.33	23.31	23.66	14.53
	100	63.80	33.42	30.98	18.30
C_{00}	25	14.69	8.68	5.80	4.86
	50	30.97	15.89	13.60	12.69
	75	44.85	21.03	19.91	13.21
	100	56.79	25.89	26.63	16.25

fragility functions, and it is often used to describe the fragilities of building components in the seismic risk analysis of buildings. An *IM*-based fragility function describes the probability that a structural component will meet or exceed a certain damage state for a given intensity of ground motion. The fragility function for the *ith* damage state for an intensity measure, *IM*, is given as:

$$Fr_i(im) = P[DM \geq d_i | IM = im] \tag{1}$$

The probability that the structural component will be in a particular damage state is given by:

$$Pr_i(im) = \begin{cases} 1 - Fr_i(im) & i = 1 \\ Fr_i(im) - Fr_{i+1}(im) & 1 \le i < N \\ Fr_i(im) & i = N \end{cases} \quad (2)$$

where, N is the total number of discrete damage states.

Fragility functions are widely used in seismic risk analysis for predicting the level of damage sustained by a component during an earthquake. There are three major approaches typically used for determining the fragility functions for structural components: (1) expert-based approach, (2) experimental or empirical approach and (3) analytical approach. In this paper, the analytical approach is used to develop fragility functions. In this approach, both the damage state capacities, Ci, and the demand generated on the component as a function of ground motion intensity, $D(im)$, are modelled analytically, and the fragility function corresponding to the i_{th} damage state can be described as follows:

$$Fr_i(im) = P\left[\frac{D(im)}{C_i} \ge 1\right] \quad (3)$$

where $Fri(im)$ is the probability that the ith damage state will be met or exceeded.

The structural capacity and seismic demand are often represented by a lognormal distribution. In this case, the probability that a particular damage state will be met or exceeded will also be described by a lognormal CDF as follows:

$$Fr_i(im) = \Phi\left[\frac{\mu_{\ln D(im)} - \mu_{\ln C_i}}{\sqrt{\beta_{D(im)}^2 + \beta_{C_i}^2}}\right] \quad (4)$$

where $\mu_{\ln D(im)}$ is mean of the natural logarithm of the seismic demand at an intensity level im, $\mu_{\ln Ci}$ is mean of the natural logarithm of the ith damage state capacity, $\beta_{D(im)}$ is the logarithmic standard deviation of the seismic demand at an intensity level im, $\beta_{\ln Ci}$ is the logarithmic standard deviation of the i_{th} damage state capacity and $\Phi[*]$ is the standard normal distribution function

Structural modelling of corroded RC bridge columns

Corrosion of reinforcement damages concrete structures in multiple ways. A full description of the corrosion and structural deterioration model used in the study can be found in Rao et al. (in press). For demonstration in this paper, a uniform reinforcement corrosion model is adopted. While the effects of highly localised pitting corrosion, localised changes in reinforcement ductility, and localised concrete deterioration and cracking are of critical concern, the simple uniform model is used to demonstrate the multi-scale integration of materials deterioration modelling with structural deterioration modelling for the creation of fragility functions.

As discussed in Rao et al. (in press), the inclusion of more complex deterioration phenomena did not substantially increase the accuracy of the structural deterioration model. However, recognising the importance of non-uniform deterioration going forward, and building from this simple framework, the authors continue to integrate more sophisticated deterioration models

that bridge multi-physics deterioration models (i.e. materials-level), as demonstrated by Michel, Geiker, Stang, and Lepech (2015) and Lepech et al. (2015), and larger scale models (i.e. building-level or system-level), as demonstrated by Rao, Lepech, and Kiremidjian (2013) that allow for detailed capture of spatially varying environmental loads, corrosion propagation and crack formation.

In order to develop fragility functions for reinforced concrete bridge columns subjected to seismic loads that have undergone deterioration due to corrosion, it is first necessary to develop a model for dynamic analysis of non-corroded columns that captures their properties and dynamic behaviour. An analytical model for a non-corroded bridge column was first developed using the Open System for Earthquake Engineering Simulation (OpenSees) software (Mazzoni, McKenna, Scott, & Fenves, 2007). OpenSees was chosen because it provides an extensive library of material models, non-linear elements and solvers that can be easily interchanged, and it has been widely adopted by the research community, making it highly conducive for the modelling and simulation of complex structural systems subject to seismic loading.

Other researchers have successfully adopted the use of lumped plasticity models for fragility analysis of corroded reinforced concrete structures, which can be more computationally efficient than fibre beam element models. However, as discussed in Rao et al. (in press), such models are not considered for this work even though they are highly computationally efficient since they do not facilitate easy integration with adjacent scale models, such as those demonstrated by Michel et al. (2015) and Lepech et al. (2015).

The column is modelled as a non-linear distributed plasticity, fibre beam-column element, which is a flexibility-based element that considers the spread of plasticity along the length of the element. The assumptions and the details of the finite element model used to capture the behaviour of the column are given in Rao (2014) and summarised in Rao et al. (in press). After the model for the non-corroded column was formulated and tested, a structural model for the corroded column was developed. To account for the effects of corrosion, the following changes were made as a function of the degree of corrosion:

- The area of the reinforcing steel is reduced.
- The strength of the cover concrete elements is reduced.
- The bond strength is modified.

Two test data-sets were used to validate the structural model for the corroded columns as described in Rao et al. (in press). The first set of tests comprised of eight columns having the same specifications that were subjected to different axial loads and varying degrees of accelerated corrosion in an experimental study conducted at the State Key Laboratory of Coastal and Offshore Engineering at Dalian University of Technology in China (Li & Gong, 2008). The second set of tests comprised of one non-corroded column and one corroded column, and are taken from a larger set of tests conducted at the Newmark Structural Engineering Research Laboratory of the Department of Civil and Environmental Engineering at University of Illinois at Urbana-Champaign (Aquino, 2002; Aquino & Hawkins, 2007). An extensive discussion of this analysis is presented in Rao et al. (in press).

Table 4. Bridge column damage states descriptions.

Damage state	Description	Repair
D1: Slight	Minor cracks	Seal cracks and paint
D2: Moderate	Shear cracks and spalling	Epoxy injection and spall repair
D3: Extensive	Bar buckling	Replace buckled bars and FRP wrap
D4: Complete	Collapse or imminent collapse	Replace column

Table 5. Bridge column damage state capacities.

	Median drift ratio		
	Column C_{60}	Column C_{80}	Column C_{00}
Damage state	Pre-1971 Era	1971 – 1990 Era	Post-1990 Era
D1: Slight	.23%	.23%	.23%
D2: Moderate	.85%	2.2%	2.20%
D3: Extensive	1.93%	5.41%	7.28%
D4: Complete	3.28%	14.07%	17.83%

Based on the validation tests, it is observed that the simplest of the analytical models that takes into consideration only the reduction in rebar cross-sectional area is able to provide the same level of accuracy as the model proposed by Berry and Eberhard (2007) for non-corroded columns. It was also found that inclusion of changes in concrete material behaviour or bond-strength deterioration did not lead to any significant improvements in the accuracy of the model. It is noted that the simplified approach was found to be unsuitable for modelling severely damaged columns that are also subject to a high axial load. However, the objective of this paper was to illustrate the general formulation and a more complex model can certainly be used with the proposed approach. Computational efficiency is also a major consideration in the development of fragility functions often trading complex analytical models for efficient analysis. Thus, the simplified model was used in the development of the time-dependent fragility functions described in the next section.

Bridge column damage states

In order to develop fragility functions for corroded columns, it is necessary to define specific damage states. Various definitions for damage states have been proposed each reflecting different degrees of observed damage (e.g. ATC13 1985; Mackie, Wong, & Stojadinovi'c, 2008; Vosooghi & Saiidi, 2010; HAZUS 2011). The definitions proposed by the California Department of Transportation's Structures Maintenance and Investigations division (CALTRANS, 2008) were chosen for this analysis. The four damage states are defined in Table 4.

Damage state capacities for corroded columns

The methods used in this paper for estimating the capacity level associated with the different damage states described in Table 4 are obtained quantitatively in terms of material and geometric properties of components using either physics-based characterisations of the mechanics of the problem or through empirical equations based on experimental tests. Different EDPs can be used to describe the damage state capacities of bridge columns.

For instance, Mackie et al. (2008) used both maximum column drift and residual drift to predict damage state capacities. Hwang, Liu, and Chiu (2001) defined damage state capacities in terms of displacement ductility, whereas use of the curvature ductility to define damage state capacities has been preferred by Choi, DesRoches, and Nielson (2004), Nielson (2005), Padgett and DesRoches (2008), Ghosh and Padgett (2010), and Ramanathan (2012). Vosooghi and Saiidi (2012) obtained damage state capacities in terms of six different EDPs, including maximum drift ratio, residual drift ratio, frequency ratio (defined as the ratio of the fundamental frequency of a column measured

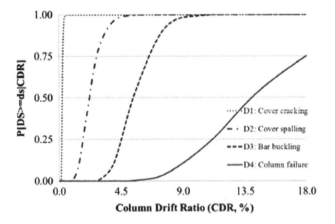

Figure 6. Capacity fragility function for the C_{80} column.

during a test to its elastic fundamental frequency), maximum longitudinal steel strain and maximum transverse steel strain. The maximum drift ratio was preferred as the EDP for damage state capacities over the curvature ductility, as the curvature ductility has limited applicability to the non-ductile pre-1971 column (C_{60}) included in this study.

For each damage state, specific analytical formations were used to correlate the column median drift ratio to observed damage. These are given in Rao (2014). The values of the median drift ratios are summarised in Table 5 for each of the three columns in their non-corroded state. The coefficients of variation used in the analysis are those proposed by Berry and Eberhard (2007), Elwood and Moehle (2005) and Mackie et al. (2008). Lognormal distributions are then used to model the capacity fragility functions using the median values given in the table. Figure 6 shows the capacity functions for the C_{80} column in its non-corroded state.

To account for the effects of corrosion, the model for the non-corroded column was used by changing the material and geometric properties of the columns. Figures 7(a) and (b) display the capacity curves for degrees of corrosion ranging from 0 to 40% of steel mass loss, in increments of 10%, for damage states 3 and 4. These figures correspond to corrosion deterioration in the C_{80} column models.

Seismic demand analysis of corroded columns

The need to model the seismic demand on a structural component probabilistically arises from the consideration that the same structural component can behave quite differently when subjected to different ground motion records of similar intensity levels. In addition, uncertainties in the geometric and material

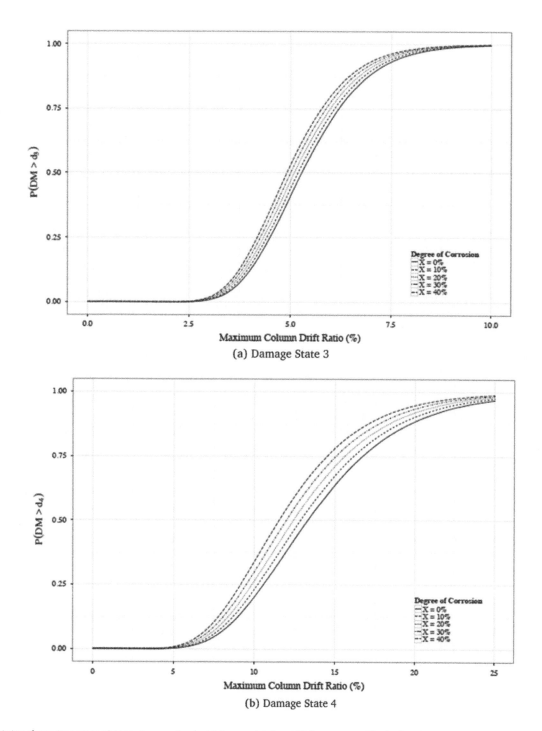

Figure 7. Variation of capacity curves with increasing mass loss for (a) damage state 3 and (b) damage states 4 for the C_{80} columns.

properties of the structural component also contribute to the uncertainty in the seismic demand, and a probabilistic description of demand can be used to capture this uncertainty.

In order to obtain the probabilistic demand, an analytical model is used to subject the structure to a suite of ground motion. A power law is then often used to predict the median demand D as a function of the ground motion intensity IM (Cornell and Krawinkler, 2000):

$$\hat{D} = aIM^b \qquad (5)$$

where the coefficients a and b are estimated through regression analysis of $\ln D$ on $\ln IM$. This method has been preferred by Choi et al. (2004), Nielson and DesRoches (2007), Padgett and DesRoches (2008) and Ghosh and Padgett (2010), amongst others, for the generation of seismic fragility functions for bridge columns.

The record-to-record dispersion of the demand at a given intensity level, $\beta_{D|IM}$, is almost always assumed to be constant across the entire range of intensity levels considered in a cloud analysis. In reality, $\beta_{D|IM}$ is not constant, but tends to increase with increasing intensity of ground motion. Assuming a constant $\beta_{D|IM}$ over the entire range of intensity levels may lead to non-conservative estimates of the risk (Jalayer, 2003). Various wide-range probabilistic seismic demand analysis techniques which allow for different $\beta_{D|IM}$ values over the range of intensity levels can be used to obtain the relationship between D and IM instead of cloud analysis.

Examples of such techniques include incremental dynamic analysis (IDA) and multiple stripe analysis (MSA). The difference between IDA and MSA is that in IDA the individual records are scaled to different intensities and the engineering demand parameter is recorded. In MSA a set of records are scaled to a particular intensity measure and all the records are used to obtain a median and a dispersion value of the engineering demand parameter at that intensity measure. In this particular application, the MSA was found to be computationally more efficient and thus used in the analysis.

Similar to the capacity computations, first the demand curves for the non-corroded columns are obtained and then column deterioration is incorporated to reflect the degree of mass loss due to corrosion. For the single column analysis presented in this paper the ground motion intensity measure selected is the peak ground acceleration (PGA). The set of 40 broad-band ground motions for rock sites provided by Baker and Cornell (2005) was used to simulate the demand on the column. As in all other previous analyses, the collapse probability is evaluated separately. Figure 8(a) shows the median, 16th and 84th percentile of maximum drift ratio as a function of PGA for the C_{60} column without corrosion. The collapse cases are shown in Figure 8(b). From these figures it can be observed that the dispersion increases with greater drift and are particularly large at PGA values greater than .5 g. It should be noted using the spectral acceleration as the IM parameter did not result in any reduction in the dispersion.

To account for the deterioration due to corrosion in the column, the following changes are made in the structural model as a function of the degree of corrosion: (a) the area of the longitudinal reinforcing steel is reduced and (b) the strength of the cover concrete elements is also reduced. The change in bond strength with corrosion was not considered in this analysis. It is hypothesised that the fundamental period of the column will

change with increased degree of corrosion. Eigenvalue analyses were performed to investigate this change and the results are summarised in Table 6.

The results of the MSA for the corroded C_{60} columns are shown in Figures 9(a)–(f). Figures 9(a), (c) and (e) correspond to the median, 16th and 84th percentile of maximum column drift ratio as a function of PGA. These figures are correspondingly for 0, 20 and 40% mass loss in the reinforcing bars. Figures 9(b), (d) and (f) are the corresponding plots for column collapse. Similar results were also obtained for the C_{80} and C_{00} columns.

Seismic fragility estimates

With the median and dispersion of damage state capacities and the median and dispersion of the demand at various intensity levels evaluated, the fragility for the four damage states for the non-corroded and corroded columns can be obtained by integrating the two curves. The probability of being or exceeding a specified damage state i is given by:

$$F_i(im) = P_{NC}(im) \times \Phi\left[\frac{\mu_{\ln D(im)|NC} - \mu_{\ln C_i}}{\sqrt{\beta_{D(im)|NC}^2 + \beta_{C_i}^2}}\right] + P_C(im) \quad (6)$$

where $\mu_{\ln D(im)|NC}$ = mean of the natural logarithm of the seismic demand at intensity level im given that there is no collapse; $\mu_{\ln C_i}$ = mean of the natural logarithm of the ith damage state capacity; β corresponds to the logarithmic standard deviations of capacity and demand; $P_{NC}(im)$ and $P_C(im)$ are, respectively, the probabilities of no collapse and collapse.

A smooth curve can now be fitted to these fragility values to get a fragility function or fragility curve. The cumulative lognormal distribution function has been found to be appropriate for modelling the fragility of a variety of structural and non-structural components (Porter, Hamburger, & Kennedy, 2007), and it is widely used in seismic risk analysis. Thus, the cumulative lognormal distribution function is used to model the damage state fragilities in this study, and the maximum likelihood method is used to estimate the parameters of the fitted cumulative lognormal distribution function. Figures 10(a)–(c) show the cumulative

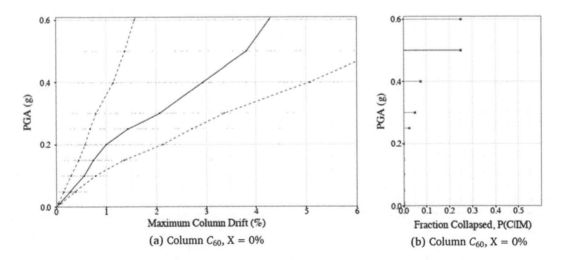

(a) Column C_{60}, X = 0%   (b) Column C_{60}, X = 0%

Figure 8. Median, 16th and 84th percentile of maximum drift for the C_{60} column in non-corroded state.

Table 6. Changes in the fundamental period of vibration of columns with increasing level of corrosion.

Column	Degree of corrosion (%)								
	0	5	10	15	20	25	30	35	40
C_{60}	.529	.530	.531	.531	.532	.533	.534	.535	.536
C_{80}	.520	.521	.522	.523	.524	.525	.526	.527	.637
C_{00}	.491	.493	.496	.499	.502	.504	.507	.510	.512

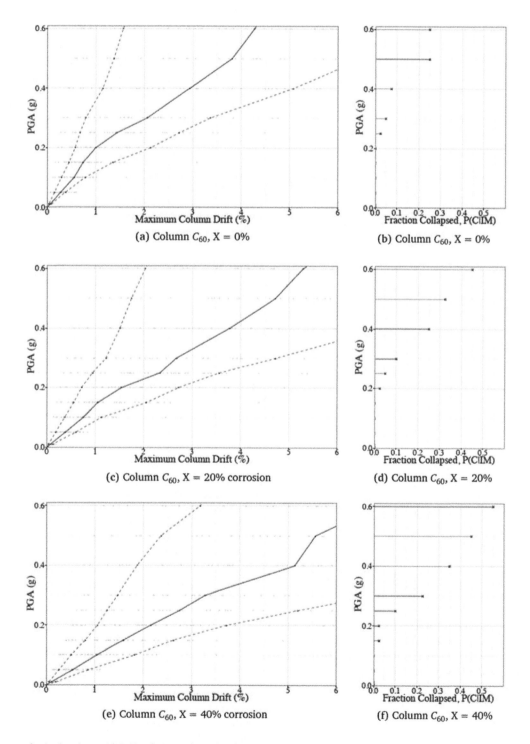

(a) Column C_{60}, X = 0%

(b) Column C_{60}, X = 0%

(c) Column C_{60}, X = 20% corrosion

(d) Column C_{60}, X = 20%

(e) Column C_{60}, X = 40% corrosion

(f) Column C_{60}, X = 40%

Figure 9. MSA curves for the C_{60} column with 0, 20 and 40% steel mass loss due to corrosion.

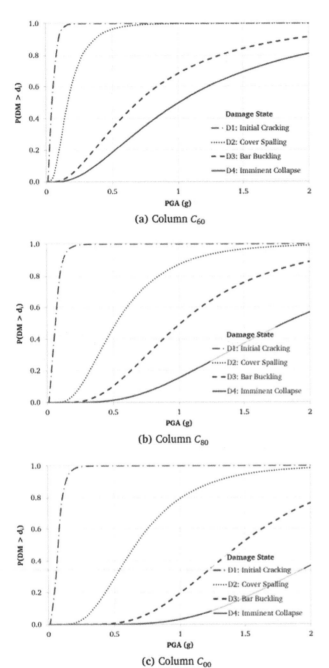

Figure 10. Fragility functions for the non-corroded (a) C_{60}, (b) C_{80} and (c) C_{00} columns.

lognormal fragility functions obtained for the three columns for the different damage without corrosion. Medians and dispersions of the fitted fragility functions for the different damage states for the three columns can be found in Rao (2014).

When the effect of corrosion is included, the fragility function is modified and the part of Equation (6) that changes is given as follows:

$$F_i(im, w) = P_{NC}(im, w) \times \Phi \left[\frac{\mu_{\ln D(im,w)|NC} - \mu_{\ln C_i(w)}}{\sqrt{\beta_{D(im,w)|NC}^2 + \beta_{C_i(w)}^2}} \right] + P_C(im, w) \tag{7}$$

where the variable w represents the degree of mass loss.

Fragility increment functions for corroded bridge columns

Mori and Ellingwood (1993) modelled the strength of a deteriorating component as time-dependent through the use of a degradation function, and used it to calculate the time-dependence of the failure-probability of the component:

$$r(t) = r \cdot g(t) \tag{8}$$

where $r(t)$ is the strength at time t, r is the initial strength and $g(t)$ is the degradation function. In their approach, corrosion was modelled using a linear degradation function, whereas sulphate attack was modelled using a parabolic function, and diffusion-controlled degradation was modelled using a square root function. Enright and Frangopol (1998a, 1998b) also used a resistance loss function to model the degradation and Choe, Gardoni, Rosowsky, and Haukaas (2008, 2009) modelled the deformation and shear failure fragilities of deteriorating bridge columns as functions of the deformation or shear demand and time t. Ghosh and Padgett (2010) proposed using a quadratic function to model the time-dependent median and dispersion of the fragility functions of a deteriorating bridge system.

The major drawback of expressing damage state fragilities of deteriorating components as functions of time is that the time-dependence of the deterioration is implicit in such functions. As such, the time-variant fragility functions derived for a deteriorating bridge column will be specific to the deterioration models used for generating the functions, as well as the particular environmental exposure conditions assumed in those deterioration models.

If a different deterioration model is required to be used, or if the environmental exposure conditions change, the entire process of generating time-variant fragility functions will need to be repeated for the changed conditions. These drawbacks can be avoided if the parameters of the damage state fragility functions (i.e. the median and dispersion) are expressed as functions of the level of deterioration w, rather than as functions of time t. Thus, in the proposed formulation, the time-dependent deterioration model is separated from the corrosion dependent fragility function. Thus, once the degree of deterioration at time t is obtained using the deterioration model of preference to the analyst, then the appropriate fragility function can be used without the need for the fragility function to be recomputed. This step represents a significant savings in computational effort.

The following deterioration model is proposed for use in the fragility functions:

$$m(w) = m_0 e^{-\alpha_m w} \tag{9}$$

$$\xi(w) = \xi_0(1 - \alpha_\xi w) \tag{10}$$

where $m(w)$ and $\xi(w)$ are, respectively, the median and dispersion of the fragility function at a level of deterioration w, m_0 and ξ_0 are the median and dispersion of the fragility function for the column in its non-corroded state, α_m defines the exponential decrement function for the median and α_ξ is the coefficient of the linear decrement function for the dispersion of the fragility function. This form of the deterioration model is introduced

Table 7. Median and dispersion for fragility functions of the C_{60} column as function of mass loss.

W(%)	D1		D2		D3		D4	
	μ	ξ	μ	ξ	μ	ξ	μ	ξ
0	.045	.550	.162	.628	.689	.777	1.005	.786
5	.04	.609	.149	.641	.626	.723	.884	.752
10	.037	.644	.141	.64	.572	.737	.764	.692
15	.035	.658	.129	.641	.546	.744	.738	.673
20	.034	.651	.120	.651	.523	.754	.706	.686
25	.030	.671	.109	.667	.488	.786	.691	.693
30	.026	.658	.100	.675	.465	.797	.680	.733
35	.024	.674	.091	.682	.429	.838	.638	.757
40	.022	.701	.084	.672	.404	.828	.596	.831

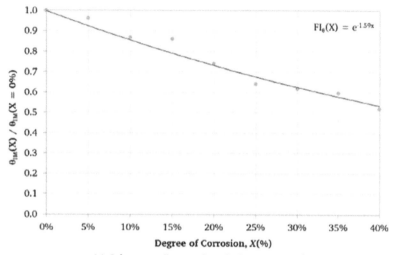

(a) Column C_{60}: Damage State 2 - Increase in median

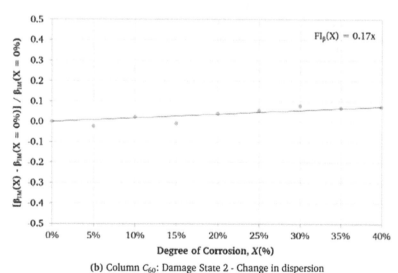

(b) Column C_{60}: Damage State 2 - Change in dispersion

Figure 11. Change in median and dispersion values for the fragility function for damage state 2 for the C_{60} column with increasing levels of corrosion.

primarily for computational efficiency. The fragility functions for deteriorating columns can then be computed as follows:

$$F_i(im, w) = \Phi \left[\frac{\ln(im) - \ln(m_0 e^{-\alpha_m w})}{\xi_0 (1 - \alpha_\xi w)} \right] \quad (11)$$

By keeping the deterioration analysis separate from the fragility modelling for deteriorated columns, we avoid the need to repeatedly compute the fragility functions each time the deterioration model or one of its parameters is changed.

Table 7 illustrates the median and dispersion of the fragility functions for C_{60} column at different levels of corrosion deterioration. Figures 11(a) and (b) depict the median and dispersion of the fragility functions for C_{60} column with increasing levels of corrosion (normalised by the median of the fragility function for the column in its non-corroded state), and the fitted fragility

Table 8. Values of the exponent of the median function (Equation 9) and the coefficient of determination for each column at different damage states.

Damage state	Column C_{60}		Column C_{80}		Column C_{00}	
	a_m	R^2	a_m	R^2	a_m	R^2
D1	1.70	.99	.64	.98	.48	.98
D2	1.60	.99	.57	.99	.64	.98
D3	1.35	.97	.53	.99	.75	.99
D4	1.43	.83	.35	.93	.58	.99

Table 9. Values of the linear coefficient of the dispersion function (Equation 10) and the coefficient of determination for each column at different damage states.

Damage state	Column C_{60}		Column C_{80}		Column C_{00}	
	a_{xi}	R^2	a_{xi}	R^2	a_{xi}	R^2
D1	.73	.24	−.17	−1.06	.12	.46
D2	.18	.22	.44	.85	.31	.79
D3	.05	.08	.50	.85	.69	.94
D4	−.18	−.55	.53	.90	.64	.92

increment function for damage state 2. The fitted curve for the median and the dispersion as functions of the degree of corrosion W in per cent are also shown on these figures. Unlike the decrease in the median with increasing corrosion, no clear trend is discernible in the case of dispersion of the fragility function. The same observation is made for all damage states of this and the remaining columns investigated in this study. Closed-form equations were developed for each damage state and for each column type (see Rao, 2014 for more details).

The coefficients α_m and α_ξ defined in Equations (9) and (10), respectively, are listed in Tables 8 and 9. The coefficient of determination, R^2 is also given in these tables. It can be observed that the coefficient of determination is very high (.98–.99) for the coefficient for the median but it is rather low for majority of the regressions for the dispersion coefficient. Figures 12(a) and (b) illustrate example fragility curves for the C_{60} column for damage states 1 and 2 as functions of increased corrosion.

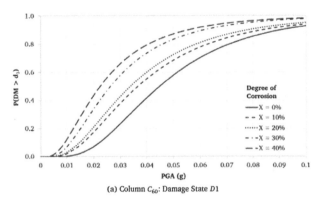

(a) Column C_{60}: Damage State $D1$

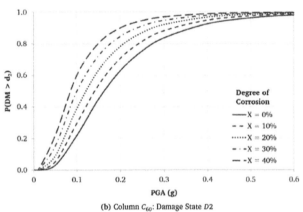

(b) Column C_{60}: Damage State $D2$

Figure 12. Fragility curves for the C_{60} column for damage states 1 and 2 as functions of increased corrosion.

These functions will be used in evaluation of the loss estimates for damage states other than collapse. Probabilities of collapse for deteriorating columns will be treated separately as the fragility functions for collapse depend on the intensity measure directly and do not include separate treatment of capacity and demand.

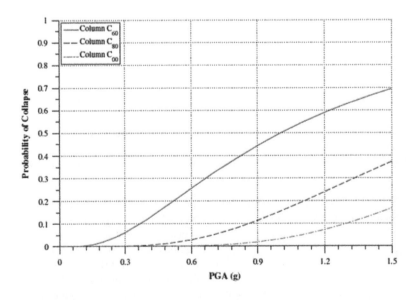

Figure 13. Probability of collapse conditioned on *IM* for columns C_{60}, C_{80} and C_{00} in the non-corroded state.

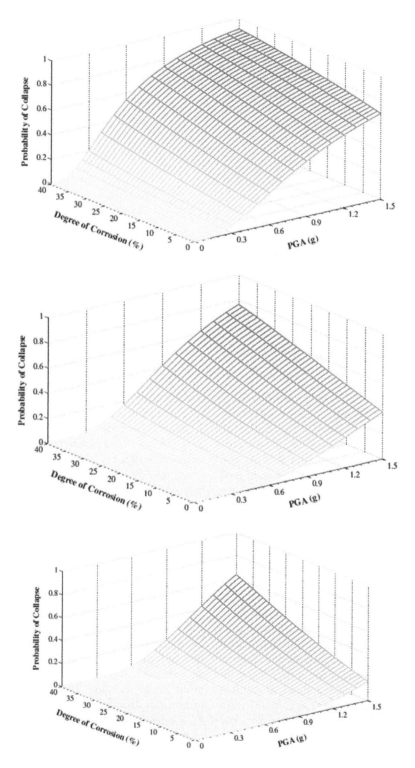

Figure 14. Probability of collapse given *IM* and *W* for column C_{60}, C_{80} and C_{00}.

Mean annual frequency of collapse for corroded

Bridge columns

The fragility function for damage state 4 gives the probability of the non-corroded column being in a state of actual or imminent collapse. It should be noted that this definition encompasses the collapse cases indicated by numerical instability, the collapse

cases corresponding to the flattening of the IDA curve, and the probability of imminent collapse indicated by the drift capacity for this damage state. The probability of collapse conditioned on the ground motion intensity, *IM*, for columns C_{60}, C_{80} and C_{00} in the non-corroded state is shown in Figure 13.

Clearly, improving bridge design standards have led to a significant decrease in the probability of collapse of bridge columns.

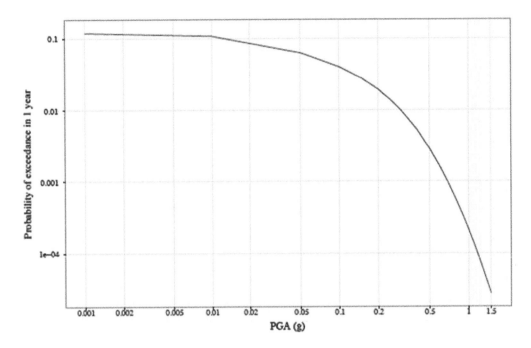

Figure 15. Seismic hazard curve for selected site in Oakland, CA.

For instance, at a PGA of .7 g, which roughly corresponds to the 5% in 50-year hazard level for Oakland, CA, the probability of collapse for the oldest of the three columns, C_{60}, is about .3, the probability of collapse for C_{80} is about .05 and the probability of collapse for C_{00} is negligibly small. At a PGA of .88 g, which roughly corresponds to the 2% in 50-year hazard level for Oakland, the probability of collapse for C_{60} is about .45, the probability of collapse for $C80$ is about .1 and the probability of collapse for C_{00} is about .02.

Similarly, the fragility function for damage state 4 computed using Equation (11) gives the probability of a deteriorating column being in a state of actual or imminent collapse. The probability of collapse conditioned on the ground motion intensity, IM, and the level of deterioration, W, for columns C_{60}, C_{80} and C_{00} is shown in Figures 14(a)–(c) as three-dimensional failure surfaces. For all three columns, we observe that increasing deterioration leads to a higher probability of collapse at any intensity level. It is interesting to note the effect of corrosion on the probability of collapse for column C_{00} at higher intensity levels: even though the column has a probability of collapse of only about .02 at a PGA of .88 g, as noted above, this value increases to almost .2 when the level of reinforcement corrosion reaches 40% steel mass loss.

Although the newer columns are often provided with an increased cover thickness compared to the older columns in order to delay the time to initiation of corrosion, the above results show that once the process of reinforcement corrosion is underway, the effect of corrosion on the seismic risk cannot be neglected at the higher seismic hazard levels, even for columns designed according to the latest bridge standards. Thus, a scenario-based seismic risk analysis investigating a large magnitude scenario event that fails to consider the effects of corrosion will end up underestimating the potential loss and impacts. Furthermore, the above results also show that the effect of corrosion on the probability of collapse is more pronounced for the older columns

at all hazard levels, underscoring the need to consider the effects of deterioration in the seismic risk assessment for such columns.

When these failure surfaces shown in Figures 14(a)–(c) are integrated with the absolute value of the derivative of the mean seismic hazard curve at the site, we obtain the mean annual frequency or mean annual rate of collapse for the three columns at varying levels of corrosion. The equation used to compute the annual rate of collapse is as follows:

$$\lambda_{\text{collapse}} = \int_{im} P[DM \geq dm_4 | im] d\lambda_{IM}(im) \tag{12}$$

Figure 15 presents the mean seismic hazard curve for the hypothetical location of the bridge columns in Oakland, CA. Figure 16 shows the mean annual frequency of collapse for the three columns computed using the above hazard curve. We make the observation that the mean annual frequency of collapse is highest for column C_{60} and lowest for column C_{00}, indicating an improvement in performance due to improvements in the bridge design standards. Even with a 40% loss of steel mass due to corrosion, the mean annual frequency of collapse for column C_{00} is below .004. On the other hand, the mean annual frequency of collapse for column C_{60} increases from about .0027 in the non-corroded state to about .0073 at a deterioration level of 40% steel mass loss, an increase of over 170%. Once again, this highlights the fact that neglecting the effects of deterioration for the older bridge columns could lead to a severe underestimation of the seismic risk.

Conclusions

In this paper, the widely used multi-stage phenomenological model is used to obtain parameters for the various components of the model. Monte Carlo simulations were used to obtain

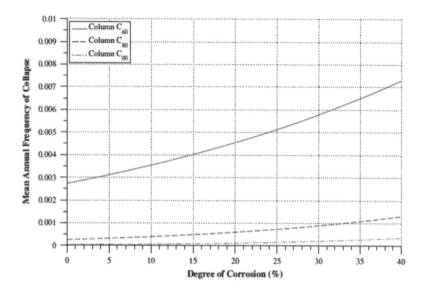

Figure 16. Mean annual frequency of collapse for columns C_{60}, C_{80} and C_{00}.

distributions for the time to corrosion initiation and the degree of corrosion for three California bridge columns designed according to different bridge standards. The three probability density functions used to represent the uncertainty in the time to corrosion initiation and the degree of corrosion include the lognormal, Weibull and Gamma distribution.

Based on the Kolmogorov–Smirnov, Cramer–von Mises and Anderson–Darling goodness of fit tests, it is shown that the lognormal distribution has the best fit for both variables. For a given column, the mean time to corrosion initiation was found to increase with increasing distance of the column from the coast; and for a given location, the mean time to corrosion initiation was found to increase with increasing cover thickness reflecting newer design criteria. The degree of corrosion was shown to be greater for columns in the splash zone than for those in the marine atmospheric zone, but the amount of corrosion was found to be smaller with the newer columns over time than with the older designs.

Deterioration due to reinforcement corrosion causes a decrease in the median capacities for each damage state. This decrease in median capacities can potentially be captured by the use of the changed geometric and material properties of the deteriorating column in prescriptive equations for predicting the capacities. Such prescriptive equations are either physics based, or empirically derived from experimental observations. Currently, there is a lack of damage state capacity models for deteriorating columns based on either approach.

An increasing amount of corrosion also leads to an increase in the median seismic demands at a given intensity level. The seismic demand generated on a deteriorating column can be assessed probabilistically by non-linear demand estimation procedures such as IDA or MSA, using a non-linear model of the column that incorporates the key structural effects of corrosion.

The decreasing damage state capacities coupled with increasing seismic demand leads to fragility functions that increase with increasing deterioration. The use of fragility increment functions based on the level of deterioration and the procedure for

estimating collapse probabilities are demonstrated using the three California highway bridge columns built in three different design eras. Improving bridge design standards have led to a significant decrease in the probability of collapse of bridge columns. For instance, at a PGA of .7 g, which roughly corresponds to the 5% in 50-year hazard level for Oakland, CA, the probability of collapse for the oldest of the three columns, C_{60}, is about .3, the probability of collapse for C_{80} is about .05 and the probability of collapse for C_{00} is negligibly small. At a PGA of .88 g, which roughly corresponds to the 2% in 50-year hazard level for Oakland, the probability of collapse for C_{60} is about .45, the probability of collapse for C_{80} is about .1 and the probability of collapse for C_{00} is about .02.

Acknowledgements

The authors also express their gratitude to Dr W. Aquino and Dr Y. Sun for providing the accelerated corrosion data.

Disclosure statement

No potential conflict of interest was reported by the authors.

Funding

The work was supported by the Division of Civil, Mechanical and Manufacturing Innovation (NSF) [grant numbers CMMI-1233694 and NEESR 1207911]. Dr Anirudh Rao was partially supported through Stanford University Leavell and Blume Fellowships.

References

Akiyama, M., Frangopol, D. M., & Matsuzaki, H. (2011). Life-cycle reliability of RC bridge piers under seismic and airborne chloride hazards. *Earthquake Engineering and Structural Dynamics, 40*, 1671–1687.

Akiyama, M., Frangopol, D. M., & Suzuki, M. (2012). Integration of the effects of airborne chlorides into reliability-based durability design of reinforced concrete structures in a marine environment. *Structure and Infrastructure Engineering, 8*, 125–134.

Applied Technology Council. (1985). *ATC-13: Earthquake damage evaluation data for California*. Washington, DC: Federal Emergency Management Agency.

Aquino, W. (2002). *Long-term performance of seismically rehabilitated corrosion-damaged columns* (PhD dissertation). University of Illinois, Urbana-Champaign.

Aquino, W., & Hawkins, N. M. (2007). Seismic retrofitting of corroded reinforced concrete columns using carbon composites. *ACI Structural Journal, 104*, 348–356.

Baker, J. W., & Cornell, C. A. (2005). *Vector-valued ground motion intensity measures for probabilistic seismic demand analysis* (Technical report 150). Stanford, CA: The John A. Blume Earthquake Engineering Center.

Berry, M. P., & Eberhard, M. O. (2007). *Performance modeling strategies for modern reinforced concrete bridge columns* (Technical report). Berkeley, CA: Pacific Earthquake Engineering Research Center, University of California.

Billah, A. H. M. M., & Alam, M. S. (2015). Seismic fragility assessment of highway bridges: A state-of-the-art review. *Structure and Infrastructure Engineering, 11*, 804–832.

Biondini, F., Palermo, A., & Toniolo, G. (2011). Seismic performance of concrete structures exposed to corrosion: Case studies of low-rise precast buildings. *Structure and Infrastructure Engineering, 7*, 109–119.

California Department of Transportation. (2008). *Caltrans memo to designers 20-4: Seismic retrofit guidelines for bridges in California*. Sacramento, CA: Author.

California Department of Transportation. (2012, November). *Corrosion guidelines* (Version 2.0. Technical report). Sacramento, CA: Author.

Choe, D.-E., Gardoni, P., Rosowsky, D. V., & Haukaas, T. (2008). Probabilistic capacity models and seismic fragility estimates for RC columns subject to corrosion. *Reliability Engineering & System Safety, 93*, 383–393.

Choe, D.-E., Gardoni, P., Rosowsky, D. V., & Haukaas, T. (2009). Seismic fragility estimates for reinforced concrete bridges subject to corrosion. *Structural Safety, 31*, 275–283.

Choi, E., DesRoches, R., & Nielson, B. G. (2004). Seismic fragility of typical bridges in moderate seismic zones. *Engineering Structures, 26*, 187–199.

Cornell, C. A., & Krawinkler, H. (2000). Progress and challenges in seismic performance assessment. *PEER Center News, 3*, 1–3.

Elwood, K. J., & Moehle, J. P. (2005). Drift capacity of reinforced concrete columns with light transverse reinforcement. *Earthquake Spectra, 21*, 71–89.

Enright, M. P., & Frangopol, D. M. (1998a). Service-life prediction of deteriorating concrete bridges. *Journal of Structural Engineering, 124*, 309.

Enright, M. P., & Frangopol, D. M. (1998b). Probabilistic analysis of resistance degradation of reinforced concrete bridge beams under corrosion. *Engineering Structures, 20*, 960–971.

FEMA. (2011). *Hazus-MH MR5 Technical manual. Multi-hazard loss estimation methodology: Earthquake model* (Technical report). Washington, DC: Author.

Frangopol, D. M., & Das, P. C. (1999). Management of bridge stocks based on future reliability and maintenance costs. In P. C. Das, D. M. Frangopol, & A. S. Nowak (Eds.), *Current and future trends in bridge design, construction and maintenance* (pp. 45–58). London. The Institution of Civil Engineers, Thomas Telford.

Frangopol, D. M., Kong, J. S., & Gharaibeh, E. S. (2000). Bridge management based on lifetime reliability and whole life costing: The next generation. In M. Ryall, G. Parke, & J. Harding (Eds.), *Bridge management 4: Inspection, maintenance, assessment and repair* (pp. 392–399). London: Institution of Civil Engineers, Thomas Telford.

Frangopol, D. M., Kong, J. S., & Gharaibeh, E. S. (2001). Reliability-based life-cycle management of highway bridges. *Journal of Computing in Civil Engineering, 15*, 27–34.

Ghosh, J., & Padgett, J. E. (2010). Aging considerations in the development of time-dependent seismic fragility curves. *Journal of Structural Engineering, 136*, 1497–1511.

Hwang, H., Liu, J. B., & Chiu, Y.-H. (2001). *Seismic fragility analysis of highway bridges* (Technical report). Memphis, TN: Center for Earthquake Research and Information, The University of Memphis.

Jalayer, F. (2003). *Direct probabilistic seismic analysis: Implementing non-linear dynamic assessments* (PhD dissertation), Stanford University, Stanford, CA.

Lepech, M., Rao, A., Kiremidjian, A., Michel, A., Stang, H., & Geiker, M. (2015). Multi-physics modeling and multi-scale deterioration modeling of reinforced concrete part II: Coupling corrosion and damage at the structural scale. In K. Dahl, H. Stang, & M. Bræstrup (Eds.), *Proceedings of 2015 fib symposium: Innovation and design*. Lausanne, Switzerland: Federation International du Beton.

Li, J., & Gong, J. (2008). Influences of rebar corrosion on seismic behavior of circular RC columns. *China Journal of Highway and Transport, 21*, 55–60.

Mackie, K. R., Wong, J.-M., & Stojadinović, B. (2008). *Integrated probabilistic performance-based evaluation of benchmark reinforced concrete bridges* (Technical report). Berkeley: PEER Center, University of California.

Mazzoni, S., McKenna, F. T., Scott, M. H., & Fenves, G. L. (2007). *Open system for earthquake engineering simulation user command-language manual*. Berkeley: PEER Center, University of California.

Michel, A., Geiker, M., Stang, H., & Lepech, M. (2015). Multi-physics modeling and multi-scale deterioration modeling of reinforced concrete part I: Coupling transport and corrosion at the material scale. In K. Dahl, H. Stang, & M. Bræstrup (Eds.), *Proceedings of 2015 fib symposium: Innovation and design*. Lausanne, Switzerland: Federation International du Beton.

Mirza, S. A., & MacGregor, J. G. (1982). Probabilistic study of strength of reinforced concrete members. *Canadian Journal of Civil Engineering, 9*, 431–448.

Mori, Y., & Ellingwood, B. R. (1993). Reliability-based service-life assessment of aging concrete structures. *Journal of Structural Engineering, 119*, 1600–1621.

Nielson, B. G. (2005). *Analytical fragility curves for highway bridges in moderate seismic zones* (PhD dissertation). Georgia Institute of Technology, Atlanta, GA.

Nielson, B., & DesRoches, R. (2007, August). Analytical seismic fragility curves for typical bridges in the central and southeastern United States. *Earthquake Spectra, 23*, 615–633.

Padgett, J. E., & DesRoches, R. (2008). Methodology for the development of analytical fragility curves for retrofitted bridges. *Earthquake Engineering & Structural Dynamics, 37*, 1157–1174.

Porter, K. A., Hamburger, R. O., & Kennedy, R. P. (2007, May). Practical development and application of fragility functions. *Proceedings of SEI structures congress* (pp. 16–19). Long Beach, CA.

Ramanathan, K. N. (2012). *Next generation seismic fragility curves for California bridges incorporating the evolution in seismic design philosophy* (PhD dissertation). Georgia Institute of Technology, Atlanta, GA.

Rao, A. (2014). *Structural deterioration and time-dependent seismic risk analysis* (PhD dissertation). Stanford University, Stanford, CA.

Rao, A. S., Lepech, M. D., & Kiremidjian, A. S. (2013, June 16–20). *Time-dependent risk assessment of deteriorating reinforced-concrete bridges for sustainable infrastructure design*. Proceedings of the 11th international conference on structural safety & reliability, New York, NY.

Rao, A. S., Lepech, M. D., Kiremidjian, A., & Sun, X.-Y. (in press). Development of structural deterioration model for structural fragilities, *Structure and Infrastructure Engineering*. Accepted for publication.

Thoft-Christensen, P. (1999). Estimation of bridge reliability distributions. In P. Das, D.M. Frangopol, & A. Novak (Eds.), *International conference on current and future trends in bridge design construction and maintenance* (pp. 15–25). Singapore: Thomas Telford.

Tuutti, K. (1982). *Corrosion of steel in concrete*. Stockholm: Swedish Cement and Concrete Research Institute.

Val, D. V., & Stewart, M. G. (2003). Life-cycle cost analysis of reinforced concrete structures in marine environments. *Structural Safety, 25*, 343–362.

Vosooghi, A., & Saiidi, M. S. (2010). *Post earthquake evaluation and emergency repair of damaged RC bridge columns using CFRP materials* (Technical report). Reno, NV: University of Nevada.

Vosooghi, A., & Saiidi, M. S. (2012). Experimental fragility curves for seismic response of reinforced concrete bridge columns. *ACI Structural Journal, 109*, 825–834.

Weyers, R. T., Fitch, M. G., Larsen, E. P., Al-Qadi, I. L., Chamberlin, W. P., & Hoffman, P. C. (1994). *Concrete bridge protection and rehabilitation: Chemical and physical techniques – service life estimates*. Washington, DC: Strategic Highway Research Program, SHRP-S-668, National Research Council.

Big data-based deterioration prediction models and infrastructure management: towards assetmetrics

Kiyoshi Kobayashi and Kiyoyuki Kaito

ABSTRACT

In the past decades, infrastructure management has been performed based on implicit knowledge, consisting experience and knowledge of professional engineers. The objective of assetmetrics is to convert such decision processes based on implicit knowledge and experience into systematic decision processes based on formal knowledge. The presented research and development policy is a practical approach that tries to create methodologies based on data obtained through daily and periodic inspections. Moreover, the authors point out that in the field of asset management, technical knowledge related to the analysis of existing data is more important than hardware technologies for obtaining new data. The authors also discuss these ideas as pertaining to the concept of big data. Finally, by presenting examples of advanced researches on assetmetrics, the authors give an overview of monitoring methodologies and comprehensive risk management in the relevant field.

1. Introduction

The deterioration of infrastructures takes place in a gradual manner. In difficult economic times, it is important to provide citizens and the users of infrastructures with an appropriate explanation of the importance of maintaining and managing infrastructures, as well as the necessity of ensuring a budget for that purpose. Of course, administrators and engineers involved in maintenance and management already inspect structures to predict their degree of deterioration, making decisions related to repair and reinforcement while taking into consideration economic aspects. The practice of asset management consists of that chain of decisions. On the other hand, the role of assetmetrics is to transform that experience-based decision process derived from implicit knowledge into a visible, systematic decision process based on formal knowledge (Kobayashi, 2010). Implicit knowledge is accumulated by individual engineers. On the other hand, formal knowledge can not only be accumulated within organisations but can also be corrected and improved to eventually become shared knowledge. Ultimately, the transformation of formal knowledge into shared knowledge makes it possible (1) to fulfil the responsibility of clarifying issues related to the management of infrastructures and (2) to share knowledge among infrastructure administrators within an organisation, facilitating technological inheritance (in other words, carrying out 'knowledge management').

Asset management is a practical science, and as such, it does not contribute to actual work unless its methodologies follow existing decision processes. On the other hand, the objective of assetmetrics is to strategically improve asset management, not by simply formalising decision processes that are currently based on implicit knowledge, but also by collecting on-site data, improving analysis processes and improving operation efficiency. Currently, because the majority of infrastructure administrators make decisions based on visual inspections and periodic inspection data, creating a methodology based on such inspection data is indispensable. Regarding these inspections, the authors have adopted, as a policy underlying their research and development work, a fully practical approach, where methodologies are created based on data that can be obtained through daily and periodic inspections. The information used for making decisions is accumulated on-site. However, in many cases, such information only exists as printed matter. Moreover, each piece of information is incomplete and does not carry sufficient significance for a statistical analysis. When incomplete information is stored on a large scale, extracting significant information from it requires analytical techniques, i.e. intelligence technology.

The common statistical analysis methods using inspection data include regression (Madanat, Bulusu, & Mahmoud, 1995) and hazard models (Mishalani & Madanat, 2002). They are compatible with the big data concept that is receiving increasing attention recently. Furthermore, development of a probability model to explain complicated deterioration phenomena of infrastructure and development of Bayesian statistics using the Markov Chain Monte Carlo (MCMC) methods in order to estimate the deterioration using inspection data are the contributing factors to the increasing application of asset management to the practical operation.

Based on an awareness of the above-mentioned issues, in the present research, we first describe the evolution of such techniques with an overview of pioneering examples, with emphasis on deterioration prediction, which has been the special focus of the authors' research. We discuss the role played by monitoring in management, propose future directions for research and development and give a perspective of comprehensive risk management that encompasses asset management. The methodology outlined in this paper is applicable specifically to pavement, bridge and sewage, although it is generally applicable to any infrastructure systems as long as their visual inspection data are available. In Section 2, the authors describe the basic standpoint of the research. In Section 3, the authors describe deterioration prediction models, which constitute a basic technique of asset-metrics, as well as related estimation methods. In Section 4, the authors mention the role of monitoring in asset management, as well as methods for analysing the monitored data. Finally, in Section 5, the authors explain the concept of comprehensive risk management as a general directive for asset management.

2. Basic standpoint of this study

2.1. Management curve for decision-making

If we interpret the basic proposition of asset management in its narrowest sense, which is to discuss management as a means to extend the lifetime of infrastructures, the first objective of asset management would be to provide an optimum repair policy (Kaito, Yasuda, Kobayashi, & Owada, 2005; Kobayashi, 2005) that achieves the minimum life cycle cost (FHWA, 1998; Frangopol, Lin, & Estes, 1997). Simply stated, the life cycle cost is obtained simply by estimating cost for the necessary maintenance and management of infrastructures and deciding upon the timing of investments. Thus, the idea of minimising the life cycle cost can be easily accepted. It is also true, however, that sceptical views exist with respect to life cycle cost evaluation. It is also the fact that there are many sceptical views concerning life cycle cost assessment because of (1) discount rate, (2) uncertainty of input information and (3) deterioration prediction technology.

As for (1), it is possible to attain optimal policy to minimise the life cycle cost without using discount rate based on the average cost method under the premise that the infrastructure is for permanent use without functional improvement. As for (2), there are many variables such as observation error of inspection data, judgment error of inspectors, life cycle cost assessment period and repair effect, and it is closely related to (3) deterioration prediction. Regarding the two constituent elements of life cycle cost (cost and investment timing), in the cost aspect, it is possible to calculate reliable numerical values based on the history and repair databases. Even if no information related to the cost is available, it is possible to carry out a relatively precise cumulative calculation in view of the maintenance method and actual damage conditions. Therefore, sceptical views related to life cycle cost evaluation predominantly point out the investment timing aspect, i.e. the low reliability of deterioration modelling (deterioration prediction methods).

Deterioration prediction methods can be roughly classified into theoretical and statistical methods. Theoretical models are those that derive predictive equations based on theoretical dynamic analyses or empirical rules after clarifying the deterioration and damage mechanism, using model experiments or the like. The deterioration process based on a mechanical approach shows a performance curve, for which the elapsed years are used for horizontal axis and structural capacity such as durability and load-bearing capacity is used for vertical axis. On the other hand, a statistical method describes statistical regularity that exists behind a large volume of data. The deterioration process based on a statistical method shows a management curve, for which discrete assessment indicators that are obtained as a result of inspection are used for vertical axis. Therefore, theoretical methods are considered advantageous for deterioration prediction from a microscopic perspective targeted on specific infrastructures and materials.

In contrast, statistical methods are more suitable for deterioration prediction from a macroscopic perspective focused on entire infrastructures. The use of theoretical or statistical methods for deterioration prediction in asset management obviously depends on the final objective. However, theoretical methods need different information and predictive equations according to the target degradation and damage conditions, and the necessary information often cannot be obtained during normal inspection operations, which inevitably results in a practical disadvantage. In contrast, statistical methods may be more suitable in practice because of their ability to provide invariant deterioration prediction (e.g. by applying Markov chain models) even if the target infrastructures suffer different types of deterioration and damage, provided that daily and regular inspections are obligatory for all infrastructures (normally visual inspections) and that the inspection results are properly evaluated as the discrete degree of soundness.

In asset management, what the administrator should know is not the design-level structural performance of infrastructures, but rather the characteristics of the deterioration that actually occurs in the facilities, as well as practical information regarding the timing for the execution of repairs and reinforcements, and the construction methods to be adopted. The results of degradation prediction based on statistical degradation prediction methods show the degree of soundness on the vertical axis. Soundness consists of discrete evaluation parameters obtained as the result of inspections (inspection data) and relate to the damage and deterioration of structures. However, it only takes into consideration superficial damage that can be noticed through visual inspection, without reflecting the actual structural performance. On the other hand, it is important to note that the definition of soundness contained in inspection manuals is not only related to the engineering meaning of the term but also expresses basic management policies in terms of repair and reinforcement. Therefore, if the number of years it takes to reach a certain level of soundness is statistically analysed, we can obtain important management information related to investment timing. For an infrastructure administrator responsible for the preparation of a maintenance and repair plan, understanding the investment timing is a central management issue. For this reason, prediction results calculated using statistical methods are not mere deterioration performance curves, but rather are management curves that determine the investment timing. Therefore, the main objective of deterioration prediction using inspection data is not to assess the soundness of structures and predict their lifetimes, but rather to provide a posteriori evaluation of the administrator's performance in terms of maintenance, repair and investment actions, which will ultimately contribute to management improvement.

2.2. Big data and infrastructure management

The concept of big data is 'to change relations between markets, organisations, and also citizens and the government by executing large scale actions that cannot be done on a small scale' (Schonberger & Cukier, 2013). It has been pointed out that the following changes in terms of data analysis occur owing to the use of big data: (1) big data encompasses nearly all of the data, (2) a sufficiently large quantity overwhelms the need for accuracy and (3) correlation is more important than causal relations (hereinafter referred to as 'concepts 1, 2, and 3 of big data').

The objective for the assetmetrics pursued by the authors is to apply them in an elaborate statistical analysis of inspection data, which have not received close attention in previous scientific studies. In infrastructure management, decisions regarding maintenance, management, repair and reinforcement must be made based on the data observed on-site. Compared to other types of inspection data, there is a large quantity of daily and periodic inspection data obtained at infrastructures, which contain a large amount of information and can thus literally be called 'big data.' Moreover, the investigative approach of using the available inspection data to obtain meaningful management information data is close to the concept of big data. It can also be said that the use of statistical methods to describe degradation in terms of probability models is justified by the fact that inspection data contain a large amount of uncertainty, and the low quality of the inspection data is compensated by the sophistication of the modelling and the amount of data. In addition, as mentioned in 2.1, the fact that we are using management curves, instead of performance curves that are determined by clarifying the causal relation of the degradation process, is also similar to the concept of big data. Visual inspection data represent just the observation of the conditions for the superficial degradation of structures. In that sense, the ideal scenario would be to actually measure and evaluate the performance and soundness of structures before making decisions.

Complementary information such as visual inspection data just contains correlations with the information that matters (such as investment timing). Thus, causal relations in the engineering sense seldom exist. In the case of infrastructure management, the correlations contain important meanings for decision-making compared to the causal relations concerning the facts. To implement infrastructure management, it becomes necessary to take actions such as coping with the hazards and risks involved in practical implementation (getting quickly to the work site and making decisions), as well as making use of every opportunity to improve the work and operations. In the practice of management, it is important to create such opportunities, and from that perspective, it is possible to say that information concerning causal relations is often not required. The final output of assetmetrics (degradation prediction results and optimum repair policies)

consists of information to assist the administrator in making decisions, rather than the decisions themselves.

There are also differences between big data in the field of infrastructure management and big data applied to other fields. The most important difference is that the data related to infrastructure inspection do not constitute open information. If we consider the social importance of infrastructures, it would be surprising whether anyone could easily access inspection data. From the infrastructure administrator's point of view, it is advantageous to provide inspection data to researchers and engineers who possess sophisticated analysis techniques and may potentially yield outputs that satisfy the needs of those actually involved in operations. Regarding this point, here we provide an additional explanation using an example involving the hidden Markov hazard model (Kobayashi, Kaito, & Nam, 2012b) developed by the authors.

A large variety of methods for infrastructure facility asset management are under development. As will be explained in 2.3, from the perspective of big data applied to asset management, management in the field of pavement work is advancing in a substantial manner, both at the practical and academic levels. One of the driving reasons is the development of a road surface-measuring vehicle, which makes it possible to obtain large amounts of inspection data. On the other hand, the vehicle performs measurements of road conditions at normal traffic speeds, and the effect of measuring errors cannot be ignored. According to the first principle of big data, it is of fundamental importance to use all the inspection data for prediction.

Using the Markov degradation model (Tsuda, Kaito, Aoki, & Kobayashi, 2006) for estimation requires information samples obtained from two inspections. If no repair is carried out, the soundness will never improve naturally by the second inspection. However, in a case where degradation prediction was carried out based on the rutting amount data measured by the road surface-measuring vehicle, soundness improvements were observed for about half the information samples. This was due to the fact that the road surface-measuring vehicle did not pass over exactly the same line in the two inspections. This kind of information sample is normally eliminated in a model-based prediction, but we cannot neglect the possibility of observation errors even in the information samples that are actually used (samples where the soundness is constant or is reduced). Under these conditions, if we simply eliminate the samples where the soundness improves, we may end up introducing a kind of prediction bias (Kobayashi, Kumada, Sato, Iwasaki, & Aoki, 2007). The hidden Markov degradation model (Kobayashi et al., 2012) is a degradation model that actually considers the influence of measuring errors, but the data used are the same as a standard Markov degradation hazard model. Even using the same inspection data, the output differs completely according to the analyst's skills. As mentioned above, the big data applied to infrastructure management will be mostly handled by a few researchers and engineers.

2.3. Intelligence technology in field of big data

Two types of technology exist: hardware and software. In the following, we try to suggest directions for future technology development targeted on assetmetrics (infrastructure management based on big data). Figure 1 contains a diagram showing

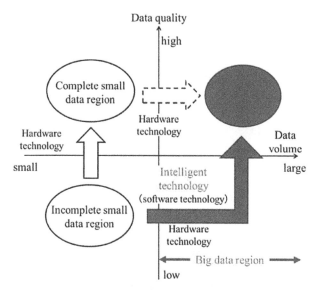

Figure 1. Concept of big data and intelligent technology.

how decision-making can be graphically divided into four areas from the viewpoints of data volume and information quality. The values along the vertical axis represent the quality of the information used in the infrastructure manager's final decision-making. The quality of the information can be improved by improving the inspection technologies used in the inspection of structures, which will result in higher quality for the obtained data. Moreover, better analysis techniques will also result in improvements, even if the inspection data remain the same. The figure shows that in the past, decisions had to be made in the 'incomplete small data' region, where the quality of the data was poor, and the volume was small. Prior to the big data era, the probability and statistics focused on incomplete small data, and the objective of research and development was to extract meaningful (high-quality) information from the data.

Two directions for technological development can be observed in this incomplete small data region: towards the complete small data region and towards the incomplete big data region. Before the popularisation of the big data concept, the target of technological development was to improve the quality of data in the small data region, i.e. to approach the complete small data region. When the goal is high accuracy and clear causal relations, despite the small amount of information used in decision-making (Schonberger & Cukier, 2013), technological development tends to prioritise hardware rather than software owing to the need to obtain new data. This trend is not limited to civil engineering.

Actually, technologies related to non-destructive inspection and sensors for monitoring have quickly advanced in the field of infrastructure management. These technologies have found practical applications such as detailed inspection technologies, providing useful information for making decisions regarding the execution of repairs in heavily damaged infrastructures and the selection of repair methods. However, detailed inspections are subject to strong restrictions in terms of cost and time, and their application scope is limited. They have contributed to the development of maintenance and management engineering focused on repair and reinforcement methods for specific infrastructures, but their application to infrastructure management for

decision-making in all infrastructures is difficult. On the other hand, in the incomplete big data region, (1) the versatility and decreasing cost of existing sensors and (2) the development of sensor network technologies have supported the transition shown in the diagram. In any case, the above-mentioned region transition requires the development of innovative hardware technologies.

It should be noted that in the incomplete big data region, which is the target of the big data concept, the amount of data is increasing but the quality is not necessarily improving. It is true that in the incomplete small data region, the amount of data and the improvement of statistical analysis methods are in a trade-off relation. In the big data region, by improving statistical analysis methods based on the knowledge and results of research accumulated in the small data region, it will be possible to extract more valuable information (information with practical utility) from the same data. Therefore, statistical methods developed in the small data region will not become useless in the big data region. As long as incomplete data are handled, it is important to improve the statistical methods in order to obtain information with higher quality (with more practical utility). In fact, technologies for modelling degradation phenomena using complex probability models are evolving. This kind of modelling was obviously already available in the past. However, what made estimation using actual inspection data possible was the dissemination of big data concept and development of Bayesian statistics using the MCMC methods. (Iba et al., 2005; Wago, 2005).

According to the general concept of big data, inspection data based on the visual inspection of infrastructures are often not available in large quantities. Thus, in the strictest sense, these situations do not characterise big data. However, in practice, a road surface-measuring vehicle is actually used for road condition inspections, just as rail inspection cars are used for railway inspections. In both cases, inspection technologies that make it possible to obtain the existing data in a more efficient way have been developed, rather than technologies related to new types of data. In addition, placing sensors in multiple scattered structures to construct a monitoring system produces a burden on each system in terms of cost and management, and increases the size of the network for collecting data. In fact, inspections using road surface-measuring vehicles and rail inspection cars are carried out by moving the inspection system and performing an inspection and a data collection simultaneously. It is expected that the introduction of such monitoring technologies will result in an explosive increase in inspection data. Decision-making for asset management does not require hardware technologies to obtain high-quality information, but rather software (intelligence technology) to analyse the huge amount of inspection-related information that is currently being accumulated (big data).

3. Evolution of degradation prediction models and Bayes' estimation

3.1. Mixed degradation process and estimation methods

The introduction of a hazard model (Lancaster, 1990) and Markov chain model resulted in considerable improvements in statistical degradation prediction methods. A large number of degradation prediction methods have been proposed for specific phenomena, and these are reported in detail in Aoki (2008) and Kaito, Aoki, and Kobayashi (2010). The development of degradation prediction models has advanced since then, such as the infrastructure degradation prediction method for individual infrastructures represented by benchmark analysis (Obama, Okada, Kaito, & Kobayashi, 2008), the competition-based approach where individual degradations are not treated as isolated degradation events but rather as competing phenomena whose degradation factors influence each other (Kobayashi, Kaito, & Nam, 2014), and the degradation prediction models are based on hierarchical (Kobayashi, Kaito, Eguchi, Ohi, & Okizuka, 2011; Nam, Kaito, Kobayashi, & Okizuka, 2012) or mixed (Matsumura, Kobayashi, Kaito, Ohi, & Yamaguchi, 2013) relations between deterioration processes.

As specific examples, the authors have worked on the relation between potholes and cracks (Nam et al., 2012), hierarchical hidden Markov degradation models that express the mixed relation between cracks and the load-bearing capacity of asphalt (Kobayashi et al., 2011), and hidden Poisson models to express the hierarchical relation between on-road fallen objects and complaints (Obama, Kaito, Kobayashi, Fukuda, & Itagaki, 2013). Figure 2 shows the concept of the Poisson hidden Markov model. A hierarchical relation exists between pavement potholes and cracks. As the rate of cracks increases, the occurrence of potholes increases accordingly. In particular, in the case of dense graded asphalt, it is known from practical experience that the development of cracks eventually results in potholes. This relation is shown in the same figure. A scale parameter related to the rate of occurrence of potholes becomes larger owing to the progress of cracks. This scale parameter is called a latent variable and is not actually measured. The expectation–maximisation (EM) algorithm (Bishop, 2010) was the most common estimation method for this kind of mixed probability model based on the formalisation of multiple degradation processes involving latent variables.

However, the advent of Bayes' estimation (Kaito, Kobayashi, Aoki, & Matsuoka, 2012; Kobayashi, Kaito, & Nam, 2012a) brought a significant improvement for more complex probability models. In the subsequent sections, details for Poisson hidden Markov model (in 3.2) and hierarchical hidden Markov model (in 3.3) are presented. Both models focus on two deterioration phenomena and examine the impact of one deterioration on the other . If the inspection data for 2 deterioration phenomena can be evaluated as the discrete soundness level, the latter model is used. If the inspection data for one deterioration phenomenon (of 2 deterioration phenomena) are the enumerated data with a focus on the number of damages, the former model is used.

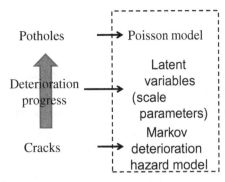

Figure 2. Concept of Poisson hidden Markov model.

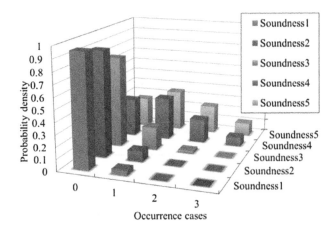

Figure 3. Example of Poisson hidden Markov model.

3.2. *Poisson hidden Markov model*

Figure 3 shows the hierarchical relation between potholes and the crack rate based on the degradation prediction results of the Poisson hidden Markov model. After formulating the pothole formation process using a Poisson model and the crack rate degradation process using a Markov degradation hazard model, we analysed inspection data measured in a specific road section. An inspection related to potholes was carried out as a daily inspection (once a day), whereas the measurement of the crack rate was conducted during regular inspections (once every 3 to 5 years). The above differences in the measurement frequencies and the asynchrony of the inspection dates were considered in the model (Miyazaki, Obama, & Kaito, 2013). Moreover, the above-mentioned hierarchical relation was expressed using latent variables (a scale parameter showing the proportional influence of the gradual reduction of soundness due to the crack rate on the pothole rate). For this complex mixed probability model involving latent variables, Bayes' estimation was used to obtain the unknown parameters. Let us look at the figure again.

For example, when the soundness with respect to the crack rate is 1 (0% crack rate), the probability that potholes occur in one or more places (occurrence rate in one year in a randomly chosen 100-m road section) is 5% or less. As the crack rate soundness advances from 2 to 3, the occurrence rate slowly increases. However, after the soundness reaches a value of 4 (a crack rate of 10%), it is possible to see that the situation changes drastically. For example, for a soundness of 5 (a crack rate of 20%), the occurrence rate exceeds 80%. The hierarchical relation discussed above can be sensed by engineers involved in daily inspections, even if it is not expressed in a quantitative form. The objective of assetmetrics is to provide a formalisation of that kind of implicit knowledge.

We will provide additional information regarding the application of this kind of formulated knowledge. In general, the criterion for the execution of asphalt repair (overlay) is based on the crack rate. Most inspection manuals suggest a crack rate decision criterion of 20%. A road surface-measuring vehicle is used to measure the crack rate every 3 to 5 years. On the other hand, detecting potholes is a routine part of the daily inspections by workers. Emergency pothole repairs are done using compounds at normal temperatures. However, in road sections where the number of potholes is especially large, it is desirable to quickly execute an overlay to ensure the users' safety. Certainly, those involved in practical operations know how to identify such road sections. Nonetheless, it should be noted that actually there is no rule for making decisions regarding the execution of an overlay based on the number of potholes (manuals do not contain such a rule). The need for an overlay must be explained every time.

An output of the current research regarding this practical issue (the difference between the data obtained through daily inspections and those used to justify decisions) is a degradation prediction method that considers the hierarchical relation between the pothole occurrence process and the progress of cracks. In concrete terms, we can understand, for example, that in road sections where the probability of occurrence of at least one pothole in one year is 70% or more, the probability of cracks is 10% or more. We are not asking for the establishment of new standards based on the frequency of potholes or the use of new types of information to be obtained through new inspection technologies. Rather, we are focusing on existing inspection data and inspection frequencies to present a new insight into their hierarchical relations by improving probability models and statistical analysis methods.

3.3. Hierarchical hidden Markov model

A similar hierarchical relation can be found between the crack rate and the asphalt load-bearing capacity. The asphalt load-bearing capacity can be evaluated by means of falling weight deflectometer (FWD) tests. However, FWD tests are not performed regularly, but only when necessary. Moreover, an FWD test requires a temporary traffic regulation. Thus, road administrators cannot perform them frequently. Moreover, it is extremely important to make good decisions regarding where these should be conducted. The hierarchical hidden Markov model is a model that expresses the hierarchical relation between the crack rate and the asphalt load-bearing capacity (i.e. the implicit knowledge that when the asphalt load-bearing capacity decreases, the crack rate rapidly increases). Figure 4 depicts an example of the prediction result. The degradation processes for the crack rate and asphalt load-bearing capacity are modelled using a five-step Markov degradation hazard model. The continuous red line in the figure corresponds to the expected degradation path for the

asphalt load-bearing capacity. It can be seen that the expected life is about 50 years.

For the crack rate, four expected degradation paths are shown. The yellow continuous line corresponds to the expected degradation path for the crack rate when the soundness of the asphalt load-bearing capacity is equal to one. This leads to the conclusion that when the load-bearing capacity exhibits a high level of soundness, the expected life in relation to the crack rate is approximately 30 years. However, the expected degradation path when the soundless level of the asphalt load-bearing capacity decreases to four (the blue continuous line in the figure) shows that the life is approximately 12 years. From this, we conclude that the expected life of the surface in terms of the crack rate can be reduced by half depending on asphalt load-bearing conditions. Therefore, when the surface life is reduced to about half the original time after multiple overlays, this can be used as a trigger for scheduling an FWD test (Kaito, Kobayashi, & Obama, 2014).

Moreover, it is known in practice that the relation between the crack rate and the asphalt load-bearing capacity is not a unidirectional degradation phenomenon where just the load-bearing capacity produces an influence on the crack rate, but rather a bidirectional one. Currently, we are developing a model that can expresses compound degradation processes with that kind of bidirectional degradation relation embedded within them (Matsumura et al., 2013). Moreover, it is also possible to model multi-layered mixed degradation processes involving the potholes, crack rate and asphalt load-bearing capacity. It is also worth noting that a large number of highway sections in Japan are several decades old and require large-scale repairs of the entire asphalt structure. Thus, decisions will be required about whether to carry out simple overlays or large-scale repairs based on life cycle cost analyses, taking into consideration their respective degradation processes. We are also developing economic analysis models that consider these issues (Kobayashi et al., 2013). Furthermore, although the statistical deterioration prediction method such as Markov deterioration hazard model is basically to predict the slow deterioration process such as ageing deterioration of infrastructure, using the chronological model and detection algorithm – as described in Section 4.2 – will enable us to detect or predict deterioration of the sudden unexpected damages.

4. Monitoring data and management

4.1. Mixed degradation process and estimation methods

If we consider asset management based on visual inspection data to be a first generation method, asset management based on monitoring data by sensors would be the second generation. Inspection and monitoring methods based on multiple sensors (hereinafter referred to as monitoring) have been proposed with the objective of detecting the damage and defects of infrastructures, or evaluating their performance. Monitoring methods can be roughly classified into: (1) those that provide absolute evaluations of the performance and damage in one pass and (2) those that provide a relative evaluation through successive evaluations. Thus far, the former type of monitoring has attracted more academic attention, and abundant research results have been obtained. However, infrastructures where such methods can be

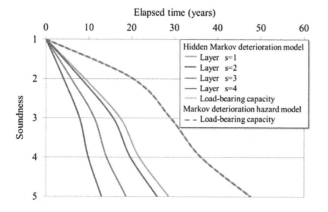

Figure 4. Hierarchical hidden Markov deterioration hazard model.

applied are restricted to special ones (in terms of the importance and degree of damage). Moreover, visual inspections constitute the current mainstream, and monitoring is needed as a way to complement them.

The latter type (successive monitoring) tries to detect changes that exceed expected ranges by successively measuring parameters that are expected to change owing to decreased performance and as a result of the occurrence and deepening of damage (abnormality detection algorithms based on time sequence analysis). For example, by detecting relative variations in vibration characteristics, it is possible to ensure an efficient allotment of technical personnel for visual inspections. The role of monitoring in infrastructure management is not to replace visual inspection but to provide support information to enable technical personnel to carry out visual inspection effectively. Additionally, from the perspective of big data, we can say that future monitoring technologies will move towards the latter type.

4.2. Time sequence model considering variance

The changes caused by big data are not restricted to a single one. Especially in the field of infrastructure management, deviations occur with respect to expected values. This is also true in the management field, as in the case of abnormality detection algorithms to support time sequence analysis. It is difficult to detect local damage in infrastructures using monitoring data related to the entire facility system (variation in the expected value in time sequences of a specific value). For local damage, it is necessary to develop algorithms using second- or higher order moments as parameters, which are expected to show larger sensitivity (than first-order moments). In the field of time sequence analysis, models such as GARCH have been developed since the 1990s to provide a formulation for temporal changes in variance, and these have especially been improved in the field of financial engineering.

Figure 5 presents a time sequence of the successively measured daily peak frequencies of acceleration in bridge joints over a period of 250 days. We can see that the expected value and variance of the frequency vary with time. Specifically, we expressed this deterioration process using Auto Regressive Moving Average model with eXogenous variables model (hereafter referred to as ARMAX model) that takes into account Xogeneous variables together with ARMA that is consisted of the autoregression and the moving average polynomials.

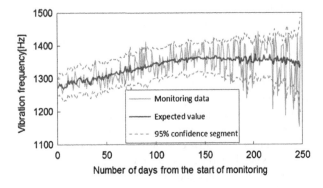

Figure 5. ARMAX-GARCH model and monitoring data.

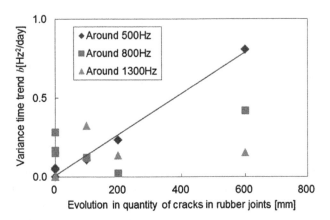

Figure 6. ARMAX-GARCH model and monitoring data.

The deterioration process is formulated using the ARMAX-GARCH model (Kobayashi, Kaito, & Kazumi, 2015) which expresses the variance of the error terms with the defined trend through GARCH (generalised autoregressive conditional heteroscedasticity model). This model obviously contains multiple unknown parameters, for which we carried out a Bayes' simultaneous estimation. We can see from the estimated results that it is possible to grasp both the variations of the expected values and their variances. In addition, in order to detect abnormalities in structures with long lifetimes such as infrastructures, it is necessary to introduce time sequence models considering long-term memory (Kobayashi, Jaafar, Ogata, & Tsukai, 2007; Tsukai & Kobayashi, 2007).

Figure 6 displays the relation between the time trend term related to variance and the proliferation of cracks in joints. Because of the small number of samples, it is impossible to ignore the possibility that the apparent relation is merely coincidental. Currently, almost no engineering knowledge has been accumulated regarding the causal relation between the variance term and the proliferation of joint cracks. However, the figure shows that a correlation exists between them, suggesting the presence of a causal relation. The concept of big data dictates that if we accumulate data from now on and confirm the significance of the correlation, it will be enough to continue monitoring using the variance term as an evaluation parameter (even though we should note that cracks are not the only cause of joint damage). As previously described, management does not consist of accumulating monitoring data. Monitoring data provide information to help engineers in their decisions. Regarding this point, the importance of monitoring is the ability to provide valuable information to engineers in the field to assist them to make decisions about the timing of detailed inspections.

5. Towards comprehensive risk management

5.1. Preventive abnormality detection system

It is known from daily work experience in pavement inspection related to preventive abnormality detection systems that recent amounts of snow and rain, as well as whether a period of continuous rain has exceeded a threshold, will contribute to higher pothole rates. Regarding pavement infrastructure management,

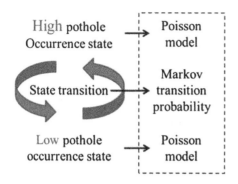

Figure 7. Markov switching model concept.

statistical prediction methods for pavement deterioration have been proposed. Moreover, in addition to planning models for the elaboration of rationalisation policies related to daily patrol and repair strategies focused on potholes (Kobayashi & Kaito, 2012; Obama, Kaito, Kobayashi, Kato, & Ikuta, 2007), life cycle cost evaluation methods linked with pavement deterioration prediction methods (Kobayashi et al., 2013) are being tested.

However, studies on existing pavement management methods (Archondo-Callao, 2008; Haas & Hudson, 1978; Madelin, 2000) have mostly considered ways to improve the average pavement performance (expected life), supposing a long stationary state. Because the occurrence of potholes engenders an immediate risk of traffic accidents, an effective measure is to introduce a flexible, management-focused scheme capable of switching from a normal management mode (called normal mode) to a high-end management mode (called abnormal mode) depending on weather conditions.

Regarding this issue, the authors are developing a decision model based on the Markov switching model (Mizutani, Kaito, & Kobayashi, 2013) (Figure 7). It is known from experience that potholes occur more often after several days of continuous rain. Moreover, in such cases, a switch occurs from normal patrol (normal mode) to interim patrol (abnormal mode). As a concrete example of interim patrol, in some cases, two patrols are carried out per day. The normal pothole generation process is modelled using a low-frequency Poisson model, whereas the pothole generation process after rain is modelled using a high-frequency Poisson model. Which of the above models is used depends on the weather, and the switching process involves the use of a Markov probability matrix. The Markov transition probability is formulated using a Markov degradation hazard model, and the explanatory variables may include such elements as the length of the continuous rain period and amount of rain.

Developing a flexible management-focused scheme capable of switching between the normal and abnormal modes depending on weather conditions would enable the construction of a more flexible patrolling scheme. Moreover, this way of thinking could be applied to other similar infrastructure management problems. In current abnormality detection (mode switching), decisions are based on whether the accumulated values of quantities such as the amount of rain exceed certain thresholds. With the assistance of probability models such as the Markov switching model, it is possible to perform status prediction. It will be possible to predict abnormalities and make decisions in a more flexible way by predicting whether the conditions are leading to a situation subject

to restrictions based on day-to-day variations in the amount of rain and environmental conditions, even if the total amount of rain at a given time is the same.

5.2. Sophistication of deterioration prediction technology

The heterogeneity in deterioration rate of infrastructure is attributable to various factors. By applying a mixed Markov deterioration hazard model, it became possible to handle the heterogeneity in deterioration rate as a heterogeneity parameter and predict the deterioration of each infrastructure. In the future, it will be possible to sophisticate deterioration prediction technology, by developing models for heterogeneity parameters and discussing them.

One is the evaluation of the management efficiency of infrastructure. For example, heterogeneity parameters can be divided into the heterogeneity in deterioration rate due to the technical problem at the initial construction stage and the heterogeneity due to managerial problems. The authors are developing a mixed Markov deterioration hazard model considering the two heterogeneities. In detail, an ideal deterioration process, excluding the effects of the deterioration due to the management environment, is described with a Markov deterioration hazard model. The degree of gap between the deterioration process that would become the best practice under given technical conditions and the current deterioration process is expressed by management efficiency parameters. Then, through the efficiency evaluation based on stochastic frontier analysis (Kumbhakar & Lovell, 2000), it is possible to quantitatively evaluate the effects of construction and management on deterioration rate.

The other is the multidimensional evaluation of infrastructure. Administrators have to evaluate the conditions of infrastructure from a multifaceted viewpoint, based on several deterioration events, but there may be some mutual interactions among deterioration events. For dealing with this problem, the authors developed a multidimensional deterioration process model considering the correlation structure among multiple deterioration events. In detail, the correlation structure among several indicators is described by expressing the progress of individual deterioration events with a mixed Markov deterioration hazard model including heterogeneity parameters and expressing the simultaneous distribution function of heterogeneity parameters with a copula (Nelsen, 1999). Mizutani, Kaito, and Kobayashi (2014) have already expressed the simultaneous distribution of heterogeneity parameters with a copula, while considering that the interaction among deterioration events can be expressed with the correlation among heterogeneities. However, in this methodology, it is hypothesised that all of the correlation structures among deterioration events are the same, and so there remains a problem that it is impossible to appropriately express the different correlations among deterioration events. Accordingly, the authors are developing a model that can flexibly express the different correlations among deterioration events based on the concept of Vine copula.

5.3. Critical infrastructure and vulnerability evaluation

It is possible to put together a comprehensive risk management method that can perform logical decisions regarding the majority of infrastructure management issues (obtaining consensus on public policies) by merging asset management, which handles daily issues and risk management (Barker & Haimes, 2009; Damnjanovic & Zhang, 2006; Zhang & Damnjanovic, 2006), which handles non-daily issues related to infrastructures. In order to do that, it is necessary to consider groups of infrastructures with complex interdependence connections as network systems to identify core infrastructures (critical infrastructure). Moreover, it is important to alleviate the risk that the entire network system will experience a functional halt owing to a domino functional stoppage of critical infrastructures.

In the field of infrastructure management, risk quantification methods such as VaR (value at risk) (McNeil, Frey, & Embrechts, 2005; Obama et al., 2007) and the real option approach (Obama, Kaito, Aoki, Kobayashi, & Fukuda, 2012), as well as risk assessment methods using fault tree analysis (Kaito, Kanaji, Sugioka, Oishi, & Matsuoka, 2013; Kaito et al., 2011), have been proposed. Moreover, the above continuous and efficient monitoring system is also indispensable from the standpoint of damage reduction for large-scale risk or checking the current situation after the event. Especially in the case of large-scale risks that may cause a functional halt of wired sensor networks, measures such as the parallel development of a wireless monitoring system (Kaito et al., 2010) are necessary.

6. Conclusions

In this paper, asset management from a big data perspective was discussed. Specifically, an overview of the statistical deterioration estimation method was presented using visual inspection data, which are assessed as the discrete soundness level. Although the pavement data were used as a case study to apply the method, this method is applicable to any infrastructure other than pavement as long as the information such as data from visual inspection or monitoring is available and can be expressed in the discrete soundness level. The reason why this method is often used in the pavement field is the availability of a large volume of data in the field.

Generally, when the soundness level is assessed in 5-scale and when the predictor variables are 2 to 3, about 1500 inspection data are needed to secure statistical reliability of a prediction model. In reality, however, only several thousands of data are available for the infrastructure such as bridges and sewage concrete pipes for which inspection method is mainly visual, while in the pavement field, road surface property-measuring vehicles are operational and the system is in place to efficiently collect large volume of data from all pavement and railways. The management standards based on the data are also in place. Therefore, it is not rare to see statistical analysis using millions of data in the pavement field, and, with the current monitoring technology level, it is possible to immediately conduct pavement management using big data.

The objective of assetmetrics is to facilitate the strategic enhancement of the management system itself as a result of the formalisation of implicit management knowledge (regarding decision-making) accumulated over many years. The role played by assetmetrics as a basic methodology for improving asset management based on such a hands-on policy is extremely important.

Disclosure statement

No potential conflict of interest was reported by the authors.

References

Aoki, K. (2008). Measuring deterioration risk of infrastructure. *Journal of Infrastructure Planning and Management, 25*, 17–35 (in Japanese).

Archondo-Callao, R. (2008). *Applying the HDM-4 model to strategic planning of road works*. Washington, DC: The World Bank Group.

Barker, K., & Haimes, Y. Y. (2009). Uncertainty analysis of interdependencies in dynamic infrastructure recovery: Applications in risk-based decision making. *Journal of Infrastructure Systems, 15*, 394–405.

Bishop, C. M. (2010). *Pattern recognition and machine learning*. New York, NY: Springer.

Damnjanovic, I., & Zhang, Z. (2006). Determination of required falling weight deflectometer testing frequency for pavement structural evaluation at the network level. *Journal of Transportation Engineering, 132*, 76–85.

FHWA (1998). *Life-cycle cost analysis in pavement design – In search of better investment decisions, Pavement Division Interim Technical Bulletin FHWA-SA-98-079, U.S.* Washington, DC: Department of Transportation, Federal Highway Administration.

Frangopol, D. M., Lin, K. Y., & Estes, A. C. (1997). Life-cycle cost design of deteriorating structures. *Journal of Structural Engineering, 123*, 1390–1401.

Haas, R. C. G., & Hudson, W. R. (1978). *Pavement management systems*. New York, NY: McGraw-Hill Book Company.

Iba, Y., Tanemura, M., Omori, Y., Wago, H., Sato, S., & Takahashi, A. (2005). *Markov Chain Monte Carlo method and its variants*. Tokyo: Iwanami (in Japanese).

Kaito, K., Aoki, K., & Kobayashi, K. (2010). Practical asset management and its perspective toward the second generation R&D. *Journal of the Japan Society of Civil Engineers, Ser. F5 (Professional Practices in Civil Engineering), 1*, 67–82 (in Japanese).

Kaito, K., Kanaji, H., Kobayashi, H., Mashima, N., Oishi, H., & Matsuoka, K. (2011). Optimum inspection policy for long span bridge based on fault tree analysis with visual inspection data. *Journal of the Japan Society of Civil Engineers F4, 67*, 74–91 (in Japanese).

Kaito, K., Kanaji, H., Sugioka, K., Oishi, H., & Matsuoka, K. (2013). Optimal inspection cycle of a long span bridge using the hybrid deterioration prediction and fault tree analysis. *Journal of the Japan Society of Civil Engineers F4, 69*, 84–101 (in Japanese).

Kaito, K., Kobayashi, K., Aoki, K., & Matsuoka, K. (2012). Hierarchical Bayesian estimation of mixed Markov hazard models. *Journal of the Japan Society of Civil Engineers D3, 68*, 255–271 (in Japanese).

Kaito, K., Kobayashi, K., & Obama, K. (2014). Investigating pavement structure deterioration with a relative evaluation model. In D. Frangopol & Y. Tsompanakis (Eds.), *Maintenance and safety of aging infrastructure*, Structures and Infrastructures Book Series, (Vol. 10, pp. 343–377). London: CRC Press.

Kaito, K., Matsuoka, K., Sakai, Y., Kawakami, Y., Arakawa, T., Kanagawa, M., & Kobayashi, K. (2010). Road-to-vehicle wireless communication monitoring aiming at application for asset management. *Journal on Applied Dynamics, Japan Society of Civil Engineers, 13*, 1017–1028 (in Japanese).

Kaito, K., Yasuda, K., Kobayashi, K., & Owada, K. (2005). Optimal maintenance strategies of bridge components with an average cost minimizing principles. *Journal of the Japan Society of Civil Engineers, 801/I-73*, 83–96 (in Japanese).

Kobayashi, K. (2005). Decentralized life-cycle cost evaluation and aggregated efficiency. *Journal of the Japan Society of Civil Engineers, 793/IV-68*, 59–71 (in Japanese).

Kobayashi, K. (2010). Practical research in civil engineering: Perspectives and methods. *Journal of the Japan Society of Civil Engineers, Ser. F5 (Professional Practices in Civil Engineering), 1*, 143–155 (in Japanese).

Kobayashi, K., Eguchi, T., Ohi, A., Aoki, K., Kaito, K., & Matsumura, Y. (2013). The optimal implementation policy of pavement inspection with deterioration uncertainty. *Journal of the Japan Society of Civil Engineers, 1*, 551–568.

Kobayashi, K., Jaafar, M. N. B., Ogata, S., & Tsukai, M. (2007). Danger warning information for transboundary hayes disaster: Regime change and long-term memory. *Journal of the Japan Society of Civil Engineers D, 63*, 478–497 (in Japanese).

Kobayashi, K., & Kaito, K. (2012). Obstacle emergence risk and road patrol policy. *Journal of Advances in Civil Engineering, 2012*, 9pp. ID 567509. doi:http://dx.doi.org/10.1155/2012/567509

Kobayashi, K., Kaito, K., Eguchi, T., Ohi, A., & Okizuka, R. (2011). A hierarchical hidden Markov deterioration model for pavement structure. *Journal of the Japan Society of Civil Engineers D3, 67*, 422–440 (in Japanese).

Kobayashi, K., Kaito, K., & Kazumi, K. (2015). Deterioration prediction of joint members based on long-term monitoring data. *European Journal of Transportation and Logistics, 4*, 5–30.

Kobayashi, K., Kaito, K., & Nam, L. T. (2012a). A Bayesian estimation method to improve deterioration prediction for infrastructure system with Markov chain model. *International Journal of Architecture, Engineering and Construction, 1*(1), 1–13.

Kobayashi, K., Kaito, K., & Nam, L. T. (2012b). A statistical deterioration forecasting method using hidden Markov model with measurement errors. *Transportation Research, Part B*, Elsevier, *46*, 544–561.

Kobayashi, K., Kaito, K., & Nam, L. T. (2014). A competing Markov model for prediction of pavement cracking process. *Transportation Research, Part B: Methodological, 68*, 345–362.

Kobayashi, K., Kumada, K., Sato, M., Iwasaki, Y., & Aoki, K. (2007). A pavement deterioration forecasting model with reference to sample dropping. *Journal of the Japan Society of Civil Engineers F, 63*(1), 1–15 (in Japanese).

Kumbhakar, S. C., & Lovell, C. A. K. (2000). *Stochastic frontier analysis*. Cambridge: Cambridge University Press.

Lancaster, T. (1990). *The econometric analysis of transition data*. Cambridge: Cambridge University Press.

Madanat, S., Bulusu, S., & Mahmoud, A. (1995). Estimation of infrastructure distress initiation and progression models. *Journal of Infrastructure Systems, 1*, 146–150.

Madelin, K. B. (2000). Developments in maintenance management in the United Kingdom. *International Journal of Pavement Engineering, 1*, 233–245.

Matsumura, Y., Kobayashi, K., Kaito, K., Ohi, A., & Yamaguchi, K. (2013). *A survey on mixed hidden Markov pavement deterioration models considering sample dropping*. Proceedings of the 46th Meeting of Infrastructure Planning and Management. Hiroshima: Japan Society of Civil Engineers (in Japanese).

McNeil, A. J., Frey, R., & Embrechts, P. (2005). *Quantitative risk management: Concepts, techniques and tools*. Princeton, NJ: Princeton University Press.

Mishalani, R. G., & Madanat, S. M. (2002). Computation of infrastructure transition probabilities using stochastic duration models. *Journal of Infrastructure Systems, 8*, 139–148.

Miyazaki, B., Obama, K., & Kaito, K. (2013). *Hidden Markov Model considering the inconsistencies of observation period*. The 13th East Asia-Pacific Conference on Structural Engineering and Construction (EASEC-13), Hokkaido.

Mizutani, D., Kaito, K. and Kobayashi, K. (2013). *Generation process of pot hole by Markov switching model*. 9th Annual Inter-University Symposium on Infrastructure Management (AISIM9), Berkeley, CA, USA.

Mizutani, D., Kaito, K., & Kobayashi, K. (2014). *Multidimensional deterioration process by copula modeling*. 10th Annual Inter-University Symposium on Infrastructure Management (AISIM10), Blacksburg, VA, USA.

Nam, L. T., Kaito, K., Kobayashi, K., & Okizuka, R. (2012). Modeling pavement deterioration processes by Poisson Hidden Markov Models. *Journal of the Japan Society of Civil Engineers F4, 68*, 62–79 (in Japanese).

Nelsen, R. B. (1999). *An introduction to copulas*. New York, NY: Springer.

Obama, K., Kaito, K., Aoki, K., Kobayashi, K., & Fukuda, T. (2012). The optimal scrapping and maintenance model of infrastructure considering deterioration process, *Journal of the Japan Society of Civil Engineers F4, 68*, 141–156 (in Japanese).

Obama, K., Kaito, K., Kobayashi, K., Fukuda, T., & Itagaki, K. (2013). Analysis of claims related to road obstacles. *Journal of the Japan Society of Civil Engineers F4, 69*, 32–46 (in Japanese).

Obama, K., Kaito, K., Kobayashi, K., Kato, T., & Ikuta, N. (2007). Road hazard risks and rational patrol policy. *Journal on Construction Management, Japan Society of Civil Engineers, 14*, 87–98 (in Japanese).

Obama, K., Okada, K., Kaito, K., & Kobayashi, K. (2008). Disaggregated hazard rates evaluation and bench-marking. *Journal of the Japan Society of Civil Engineers A, 64*, 857–874 (in Japanese).

Schonberger, V. M., & Cukier, K. (2013). *Big data: A revolution that will transform how we live, work, and think*. Boston, MA: Eamon Dolan/ Houghton Mifflin Harcourt.

Tsuda, Y., Kaito, K., Aoki, K., & Kobayashi, K. (2006). Estimating Markovian transition probabilities for bridge deterioration forecasting. *Journal of Structural Engineering and Earthquake Engineering, 23*, 241s–256s.

Tsukai, M., & Kobayashi, K. (2007). Measuring infrastructure productivity with reference to long persistent memory. *Journal of the Japan Society of Civil Engineers D, 63*, 255–274 (in Japanese).

Wago, H. (2005). *Bayes econometric analysis, Markov Chain Monte Carlo Method and its applications*. Tokyo: Toyo Keizai Shinposha (in Japanese).

Zhang, Z., & Damnjanovic, I. (2006). Quantification of risk cost associated with short-term warranty-based specifications for pavements. *Transportation Research Record: Journal of the Transportation Research Board, 1946*, 3–11.

Time-variant redundancy and failure times of deteriorating concrete structures considering multiple limit states

Fabio Biondini [iD] and Dan M. Frangopol

ABSTRACT

Structural redundancy and load redistribution capacity are desirable features to ensure suitable system performance under accidental actions and extreme events. For deteriorating structures, these features must be evaluated over time to account for the modification of the redistribution mechanisms due to damage processes. In particular, the identification of the local failure modes and prediction of their occurrence in time is necessary in order to maintain a suitable level of system performance and to avoid collapse. In fact, repairable local failures can be considered as a warning of damage propagation and possible occurrence of more severe and not repairable failures. In this paper, failure loads and failure times of concrete structures exposed to corrosion are investigated and life-cycle performance indicators, related to redundancy and elapsed times between sequential failures, are proposed. The effects of the damage process on the structural performance are evaluated based on a methodology for life-cycle assessment of concrete structures exposed to diffusive attack from environmental aggressive agents. The uncertainties involved are taken into account. The proposed approach is illustrated using two applicative examples: a reinforced concrete frame building and a reinforced concrete bridge deck under corrosion. The results demonstrate that both failure loads and failure times can provide relevant information to plan maintenance actions and repair interventions on deteriorating structures in order to ensure suitable levels of structural performance and functionality during their entire life-cycle.

Introduction

Structural reliability and durability of civil engineering structures and infrastructure facilities are essential to the economic growth and sustainable development of countries. However, aging, fatigue and deterioration processes due to aggressive chemical attacks and other physical damage mechanisms can seriously affect structure and infrastructure systems and lead over time to unsatisfactory structural performance (Clifton & Knab, 1989; Ellingwood, 2005). The economic impact of these processes is extremely relevant (ASCE, 2013; NCHRP, 2006) and emphasises the need of a rational approach to life-cycle design, assessment and maintenance of deteriorating structures under uncertainty based on suitable reliability-based life-cycle performance indicators (Biondini & Frangopol, 2014; Frangopol & Ellingwood, 2010; Saydam & Frangopol, 2011; Zhu & Frangopol, 2012). This need involves a major challenge to the field of structural engineering, since the classical time-invariant structural design criteria and methodologies need to be revised to account for a proper modelling of the structural system over its entire life-cycle by taking into account the effects of deterioration processes, time-variant loadings, and maintenance and repair interventions under uncertainty (Biondini & Frangopol, 2008a, 2016; Frangopol, 2011; Frangopol & Soliman, 2016).

In structural design the level of performance is usually specified with reference to structural reliability. However, when aging and deterioration are considered, the evaluation of the system performance under uncertainty should account for additional probabilistic indicators aimed at providing a comprehensive description of the life-cycle structural resources (Barone & Frangopol, 2014a, 2014b; Biondini & Frangopol, 2014, 2016; Frangopol & Saydam, 2014). The availability of stress redistribution mechanisms and the ability to mitigate the disproportionate effects of sudden damage under accidental actions, abnormal loads and extreme events, are often investigated in terms of structural redundancy (Bertero & Bertero, 1999; Biondini, Frangopol, & Restelli, 2008; Frangopol & Curley, 1987; Frangopol, Iizuka, & Yoshida, 1992; Frangopol & Nakib, 1991; Fu & Frangopol, 1990; Ghosn, Moses, & Frangopol, 2010; Hendawi & Frangopol, 1994; Husain & Tsopelas, 2004; Paliou, Shinozuka, & Chen, 1990; Pandey & Barai, 1997; Schafer & Bajpai, 2005; Zhu & Frangopol, 2013, 2015), structural vulnerability and robustness (Agarwal, Blockley, & Woodman, 2003; Baker, Schubert, & Faber, 2008; Biondini, Frangopol, & Restelli, 2008; Biondini & Restelli, 2008; Ellingwood, 2006; Ellingwood & Dusenberry, 2005; Ghosn et al., 2010; Lind, 1995; Lu, Yu, Woodman, & Blockley, 1999; Starossek & Haberland, 2011), and seismic resilience (Bocchini

& Frangopol, 2012a, 2012b; Bruneau et al., 2003; Chang & Shinozuka, 2004; Cimellaro, Reinhorn, & Bruneau, 2010; Decò, Bocchini, & Frangopol, 2013).

However, depending on the damage propagation mechanism, continuous deterioration may also involve alternate load redistribution paths and disproportionate effects, which can affect over time structural reliability and other performance indicators, including redundancy, robustness, resilience, and sustainability (Biondini, 2009; Biondini & Frangopol, 2014; Biondini, Camnasio, & Titi, 2015; Biondini, Frangopol, & Restelli, 2008; Biondini & Restelli, 2008; Decò, Frangopol, & Okasha, 2011; Enright & Frangopol, 1999; Frangopol & Bocchini, 2011; Furuta, Kameda, Fukuda, & Frangopol, 2004; Okasha & Frangopol, 2009, 2010; Sabatino, Frangopol, & Dong, 2015; Zhu & Frangopol, 2012, 2013). The effects of continuous deterioration can be particularly relevant for concrete structures exposed to the diffusive attack from aggressive agents, such as chlorides and sulfates, which may involve corrosion of steel reinforcement and deterioration of concrete (Bertolini, Elsener, Pedeferri, & Polder, 2004; CEB, 1992).

For reinforced concrete (RC) structures under corrosion the identification of the local failure modes and of their occurrence in time provides useful information in order to maintain a suitable level of system performance and to avoid collapse over their lifetime. In fact, repairable local failures can be considered as a warning of damage propagation and possible occurrence of more severe and not repairable failures (Mori & Ellingwood, 1994). Structural redundancy is a key performance indicator to this purpose, since it measures the ability of the system to redistribute among its active members the load which can no longer be sustained by other damaged members after the occurrence of a local failure (Biondini, Frangopol, & Restelli, 2008; Frangopol & Curley, 1987; Frangopol et al., 1992). However, this indicator refers to a prescribed point in time and does not provide a direct measure of the failure rate, which depends on the damage scenario and damage propagation mechanism (Biondini & Frangopol, 2008b, 2014).

Failure times and time intervals between subsequent failures, or elapsed time between failures, should be computed to provide complete information about the available resources after occurrence of local failures (Biondini, 2012; Biondini & Frangopol, 2014). In fact, the elapsed time between subsequent failures can be considered as a measure of system redundancy in terms of rapidity of evacuation and/or ability of the system to be repaired right after a critical damage state is reached. More specifically, the identification of all the local failure modes up to collapse and their occurrence in time could be helpful to plan emergency procedures, as well as maintenance and repair interventions to ensure suitable levels of life-cycle system performance and functionality.

In the following, failure loads and failure times of concrete structures under corrosion are investigated. Criteria and methods for the definition of life-cycle performance indicators related to redundancy and elapsed times between subsequent failures are proposed. The effects of the damage process on the structural performance are evaluated by using a proper methodology for life-cycle assessment of concrete structures exposed to diffusive attack from environmental aggressive agents (Biondini,

Bontempi, Frangopol, & Malerba, 2004a, 2006). The uncertainties in the material and geometrical properties, in the physical models of deterioration processes, and in the mechanical and environmental stressors, are taken into account in probabilistic terms. The proposed approach is illustrated through the assessment of structural redundancy and elapsed time between failures of a RC frame building and a RC bridge deck under corrosion. The goal is to show that both failure loads and failure times may provide important information to protect, maintain, restore and/or improve the life-cycle structural resources of deteriorating concrete structures.

Time-variant failure loads and failure times

A failure of a system is generally associated with the violation of one or more limit states. Focusing on RC frame systems, limit states of interest may be the occurrence at the material level of local failures associated to cracking of concrete and/or yielding of steel reinforcement, which represent warnings for initiation of damage propagation, as well as attainment of failures associated with the ultimate capacity of critical cross-sections and/or system collapse (Malerba, 1998).

Time-variant failure loads and redundancy

Denoting $\lambda \geq 0$ a scalar load multiplier, the limit states associated to the occurrence of a series of sequential failures $k = 1, 2, \ldots$ can be identified by the corresponding failure load multiplier λ_k (Biondini, 2012). Since the structural performance of RC structures deteriorates over time, the functions $\lambda_k = \lambda_k(t)$ need to be evaluated by means of structural analyses taking into account the effects of the damage process (Biondini et al., 2004a).

The ability of the system to redistribute the load after the failure $k = i$ up to the failure $k = j$ depends on the reserve load carrying capacity associated to the failure load multipliers $\lambda_i = \lambda_i(t)$ and $\lambda_j = \lambda_j(t)$:

$$\Delta \lambda_{ij}(t) = \lambda_j(t) - \lambda_i(t) \geq 0 \tag{1}$$

Therefore, the following quantity can be assumed as time-variant measure of redundancy between subsequent failures:

$$\Lambda_{ij}(t) = \Lambda(\lambda_i, \lambda_j) = \frac{\lambda_j(t) - \lambda_i(t)}{\lambda_j(t)} \tag{2}$$

The redundancy factor $\Lambda_{ij} = \Lambda_{ij}(t)$ can assume values in the range $[0;1]$. It is zero when there is no reserve of load capacity between the failures i and j ($\lambda_i = \lambda_j$), and tends to unity when the failure load capacity λ_i is negligible with respect to λ_j ($\lambda_i << \lambda_j$).

It is worth noting that the classical measure of redundancy refers to the ability of the system to redistribute the load after the occurrence of the first local failure, reached for $\lambda_1 = \lambda_1(t)$, up to structural collapse, reached for a collapse load multiplier $\lambda_c = \lambda_c(t)$. For the sake of brevity, the time-variant redundancy factor between the first failure and collapse is denoted $\Lambda = \Lambda(t)$ (Biondini & Frangopol, 2014):

$$\Lambda(t) = \Lambda(\lambda_1, \lambda_c) = \frac{\lambda_c(t) - \lambda_1(t)}{\lambda_c(t)} \tag{3}$$

It is also noted that redundancy is often associated with the degree of static indeterminacy of the structural system. However, it has been demonstrated that the degree of static indeterminacy is not a consistent measure for structural redundancy (Biondini & Frangopol, 2014; Biondini, Frangopol, & Restelli, 2008; Frangopol & Curley, 1987). In fact, structural redundancy depends on many factors, such as structural topology, member sizes, material properties, applied loads and load sequence, among others (Frangopol & Curley, 1987; Frangopol & Klisinski, 1989; Frangopol & Nakib, 1991). Moreover, the role of these factors may change over time due to structural deterioration, both in deterministic and probabilistic terms (Biondini, 2009; Biondini & Frangopol, 2014; Okasha & Frangopol, 2009).

Failure times and elapsed time between failures

Structural redundancy refers to a prescribed point in time and does not provide information on the failure sequence and failure rate over the structural lifetime. Failure times should be computed to this purpose and the time interval between subsequent failures, or the elapsed time between failures, could represent an effective indicator of the damage tolerance of the system and its ability to be repaired after local failures.

The failure times T_k associated to the occurrence of sequential failures $k = 1,2,\ldots$ can be evaluated by comparing the time-variant failure load multipliers $\lambda_k = \lambda_k(t)$ to prescribed time-variant target functions $\lambda_k^* = \lambda_k^*(t)$ as follows (Biondini, 2012):

$$T_k = \min \left\{ \, t \mid \lambda_k(t) < \lambda_k^*(t) \right\} \qquad (4)$$

After a local failure $k = i$, the ability of the system to delay the failure $k = j$ depends on the elapsed time between these failures occurring at times T_i and T_j (Figure 1):

$$\Delta T_{ij} = T_j - T_i \geq 0 \qquad (5)$$

For the sake of brevity, the elapsed time between the first failure, occurring for $\lambda_1 = \lambda_1(t)$ at time T_1, and the structural collapse, reached for $\lambda_c = \lambda_c(t)$ at time T_c, is denoted ΔT (Biondini, 2012; Biondini & Frangopol, 2014):

$$\Delta T = T_c - T_1 \qquad (6)$$

This is an important performance indicator, since it provides the residual lifetime after the first damage warning, for example associated with the formation of a plastic hinge, and identifies the time to global failure due to the activation of a set of plastic hinges leading to structural collapse (Biondini & Frangopol, 2008b).

As mentioned previously, the elapsed time between failures can also be considered as a measure of system redundancy in terms of rapidity of evacuation and/or ability of the system to be repaired right after a critical damage. However, even though they are related concepts, elapsed time between failures and structural redundancy refer to different system resources.

Role of the uncertainties

The geometrical and material properties of the structural systems, the mechanical and environmental stressors, and the parameters of the deterioration processes are always uncertain. Consequently, life-cycle prediction models have to be formulated in probabilistic terms and all parameters of the model have to be considered as random variables or processes. Therefore, the time evolution of the failure loads $\lambda_k = \lambda_k(t)$ and the corresponding failure times T_k, with $k = 1,2,\ldots$, are also random variables or processes, as shown in Figure 2 where uncertainties are associated with initial load capacity, damage initiation, deterioration rate, load capacity after maintenance/repair interventions, and failure time without or with maintenance/repair (Frangopol, 2011; Biondini & Frangopol, 2016). Therefore, a lifetime probabilistic analysis is necessary to investigate the time-variant effects of uncertainty on both redundancy factors $\Lambda_{ij} = \Lambda_{ij}(t)$ and elapsed time intervals ΔT_{ij} between the failures i and j.

Deterioration modelling: a review

A life-cycle probabilistic-oriented approach to design and assessment of structural systems must be based on a reliable and effective modelling of structural deterioration mechanisms. Deterioration models could be developed on empirical bases, as it is generally necessary for rate-controlled damage processes, or founded on mathematical descriptions of the underlying physical

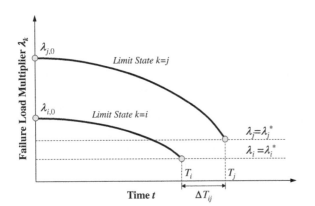

Figure 1. Time-variant failure load multipliers λ_i and λ_j, failure times T_i and T_j associated with the two limit states $\lambda_i = \lambda_i^*$ and $\lambda_j = \lambda_j^*$, and elapsed time ΔT_{ij} between these two sequential failures.

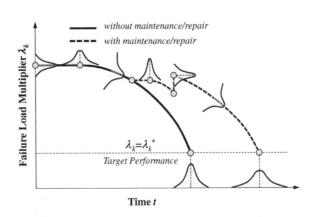

Figure 2. Time-variant failure load multiplier with uncertainties associated with initial load capacity, damage initiation, deterioration rate, load capacity after maintenance and repair interventions, and failure time without or with maintenance/repair.

mechanisms, as it is often feasible for diffusion-controlled damage processes (Ellingwood, 2005).

The latter is the case of interest for RC structures exposed to chloride ingress, where damage induced by the diffusive attack may involve corrosion of steel reinforcement and deterioration of concrete (Bertolini et al., 2004; CEB, 1992). The modelling of these processes should account for both the diffusion process and the related mechanical damage, as well as for the coupling effects of diffusion, damage and structural behaviour.

Diffusion process

The diffusion of chemical components in solids can be described by the Fick's laws which, in the case of a single component diffusion in isotropic, homogeneous and time-invariant media, can be reduced to the following second order linear partial differential equation (Glicksman, 2000):

$$D\nabla^2 C = \frac{\partial C}{\partial t} \qquad (7)$$

where D is the diffusivity coefficient of the medium, $C = C(\mathbf{x}, t)$ is the concentration of the chemical component at point $\mathbf{x} = (x, y, z)$ and time t, $\nabla C = \mathbf{grad}\, C(\mathbf{x}, t)$ and $\nabla^2 = \nabla \cdot \nabla$.

For one-dimensional diffusion, the Fick's equation is amenable to be solved analytically. This analytical solution is a convenient mathematical tool for practical applications (fib, 2006). However, the actual diffusion processes in concrete structures are generally characterised by two- or three-dimensional patterns of concentration gradients. For this reason, a numerical solution of the Fick's diffusion equation may be necessary for accurate life-cycle assessment of corroding RC structures (Titi & Biondini, 2016).

In this study, the diffusion equation is solved numerically by using cellular automata (Wolfram, 1994). With reference to a regular uniform grid of cells in two dimensions, the Fick's model can be reproduced at the cross-sectional level by the following evolutionary rule (Biondini et al., 2004a):

$$C_{ij}^{k+1} = \phi_0 C_{ij}^k + \frac{1-\phi_0}{4}(C_{i,j-1}^k + C_{i,j+1}^k + C_{i-1,j}^k + C_{i+1,j}^k) \quad (8)$$

where the discrete variable $C_{ij}^k = C(\mathbf{x}_{ij}, t_k)$ represents the concentration of the component in the cell (i, j) at point $\mathbf{x}_{ij} = (y_j, z_i)$ and time t_k, and $\phi_0 \in [0;1]$ is a suitable evolutionary coefficient. In particular, to regulate the process according to a given diffusivity D, the grid dimension Δx and the time step Δt of the automaton must satisfy the following relationship:

$$D = \frac{1-\phi_0}{4}\frac{\Delta x^2}{\Delta t} \qquad (9)$$

A proof is given in Biondini, Frangopol, and Malerba (2008).

The deterministic value $\phi_0 = 1/2$ usually leads to a good accuracy of the automaton. However, the local stochastic effects in the mass transfer can be taken into account by assuming ϕ_0 as random variable. The stochastic model also allows to simulate the interaction between diffusion process and mechanical behaviour of the damaged structure. Further details can be found in Biondini et al. (2004a).

Corrosion process

The most relevant effect of corrosion in concrete structures is the reduction of the cross-section of the reinforcing steel bars. The area $A_s = A_s(t)$ of a corroded bar can be represented as follows (Biondini et al., 2004a):

$$A_s(t) = [1 - \delta_s(t)]\, A_{s0} \qquad (10)$$

where A_{s0} is the area of the undamaged steel bar and $\delta_s = \delta_s(t)$ is a dimensionless damage index which provides a measure of cross-section reduction in the range $[0; 1]$.

Effects of corrosion are not limited to damage of reinforcing steel bars. In fact, the formation of oxidation products may lead to propagation of longitudinal splitting cracks and concrete cover spalling (Al-Harthy, Stewart, & Mullard, 2011; Cabrera, 1996; Vidal, Castel, & François, 2004). In this study, the local deterioration of concrete is modelled by means of a degradation law of the effective resistant area of concrete matrix $A_c = A_c(t)$ (Biondini et al., 2004a):

$$A_c(t) = [1 - \delta_c(t)]\, A_{c0} \qquad (11)$$

where A_{c0} is the area of undamaged concrete and $\delta_c = \delta_c(t)$ is a dimensionless damage index which provides a measure of concrete deterioration in the range $[0; 1]$. However, it may be not straightforward to relate the damage function $\delta_c = \delta_c(t)$ to the amount of steel mass loss. For this reason, in some cases, it could be more convenient to model the concrete deterioration due to splitting cracks and cover spalling through a reduction of the concrete compression strength (see Biondini & Vergani, 2015).

Additional effects of corrosion may occur depending on the type of corrosion mechanisms, i.e. uniform corrosion, localised (pitting) corrosion, or mixed type of corrosion (Stewart, 2009; Zhang, Castel, & François, 2010). As an example, depending on the amount of steel mass loss, non uniform corrosion may involve a remarkable reduction of steel ductility (Almusallam, 2001; Apostolopoulos & Papadakis, 2008) and a limited reduction of steel strength (Du, Clark, & Chan, 2005). Further information for a proper modelling of these effects can be found in Biondini and Vergani (2015). In this study, the effects of uniform corrosion only are investigated.

Damage rates

For diffusion-controlled damage processes, the deterioration rate depends on the time-variant concentration of the diffusive chemical components. In such processes, damage induced by mechanical loading interacts with the environmental factors and accelerates both diffusion and deterioration. Therefore, the dependence of the deterioration rate on the concentration of the diffusive agent is generally complex, and the available information about environmental factors and material characteristics is usually not sufficient for a detailed modelling. Despite the complexity of the problem at the microscopic level, simple coupling models can often be successfully adopted at the macroscopic level in order to reliably predict the time evolution of structural performance (Biondini & Frangopol, 2008b; Biondini, Frangopol, & Malerba, 2008; Biondini et al., 2004a).

Based on available data for sulfate and chloride attacks (Pastore & Pedeferri, 1994) and correlation between chloride content and corrosion current density in concrete (Bertolini et al., 2004; Liu & Weyers, 1998; Thoft-Christensen, 1998), a linear relationship between rate of corrosion in the range 0–200 μm/year and chloride content in the range 0–3% by weight of cement could be reasonable for structures exposed to severe environmental conditions. In this study, the time-variant damage indices $\delta_c = \delta_c(\mathbf{x}, t)$ and $\delta_s = \delta_s(\mathbf{x}, t)$ are related to the diffusion process by assuming a linear relationship between the rate of damage and the mass concentration $C = C(\mathbf{x}, t)$ of the aggressive agent:

$$\frac{\partial \delta_c(\mathbf{x}, t)}{\partial t} = \frac{C(\mathbf{x}, t)}{C_c \Delta t_c} = q_c C(\mathbf{x}, t) \qquad (12)$$

$$\frac{\partial \delta_s(\mathbf{x}, t)}{\partial t} = \frac{C(\mathbf{x}, t)}{C_s \Delta t_s} = q_s C(\mathbf{x}, t) \qquad (13)$$

where C_c and C_s are the values of constant concentration leading to a complete damage of the materials after the time periods Δt_c and Δt_s, respectively. The damage rate coefficients $q_c = (C_c \Delta t_c)^{-1}$ and $q_s = (C_s \Delta t_s)^{-1}$ depend on both the type of corrosion mechanism and corrosion penetration rate. The initial conditions $\delta_c(\mathbf{x}, t_{cr}) = \delta_s(\mathbf{x}, t_{cr}) = 0$ with $t_{cr} = \min\{t \mid C(\mathbf{x}, t) \geq C_{cr}\}$ are assumed, where C_{cr} is a critical threshold of concentration (Biondini et al., 2004a).

Structural analysis considering time effects

The lifetime structural performance is evaluated by means of structural analysis considering time-variant parameters (Biondini & Vergani, 2015; Biondini et al., 2004a). The formulation is based on the general criteria and methods for nonlinear analysis of concrete structures (Malerba, 1998). At cross-sectional level, the vector of the stress resultants (axial force N and bending moments M_z and M_y):

$$\mathbf{r} = \mathbf{r}(t) = \begin{bmatrix} N & M_z & M_y \end{bmatrix}^T \qquad (14)$$

and the vector of the global strains (axial elongation ε_0 and bending curvatures χ_z and χ_y):

$$\mathbf{e} = \mathbf{e}(t) = [\varepsilon_0 \ \chi_z \ \chi_y]^T \qquad (15)$$

are related, at each time instant t, as follows:

$$\mathbf{r}(t) = \mathbf{H}(t) \mathbf{e}(t) \qquad (16)$$

The time-variant stiffness matrix $\mathbf{H} = \mathbf{H}(t)$ of the RC cross-section under corrosion is derived by integration over the composite area of the materials, or by assembling the contributions of both concrete $\mathbf{H}_c = \mathbf{H}_c(t)$ and steel $\mathbf{H}_s = \mathbf{H}_s(t)$:

$$\mathbf{H}(t) = \mathbf{H}_c(t) + \mathbf{H}_s(t) \qquad (17)$$

$$\mathbf{H}_c(t) = \int_{A_c(x)} E_c(y, z, t) \, \mathbf{B}(y, z) \, [1 - \delta_c(y, z, t)] \, dA \qquad (18)$$

$$\mathbf{H}_s(t) = \sum_m E_{sm}(t) \, \mathbf{B}_m [1 - \delta_{sm}(t)] \, A_{sm} \qquad (19)$$

where the symbol m refers to the mth reinforcing bar located at point (y_m, z_m) in the centroidal principal reference system (y, z) of the cross-section, $E_c = E_c(y, z, t)$ and $E_{sm} = E_{sm}(t)$ are the moduli of the materials, $\mathbf{B}(y, z) = \mathbf{b}(y, z)^T \mathbf{b}(y, z)$ is a linear operator matrix, and $\mathbf{b}(y, z) = [1 \ -y \ z]$.

It is worth noting that the vectors $\mathbf{r}$ and $\mathbf{e}$ have to be considered as total or incremental quantities based on the nature of the stiffness matrix $\mathbf{H}$, which depends on the type of formulation adopted (i.e. secant or tangent) for the generalised moduli of the materials associated with the stress–strain nonlinear constitutive laws.

The proposed cross-sectional formulation can be extended to formulate the characteristics of RC beam finite elements for time-variant nonlinear and limit analysis of concrete structures under corrosion. Details can be found in Biondini et al. (2004a), Biondini & Frangopol (2008b), Biondini and Vergani (2015).

Applications

The proposed approach is applied to the probabilistic assessment of structural redundancy and elapsed time between failures of a RC frame and a RC bridge deck under corrosion.

Constitutive laws of the materials

The constitutive behaviour of the materials is described in terms of stress–strain nonlinear relationships. For concrete, the Saenz's law in compression and a bilinear elastic-plastic model in tension are assumed, with: compression strength f_c; tension strength $f_{ct} = .25 f_c^{2/3}$; initial modulus $E_{c0} = 9500 f_c^{1/3}$; peak strain in compression $\varepsilon_{c0} = .20\%$; strain limit in compression $\varepsilon_{cu} = .35\%$; strain limit in tension $\varepsilon_{ctu} = 2f_{ct}/E_{c0}$. For steel, a bilinear elastic-plastic model in both tension and compression is assumed, with yielding strength f_{sy}, elastic modulus $E_s = 210$ GPa, and strain limit $\varepsilon_{su} = 1.00\%$ associated with bond failure. In this way, the constitutive laws are completely defined by the material strengths f_c and f_{sy}.

Probabilistic modelling

The probabilistic analysis accounts for the uncertainty in both the geometrical and mechanical characteristics of the structural systems and in the parameters which define the deterioration processes. At cross-sectional level, the probabilistic model of the mechanical behaviour, diffusion process and damage propagation mechanism considers as random variables the material strengths f_c and f_{sy} of concrete and steel, respectively, the coordinates (y_p, z_p) of each nodal point $p = 1, 2, \ldots$, of the cross-section, the coordinates (y_m, z_m) and diameter $\varnothing_m$ of each steel bar $m = 1, 2, \ldots$, the diffusivity coefficient D, and the damage rate coefficients q_c and q_s. Nominal values are assumed as mean values. The probabilistic distributions and coefficients of variation are listed in Table 1 (Biondini et al., 2006; Sudret, 2008; Vismann & Zilch, 1995). The input random variables are uncorrelated to emphasise the effects of the lack of correlation on the investigated output random variables (Harr, 1996). Moreover, high values of the coefficient of

Table 1. Probability distributions and coefficients of variation (nominal values are assumed as mean values μ).

Random variable ($t = 0$)	Type	C.o.V.
Concrete strength, f_c	Lognormal	5 MPa/μ
Steel strength, f_{sy}	Lognormal	30 MPa/μ
Coordinates of nodal points, (y_p, z_p)	Normal	5 mm/μ
Coordinates of steel bars, (y_m, z_m)	Normal	5 mm/μ
Diameter of steel bars, $\varnothing_m$	Normal*	.10
Diffusivity, D	Normal*	.10
Concrete damage rate, $q_c = (C_c \Delta t_c)^{-1}$	Normal*	.30
Steel damage rate, $q_s = (C_s \Delta t_s)^{-1}$	Normal*	.30

*Truncated distributions with non negative outcomes.

variation are adopted for the random variables which mainly influence the time-variant uncertainty of the corrosion damage, such as the steel bar diameter and damage rates.

The lifetime probabilistic analysis is carried out by Monte Carlo simulation. The required accuracy of the simulation process is achieved through a posteriori estimation of the goodness of the sample size based on a monitoring of the time-variant statistical parameters of the random variables under investigation.

RC frame

The lifetime structural performance of the RC frame shown in Figure 3 is investigated in terms of redundancy and elapsed time between failures. The nominal material strengths are $f_c = 40$ MPa for concrete in compression and $f_{sy} = 500$ MPa for the yield of reinforcing steel. The frame is subjected to a dead load $q = 32$ kN/m applied on the beam and a live load λF acting at top of the columns, with $F = 100$ kN.

The frame system is designed with cross-sectional stiffness and bending strength capacities much larger in the beam than in the columns. Moreover, shear failures are avoided over the lifetime by a proper capacity design (Celarec, Vamvatsikos, &

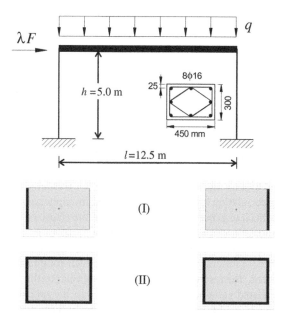

Figure 3. Reinforced concrete frame: geometry, structural scheme, cross-section of the columns, loading condition, grid of the diffusion model, and exposure scenarios with (I) columns exposed on one side and (II) columns exposed on four sides.

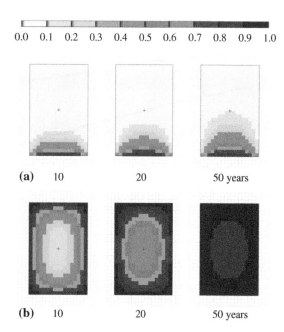

Figure 4. Maps of concentration $C(\mathbf{x}, t)/C_0$ of the aggressive agent after 10, 20, and 50 years from the initial time of diffusion penetration (nominal frame system): (a) scenario (I) with exposure on one side; (b) scenario (II) with exposure on four sides.

Dolšek, 2011; Titi & Biondini, 2014). In this way, a shear-type behaviour can be assumed, with the critical regions where plastic hinges are expected to occur located at the ends of the columns.

The structure is subjected to the diffusive attack from an aggressive agent located on the external surfaces of the columns with concentration C_0. The two exposure scenarios shown in Figure 3 are considered, with (I) columns exposed on the outermost side only, or (II) columns exposed on the four sides. A nominal diffusivity coefficient $D = 10^{-11}$ m²/sec is assumed. Figure 4 shows the deterministic maps of concentration $C(\mathbf{x}, t)/C_0$ for the two investigated exposure scenarios after 10, 20, and 50 years from the initial time of diffusion penetration.

The corrosion damage induced by diffusion is evaluated by taking the stochastic effects in the mass transfer into account (Biondini et al., 2004a). Corrosion of steel bars with no deterioration of concrete is assumed, with nominal damage parameters $C_s = C_0$, $\Delta t_s = 50$ years, and $C_{cr} = 0$. This model reproduces a deterioration process with severe corrosion of steel, as may occur for carbonated or heavily chloride-contaminated concrete and high relative humidity, conditions under which the corrosion rate can reach values above 100 μm/year (Bertolini et al., 2004).

Figure 5 shows the evolution over a 50-year lifetime of the failure load multipliers $\lambda_1 = \lambda_1(t)$ and $\lambda_c = \lambda_c(t)$ associated to the reaching of first local yielding of steel reinforcement and structural collapse of the frame system, respectively. The failure loads are computed at each time instant under the hypotheses of linear elastic behaviour up to first local yielding, and perfect plasticity at collapse.

The time evolution of the redundancy factor $\Lambda = \Lambda(t)$ of the frame system for the two investigated scenarios is shown in Figure 6. It is noted that for case (I) redundancy increases over time, even if the structural performance in terms of load capacity decreases. This is because the bending strength of the critical cross-sections corroded on the compression side deteriorates

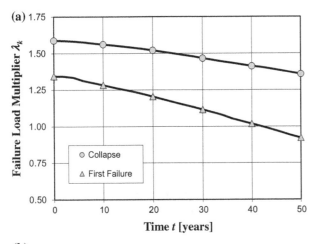

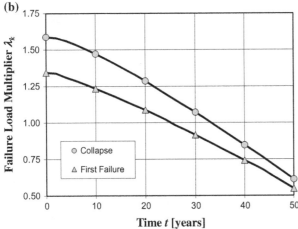

Figure 5. Time evolution of the load multipliers λ_1 and λ_c associated with the reaching of first local yielding of steel reinforcement and collapse, respectively: (a) scenario (I) with exposure on one side; (b) scenario (II) with exposure on all sides.

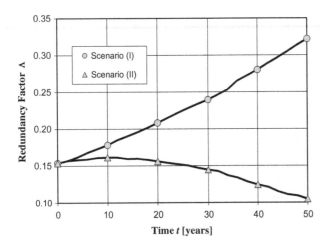

Figure 6. Time evolution of the redundancy factor Λ for scenario (I) with exposure on one side, and scenario (II) with exposure on all sides.

more slowly compared to the cross-sections corroded on the tension side. Therefore, the collapse load multiplier λ_c, which depends on the bending strengths of all critical cross-sections, has a lower deterioration rate than the load multiplier at first yielding λ_1, which is associated with the failure of a single

Table 2. Failure times T_1 and T_c and elapsed time ΔT associated to different target load values λ^*.

λ^*	T_1 [years]	T_c [years]	ΔT [years]
.70	41.4	46.4	5.0
.80	36.2	42.1	5.9
.90	30.8	37.7	6.9
1.00	25.1	33.2	8.1
1.10	19.2	28.6	9.4
1.20	12.4	24.0	11.6
1.30	4.9	19.3	14.5

cross-section. On the contrary, redundancy mainly decreases over time for case (II). Therefore, case (II) is the worst damage scenario for structural redundancy.

The time evolution of the failure loads (Figure 5) indicates that the exposure scenario (II) may be critical with respect to structural collapse. Therefore, for this scenario it is of interest the assessment of the failure times T_1 and T_c associated to the occurrence of the first yielding and collapse, respectively, as well as the elapsed time $\Delta T = T_c - T_1$ between such failures. Failure times and elapsed times associated to different target values $\lambda^* = \lambda_1^* = \lambda_c^*$ are listed in Table 2 for the nominal system. These values indicate that after local failures a significant rapidity of repair may be required under severe exposures. Moreover, it can be noticed that the failure times decrease as the target load multiplier increases. The elapsed time between failures shows instead an opposite trend. Therefore, the availability of a larger reserve of load capacity with respect to the design target is beneficial to delay the occurrence of failures, but it may require prompter repair actions after local failures occur.

The effects of the uncertainty are investigated based on the probabilistic information given in Table 1. The two sets of random variables associated to each column are preliminarily assumed as uncorrelated to emphasise the effects of the uncertainty. Figure 7 shows the probability mass functions (PMFs) of the failure times T_1 and T_c (Figure 7(a)) and elapsed time ΔT (Figure 7(b)) for two deterministic values of the target load multiplier, $\lambda^* = 1.00$ and $\lambda^* = .75$, based on a sample of 2000 Monte Carlo realisations. For $\lambda^* = 1.00$ the failure times T_1 and T_c are characterised by mean and standard deviation values lower than the values obtained for $\lambda^* = .75$. On the contrary, the mean value of the elapsed time ΔT is higher for $\lambda^* = 1.00$ than for $\lambda^* = .75$, with a small difference in terms of dispersion. These results confirm that a suitable reserve of load capacity with respect to the design target allows to delay the possible occurrence of failure events, but it demands for higher promptness and rapidity in the recovery actions.

It is worth noting that the effects of randomness on the reserve load capacity $\Delta\lambda = \lambda_c - \lambda_1$ lead to mean values of the elapsed time ΔT higher than the nominal deterministic values. Moreover, the strong correlation between the failure load multipliers λ_1 and λ_c is beneficial to achieve a lower variance of the elapsed time ΔT than the variance of the failure times T_1 or T_c.

The influence of correlation is also studied by assuming the two sets of random variables associated to each column as fully correlated. The results lead to conclusions similar to the case of uncorrelated variables, with small changes in the probabilistic parameters of the investigated performance indicators. As an example, the mean and standard deviation values obtained for the elapsed time ΔT are $\mu = 11.1$ years and $\sigma = 4.2$ years for $\lambda^* = 1.00$, and $\mu = 7.3$ years and $\sigma = 2.6$ years for $\lambda^* = .75$.

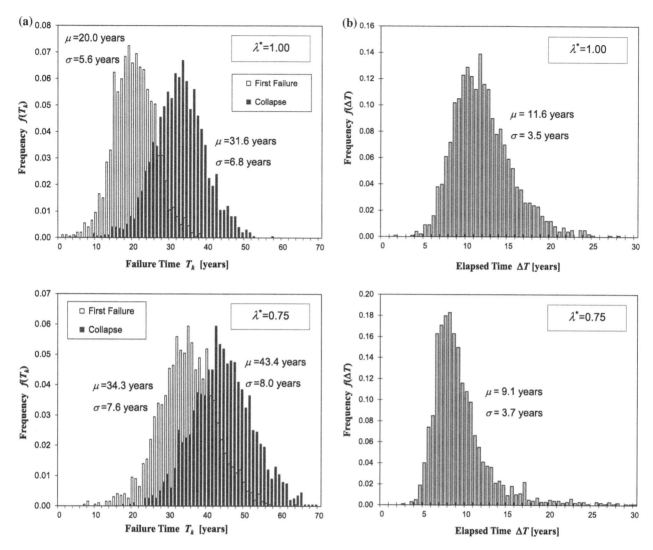

Figure 7. PMFs of (a) failure times T_1 and T_c and (b) elapsed time between failures ΔT for two values of the target load multiplier, $\lambda^* = 1.00$ and $\lambda^* = .75$, under scenario (II) with exposure on all sides.

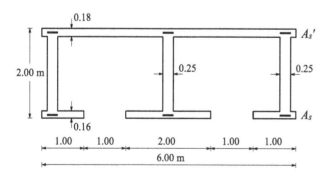

Figure 8. Reinforced concrete bridge deck: geometry of the cross-section and main steel reinforcement $A_s' = 48\varnothing28$ mm and $A_s = 21\varnothing28$ mm (additional reinforcing steel bars: $130\varnothing8$ mm in the top slab; $60\varnothing8$ mm in the bottom slab).

RC bridge deck

The lifetime structural performance of a RC bridge deck is investigated at cross-sectional level in terms of redundancy and elapsed time between failures. The geometry of the concrete cross-section

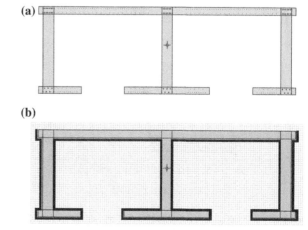

Figure 9. (a) Structural model of the bridge deck cross-section with indication of the steel reinforcement (259 steel bars: $48\varnothing28$ mm + $130\varnothing8$ mm in the top slab; $21\varnothing28$ mm + $60\varnothing8$ mm in the bottom slab). (b) Grid of the diffusion model and exposure scenario.

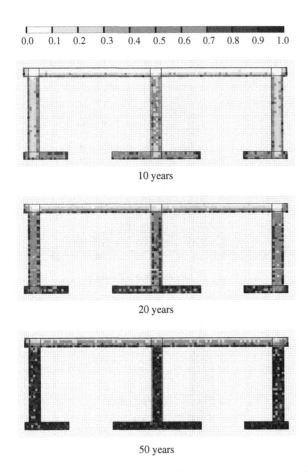

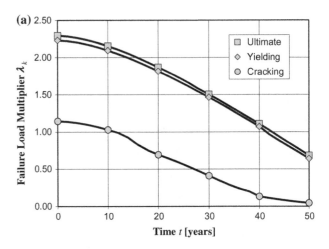

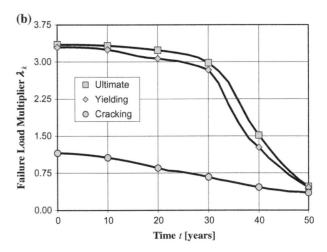

Figure 10. Maps of concentration $C(\mathbf{x}, t)/C_0$ of the aggressive agent after 10, 20, and 50 years from the initial time of diffusion penetration (nominal bridge deck under stochastic diffusion).

Figure 12. Time evolution of the failure load multipliers λ_k, with $k = 1,2,3$, at concrete first cracking ($k = 1$), steel first yielding ($k = 2$), and cross-section ultimate capacity ($k = 3$): (a) positive and (b) negative bending moment.

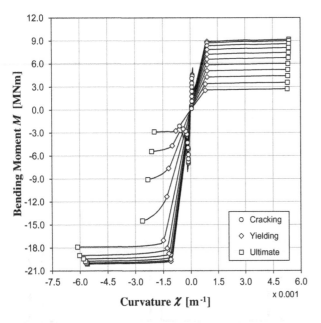

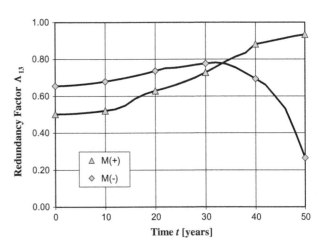

Figure 11. Time evolution of the nominal bending moment M vs. curvature χ diagrams over a 50-year lifetime ($\Delta t = 5$ years), with indication of the points associated to first cracking of concrete, first yielding of steel reinforcement, and ultimate flexural capacity of the cross-section.

Figure 13. Time evolution of the nominal redundancy factor Λ_{13} between the states (1) and (3) associated to first concrete cracking and cross-section ultimate capacity, respectively, for positive and negative bending moment.

Table 3. Failure times T_1, T_2, T_3, and elapsed times between failures ΔT_{12}, ΔT_{23} [years]: (1) concrete first cracking; (2) steel first yielding and (3) cross-section ultimate capacity.

$\lambda^* = 1.0$	T_1	T_2	T_3	ΔT_{12}	ΔT_{23}
M^+	11.5	41.4	42.6	29.9	1.2
M^-	13.7	42.6	44.1	28.9	1.5

and the location of the main steel reinforcement in the top and bottom slabs are shown in Figure 8. The nominal dimensions are: width = 6.00 m; depth = 2.00 m; web thickness = .25 m; top slab thickness = .18 m; bottom slab thickness = .16 m. The steel reinforcement located in the top slab consists of 48 bars with nominal diameter $\varnothing$ = 28 mm, and 130 bars with $\varnothing$ = 8 mm. The steel reinforcement located in the bottom slab consists of 21 bars with $\varnothing$ = 28 mm and 60 bars with $\varnothing$ = 8 mm. Figure 9(a) shows the structural model of the cross-section, with detailed location of the steel bars. The nominal material strengths are f_c = 30 MPa for concrete in compression and f_{sy} = 300 MPa for the yield of reinforcing steel.

The bridge deck cross-section is subjected to the diffusive attack from an aggressive agent located with concentration C_0 on the external surface exposed to the atmosphere. The diffusion model and the exposure scenario are shown Figure 9(b). A nominal diffusivity coefficient $D = 10^{-11}$ m²/sec is assumed. Figure 10 shows the stochastic maps of concentration $C(\mathbf{x}, t)/C_0$ after 10, 20, and 50 years from the initial time of diffusion penetration.

A severe corrosion damage scenario is assumed, with nominal parameters $C_c = C_s = C_0$, $\Delta t_c = 25$ years, $\Delta t_s = 50$ years, and $C_{cr} = 0$. The mechanical damage induced by diffusion over a 50-year lifetime is shown in Figure 11 in terms of nominal bending moment M vs. curvature χ diagrams computed by assuming the bridge deck axially unloaded.

The results shown in Figure 11 indicate that damage significantly affect the flexural performance of the cross-section, both for positive and negative bending moments. Deterioration of structural performance is mainly due to the severe exposure of the bottom slab. For positive bending moment, the corrosion of the reinforcing steel bars in tension located in the bottom slab leads to a progressive bending strength deterioration over the lifetime, with no significant changes in the curvature ductility. For negative bending moment, the lower corrosion rate of the reinforcing steel bars in tension located in the top slab involves

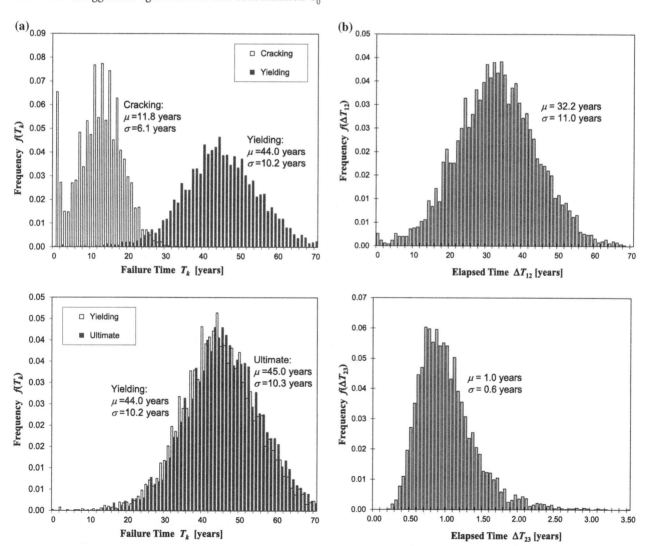

Figure 14. PMFs of (a) failure times T_1, T_2, T_3, and (b) elapsed times between failures ΔT_{12} and ΔT_{23} associated to the occurrence of the sequential limit states of (1) concrete first cracking, (2) steel first yielding and (3) cross-section ultimate capacity, for positive bending moment.

a limited reduction of bending strength over the first years of lifetime. However, after about 30 years of lifetime the severe deterioration of concrete in compression in the bottom slab causes a remarkable progressive decrease of structural performance in terms of both bending strength and curvature ductility.

At cross-sectional level, limit states of interest are the occurrence of local failures associated to cracking of concrete and yielding of steel reinforcement, which are warnings for initiation of damage propagation, as well as the attainment of the ultimate flexural capacity of the cross-section defined by the strain limits $\varepsilon_c = -\varepsilon_{cu}$ and/or $|\varepsilon_s| = \varepsilon_{su}$. The points associated to the limit states of (1) concrete first cracking, (2) steel first yielding, and (3) cross-section ultimate capacity, are indicated on the capacity curves shown in Figure 11. The corresponding time evolution of the $k = 1,2,3$, failure load multipliers $\lambda_k = \lambda_k(t)$, computed for the design values $M^+ = 4$ MNm and $M^- = -6$ MNm of the bending moments λM^+ and λM^-, is shown in Figure 12.

The reserve of load capacity after concrete cracking ensures a suitable level of overall structural redundancy for both positive and negative bending moment, as shown in Figure 13 in terms of redundancy factor $\Lambda_{13} = \Lambda_{13}(t)$ between (1) cracking and (3) ultimate capacity of the cross-section. For positive bending moment, redundancy increases continuously over time, even if the structural performance in terms of load capacity decreases. For negative bending moment, redundancy exhibits a moderate increase during the first period of exposure to damage, and rapidly decreases after about 30 years of lifetime. These results indicate that corrosion of steel reinforcement in tension, even though it involves a reduction of load capacity, may have beneficial effects in terms of structural redundancy. Contrary, the effects of deterioration of concrete in compression are generally detrimental to structural redundancy.

The reserve of load capacity after steel yielding is instead very limited and does allow for significant redundancy in between yielding and ultimate states. In this case the elapsed time between failures provides useful information about the available time to repair after a local yielding occurs.

With reference to a target load multiplier $\lambda^* = 1.0$, the failure times T_1, T_2, T_3, and the related elapsed times between failures ΔT_{12}, ΔT_{23}, associated to the occurrence of the sequential limit states of (1) concrete first cracking, (2) steel first yielding, and (3) cross-section ultimate capacity, are listed in Table 3 for the nominal system under positive and negative bending moments.

The uncertainty effects on these performance indicators are investigated also in probabilistic terms based on the probabilistic information given in Table 1. Based on 5000 Monte Carlo simulations, Figure 14 shows the PMFs of the failure times T_1, T_2, T_3 (Figure 14(a)), and elapsed time between failures ΔT_{12}, ΔT_{23} (Figure 14(b)) for the case of positive bending moment.

The deterministic and probabilistic results confirm that a remarkable rapidity of repair may be required after occurrence of a severe local failure, such as yielding of steel reinforcement. Moreover, failure loads and failure times associated to concrete cracking or other minor local failure events could provide warnings of more severe future damage states or critical threats and to support in this way the decision-making process for maintenance and repair planning.

Conclusions

Failure loads and failure times of deteriorating RC structures have been investigated. Life-cycle performance indicators, related to time-variant structural redundancy and elapsed times between sequential failures occurring under continuous deterioration processes, have been formulated. The effects of the damage process on the structural performance have been evaluated by considering uncertainties based a methodology for life-cycle assessment of concrete structures exposed to the diffusive attack from environmental aggressive agents. The proposed approach has been applied to the assessment of structural redundancy and elapsed time between failures of a RC frame and a RC bridge deck under corrosion.

The results show that the prediction of the local and global failure modes and of their occurrence in time provides useful information on the remaining life-cycle of deteriorating RC structures. In fact, after local failures occur, a very fast repair may be required under a severe exposure scenario to prevent structural collapse. Failure times and time intervals between subsequent failures must be computed for this purpose, since other damage-tolerance performance indicators, such as structural redundancy, do not provide a direct measure of the failure rate.

Therefore, failure times and elapsed time between failures are important performance indicators to be used jointly with other performance measures, such as reliability, redundancy, robustness, resilience, and sustainability for a rational approach to life-cycle design, assessment and maintenance of deteriorating structure and infrastructure systems. This approach is clearly more demanding than standard time-invariant design procedures, since it involves the modelling of complex deterioration processes and the evaluation of several performance indicators over the structural lifetime. For this reason, reliable deterioration modelling of materials and structural components and computationally efficient structural analysis procedures considering time effects, as those presented in this paper, are essential to a robust prediction of the time-variant structural performance and to support and advance the civil engineering profession in this field.

However, it is worth mentioning that deterioration models are generally very sensitive to change in the probabilistic parameters of the input random variables and their robust validation and accurate calibration are difficult tasks to be performed due to the limited availability of data. Further efforts aimed at gathering new data from both existing structures and experimental tests are crucial for a successful calibration and implementation in practice of the presented approach. Also efforts in the modelling of nonlinear structures using finite elements with time-variant properties (Biondini & Vergani, 2015; Biondini et al., 2004a), probabilistic finite element analysis (Biondini, Bontempi, Frangopol, & Malerba, 2004b; Biondini & Frangopol, 2008b; Teigen, Frangopol, Sture, & Felippa, 1991a, 1991b), reliability-based inspections (Onoufriou & Frangopol, 2002), probabilistic importance assessment of structural components (Gharaibeh, Frangopol, & Onoufriou, 2002), and developing improved models for cost and risk estimation of maintenance actions (Frangopol & Kong, 2001; Saydam & Frangopol, 2015) are necessary to ensure protection of civil infrastructure facilities over time at minimum life-cycle cost.

Disclosure statement

No potential conflict of interest was reported by the authors.

ORCID

Fabio Biondini (iD) http://orcid.org/0000-0003-1142-6261

References

Agarwal, J., Blockley, D. I., & Woodman, N. J. (2003). Vulnerability of structural systems. *Structural Safety, 25*, 263–286.

Al-Harthy, A. S., Stewart, M. G., & Mullard, J. (2011). Concrete cover cracking caused by steel reinforcement corrosion. *Magazine of Concrete Research, 63*, 655–667.

Almusallam, A. A. (2001). Effect of degree of corrosion on the properties of reinforcing steel bars. *Construction and Building Materials, 15*, 361–368.

Apostolopoulos, C. A., & Papadakis, V. G. (2008). Consequences of steel corrosion on the ductility properties of reinforcement bar. *Construction and Building Materials, 22*, 2316–2324.

ASCE. (2013, March). *Report card for America's infrastructure*. Reston, VA: American Society of Civil Engineers.

Baker, J. W., Schubert, M., & Faber, M. H. (2008). On the assessment of robustness. *Structural Safety, 30*, 253–267.

Barone, G., & Frangopol, D. M. (2014a). Life-cycle maintenance of deteriorating structures by multi-objective optimization involving reliability, risk, availability, hazard and cost. *Structural Safety, 48*, 40–50.

Barone, G., & Frangopol, D. M. (2014b). Reliability, risk and lifetime distributions as performance indicators for life-cycle maintenance of deteriorating structures. *Reliability Engineering & System Safety, 123*, 21–37.

Bertero, R. D., & Bertero, V. V. (1999). Redundancy in earthquake-resistant design. *Journal of Structural Engineering, 125*, 81–88.

Bertolini, L., Elsener, B., Pedeferri, P., & Polder, R. (2004). *Corrosion of steel in concrete*. Weinheim: Wiley-VCH.

Biondini, F. (2009). *A measure of lifetime structural robustness*. Proceedings of the SEI/ASCE Structures Congress, Austin, TX, USA, April 30–May 2. In L. Griffis, T. Helwig, M. Waggoner, & M. Hoit (Eds.), *Structures Congress 2009*. ASCE, CD-ROM.

Biondini, F. (2012). Discussion: Time-variant redundancy of ship structures, by Decò, A., Frangopol, D.M., & Okasha, N.M. *Society of Naval Architects and Marine Engineers Transactions, 119*, 40.

Biondini, F., Bontempi, F., Frangopol, D. M., & Malerba, P. G. (2004a). Cellular automata approach to durability analysis of concrete structures in aggressive environments. *Journal of Structural Engineering, 130*, 1724–1737.

Biondini, F., Bontempi, F., Frangopol, D. M., & Malerba, P. G. (2004b). Reliability of material and geometrically nonlinear reinforced and prestressed concrete structures. *Computers & Structures, 82*, 1021–1031.

Biondini, F., Bontempi, F., Frangopol, D. M., & Malerba, P. G. (2006). Probabilistic service life assessment and maintenance planning of concrete structures. *Journal of Structural Engineering, 132*, 810–825.

Biondini, F., Camnasio, E., & Titi, A. (2015). Seismic resilience of concrete structures under corrosion. *Earthquake Engineering and Structural Dynamics, 44*, 2445–2466.

Biondini, F., & Frangopol, D. M. (Eds.). (2008a). *Life-cycle civil engineering*. Boca Raton, FL, London, New York, NY, Leiden: CRC Press, Taylor & Francis Group, A.A. Balkema.

Biondini, F., & Frangopol, D. M. (2008b). Probabilistic limit analysis and lifetime prediction of concrete structures. *Structure and Infrastructure Engineering, 4*, 399–412.

Biondini, F., & Frangopol, D. M. (2014). Time-variant robustness of aging structures. Chapter 6. In D. M. Frangopol & Y. Tsompanakis (Eds.), *Maintenance and safety of aging infrastructure* (pp. 163–200). London:CRC Press, Taylor & Francis Group.

Biondini, F., & Frangopol, D. M. (2016). Life-cycle performance of deteriorating structural systems under uncertainty: Review. *Journal of Structural Engineering*. doi:10.1061/(ASCE)ST.1943-541X.0001544

Biondini, F., Frangopol, D. M., & Malerba, P. G. (2008). Uncertainty effects on lifetime structural performance of cable-stayed bridges. *Probabilistic Engineering Mechanics, 23*, 509–522.

Biondini, F., Frangopol, D. M., & Restelli, S. (2008). *On structural robustness, redundancy and static indeterminacy*. Proceedings of the SEI/ASCE 2008Structures Congress, Vancouver, BC, Canada, April 24–26, 2008. ASCE, CD-ROM.

Biondini, F., & Restelli, S. (2008). *Damage propagation and structural robustness*. First International Symposium on Life-Cycle Civil Engineering (IALCCE'08), Varenna, Italy, June 10–14. In F. Biondini & D. M. Frangopol (Eds.), *Life-cycle civil engineering* (pp. 565–570). CRC Press, Taylor & Francis Group.

Biondini, F., & Vergani, M. (2015). Deteriorating beam finite element for nonlinear analysis of concrete structures under corrosion. *Structure and Infrastructure Engineering, 11*, 519–532.

Bocchini, P., & Frangopol, D. M. (2012a). Optimal resilience- and cost-based postdisaster intervention prioritization for bridges along a highway segment. *Journal of Bridge Engineering, 17*, 117–129.

Bocchini, P., & Frangopol, D. M. (2012b). Restoration of bridge networks after an earthquake: Multicriteria intervention optimization. *Earthquake Spectra, 28*, 426–455.

Bruneau, M., Chang, S. E., Eguchi, R. T., Lee, G. C., O'Rourke, T. D., Reinhorn, A. M., Shinozuka, M., Tierney, K., Wallace, W. A., & Wintefeldt, D. V. (2003). A framework to quantitatively assess and enhance the seismic resilience of communities. *Earthquake Spectra, 19*, 733–752.

Cabrera, J. G. (1996). Deterioration of concrete due to reinforcement steel corrosion. *Cement and Concrete Composites, 18*, 47–59.

CEB. (1992). *Durable concrete structures – Design guide*. CEB Bulletin d'Information No. 183, Comité Euro-International du Béton. London: Thomas Telford.

Celarec, D., Vamvatsikos, D., & Dolšek, M. (2011). Simplified estimation of seismic risk for reinforced concrete buildings with consideration of corrosion over time. *Bulletin of Earthquake Engineering, 9*, 1137–1155.

Chang, S. E., & Shinozuka, M. (2004). Measuring improvements in the disaster resilience of communities. *Earthquake Spectra, 20*, 739–755.

Cimellaro, G. P., Reinhorn, A. M., & Bruneau, M. (2010). Framework for analytical quantification of disaster resilience. *Engineering Structures, 32*, 3639–3649.

Clifton, J. R., & Knab, L. I. (1989). *Service life of concrete*. NUREG/CR-5466. Washington, DC: U.S. Nuclear Regulatory Commission.

Decò, A., Frangopol, D. M., & Okasha, N. M. (2011). Time-variant redundancy of ship structures. *SNAME Journal of Ship Research, 55*, 208–219.

Decò, A., Bocchini, P., & Frangopol, D. M. (2013). A probabilistic approach for the prediction of seismic resilience of bridges. *Earthquake Engineering and Structural Dynamics, 42*, 1469–1487.

Du, Y. G., Clark, L. A., & Chan, A. H. C. (2005). Residual capacity of corroded reinforcing bars. *Magazine of Concrete Research, 57*, 135–147.

Ellingwood, B. R. (2005). Risk-informed condition assessment of civil infrastructure: State of practice and research issues. *Structure and Infrastructure Engineering, 1*, 7–18.

Ellingwood, B. R. (2006). Mitigating risk from abnormal loads and progressive collapse. *Journal of Performance of Constructed Facilities, 20*, 315–323.

Ellingwood, B. R., & Dusenberry, D. O. (2005). Building design for abnormal loads and progressive collapse. *Computer-Aided Civil and Infrastructure Engineering, 20*, 194–205.

Enright, M. P., & Frangopol, D. M. (1999). Reliability-based condition assessment of deteriorating concrete bridges considering load redistribution. *Structural Safety, 21*, 159–195.

fib (2006). *Model code for service life design*. Bulletin No. 34. Lausanne: Fédération internationale du béton/International Federation for Structural Concrete.

Frangopol, D. M. (2011). Life-cycle performance, management, and optimisation of structural systems under uncertainty: Accomplishments and challenges. *Structure and Infrastructure Engineering, 7*, 389–413.

Frangopol, D. M., & Bocchini, P. (2011). *Resilience as optimization criterion for the bridge rehabilitation of a transportation network subject to earthquake*. Proceedings of the SEI/ASCE Structures Congress, Las Vegas, NV, USA, April 14–16. In D. Ames, T. L. Droessler, & M. Hoit (Eds.), *Structures Congress 2011* (pp. 2044–2055). ASCE, CD-ROM.

Frangopol, D. M., & Curley, J. P. (1987). Effects of damage and redundancy on structural reliability. *Journal of Structural Engineering, 113*, 1533–1549.

Frangopol, D. M., & Ellingwood, B. R. (2010). Life-cycle performance, safety, reliability and risk of structural systems, Editorial. *Structure Magazine*. Chicago, IL: National Council of Structural Engineering Associations, NCSEA.

Frangopol, D. M., Iizuka, M., & Yoshida, K. (1992). Redundancy measures for design and evaluation of structural systems. *Journal of Offshore Mechanics and Arctic Engineering, 114*, 285–290.

Frangopol, D. M., & Klisinski, M. (1989). Weight-strength-redundancy interaction in optimum design of three-dimensional brittle-ductile trusses. *Computers and Structures, 31*, 775–787.

Frangopol, D. M., & Kong, J. S. (2001). Expected maintenance cost of deteriorating civil infrastructures. In D. M. Frangopol & H. Furuta (Eds.), *Life-cycle cost analysis and design of civil infrastructure systems* (pp. 22–47). Reston, VA: ASCE.

Frangopol, D. M., & Nakib, R. (1991). Redundancy in highway bridges. *AISC Engineering Journal, 28*, 45–50.

Frangopol, D. M., & Saydam, D. (2014). Structural performance indicators for bridges. Chapter 9. In W.-F. Chen & L. Duan (Eds.), *Bridge engineering handbook – Second Edition, Vol. 1 – Fundamentals* (pp. 185–205). Boca Raton, FL, London: CRC Press, Taylor & Francis Group.

Frangopol, D. M., & Soliman, M. (2016). Life-cycle of structural systems: recent achievements and future directions. *Structure and Infrastructure Engineering, 12*(1), 1–20.

Fu, G., & Frangopol, D. M. (1990). Balancing weight, system reliability and redundancy in a multiobjective optimization framework. *Structural Safety, 7*, 165–175.

Furuta, H., Kameda, T., Fukuda, Y., & Frangopol, D. M. (2004). Life-cycle cost analysis for infrastructure systems: Life-cycle cost vs. safety level vs. service life. Keynote Paper. In D. M. Frangopol, E. Brühwiler, M. H. Faber, & B. Adey (Eds.), *Life-cycle performance of deteriorating structures: Assessment, design and management* (pp. 19–25). Reston, VA: ASCE.

Gharaibeh, E. S., Frangopol, D. M., & Onoufriou, T. (2002). Reliability-based importance assessment of structural members with applications to complex structures. *Computers & Structures, 80*, 1111–1131.

Ghosn, M., Moses, F., & Frangopol, D. M. (2010). Redundancy and robustness of highway bridge superstructures and substructures. *Structure and Infrastructure Engineering, 6*, 257–278.

Glicksman, M. E. (2000). *Diffusion in solids*. New York, NY: Wiley.

Harr, M. E. (1996). *Reliability-based design in civil engineering*. Mineola, NY: Dover.

Hendawi, S., & Frangopol, D. M. (1994). System reliability and redundancy in structural design and evaluation. *Structural Safety, 16*, 47–71.

Husain, M., & Tsopelas, P. (2004). Measures of structural redundancy in reinforced concrete buildings. I: Redundancy indices. *Journal of Structural Engineering, 130*, 1651–1658.

Lind, N. C. (1995). A measure of vulnerability and damage tolerance. *Reliability Engineering & System Safety, 48*(1), 1–6.

Liu, Y., & Weyers, R. E. (1998). Modeling the dynamic corrosion process in chloride contaminated concrete structures. *Cement and Concrete Research, 28*, 365–379.

Lu, Z., Yu, Y., Woodman, N. J., & Blockley, D. I. (1999). A theory of structural vulnerability. *The structural engineer, The Institution of Structural Engineers, 77*, 17–24.

Malerba, P. G. (Ed.). (1998). *Analisi limite e non lineare di structure in calcestruzzo armato* [Limit and nonlinear analysis of reinforced concrete structures]. Udine: International Centre for Mechanical Sciences (CISM). (In Italian).

Mori, Y., & Ellingwood, B. R. (1994). Maintaining reliability of concrete structures. I: Role of inspection/repair. *Journal of Structural Engineering, 120*, 824–845.

NCHRP (2006). *Manual on service life of corrosion-damaged reinforced concrete bridge superstructure elements* (National Cooperative Highway Research Program, Report 558). Washington, DC: Transportation Research Board.

Okasha, N. M., & Frangopol, D. M. (2009). Lifetime-oriented multi-objective optimization of structural maintenance considering system reliability, redundancy and life-cycle cost using GA. *Structural Safety, 31*, 460–474.

Okasha, N. M., & Frangopol, D. M. (2010). Time-variant redundancy of structural systems. *Structure and Infrastructure Engineering, 6*, 279–301.

Onoufriou, T., & Frangopol, D. M. (2002). Reliability-based inspection optimization of complex structures: A brief retrospective. *Computers & Structures, 80*, 1133–1144.

Paliou, C., Shinozuka, M., & Chen, Y. (1990). Reliability and redundancy of offshore structures. *Journal of Engineering Mechanics, 116*, 359–378.

Pandey, P., & Barai, S. (1997). Structural sensitivity as a measure of redundancy. *Journal of Structural Engineering, 123*, 360–364.

Pastore, T., & Pedeferri, P. (1994). La corrosione e la protezione delle opere metalliche esposte all'atmosfera [Corrosion and protection of metallic structures exposed to the atmosphere]. *L'edilizia, 1994*, 75–92. (In Italian).

Sabatino, S., Frangopol, D. M., & Dong, Y. (2015). Sustainability-informed maintenance optimization of highway bridges considering multi-attribute utility and risk attitude. *Engineering Structures, 102*, 310–321.

Saydam, D., & Frangopol, D. M. (2011). Time-dependent performance indicators of damaged bridge superstructures. *Engineering Structures, 33*, 2458–2471.

Saydam, D., & Frangopol, D. M. (2015). Risk-based maintenance optimization of deteriorating bridges. *Journal of Structural Engineering, 141*(4), 04014120, 1–10.

Schafer, B. W., & Bajpai, P. (2005). Stability degradation and redundancy in damaged structures. *Engineering Structures, 27*, 1642–1651.

Starossek, U., & Haberland, M. (2011). Approaches to measures of structural robustness. *Structure and Infrastructure Engineering, 7*, 625–631.

Stewart, M. G. (2009). Mechanical behaviour of pitting corrosion of flexural and shear reinforcement and its effect on structural reliability of corroding RC beams. *Structural Safety, 31*, 19–30.

Sudret, B. (2008). Probabilistic models for the extent of damage in degrading reinforced concrete structures. *Reliability Engineering and System Safety, 93*, 410–422.

Teigen, J. G., Frangopol, D. M., Sture, S., & Felippa, C. A. (1991a). Probabilistic FEM for nonlinear concrete structures. I: Theory. *Journal of Structural Engineering, 117*, 2674–2689.

Teigen, J. G., Frangopol, D. M., Sture, S., & Felippa, C. A. (1991b). Probabilistic FEM for nonlinear concrete structures. II: Applications. *Journal of Structural Engineering, ASCE, 117*, 2690–2707.

Thoft-Christensen, P. (1998). Assessment of the reliability profiles for concrete bridges. *Engineering Structures, 20*, 1004–1009.

Titi, A., & Biondini, F. (2014). Probabilistic seismic assessment of multistory precast concrete frames exposed to corrosion. *Bulletin of Earthquake Engineering, 12*, 2665–2681.

Titi, A., & Biondini, F. (2016). On the accuracy of diffusion models for life-cycle assessment of concrete structures. *Structure and Infrastructure Engineering, 12*, 1202–1215.

Vidal, T., Castel, A., & François, R. (2004). Analyzing crack width to predict corrosion in reinforced concrete. *Cement and Concrete Research, 34*, 165–174.

Vismann, U., & Zilch, K. (1995). Nonlinear analysis and safety evaluation by finite element reliability method. In: *New developments in non-linear analysis method*, CEB Bulletin d'Information No. 229 (pp. 49–73). Lausanne: Comité Euro-International du Béton.

Wolfram, S. (1994). *Cellular automata and complexity – Collected papers*. Reading, MA: Addison-Wesley.

Zhang, R., Castel, A., & François, R. (2010). Concrete cover cracking with reinforcement corrosion of RC beam during chloride-induced corrosion process. *Cement and Concrete Research, 40*, 415–425.

Zhu, B., & Frangopol, D. M. (2012). Reliability, redundancy and risk as performance indicators of structural systems during their life-cycle. *Engineering Structures, 41*, 34–49.

Zhu, B., & Frangopol, D. M. (2013). Risk-based approach for optimum maintenance of bridges under traffic and earthquake loads. *Journal of Structural Engineering, 139*, 422–434.

Zhu, B., & Frangopol, D. M. (2015). Effects of post-failure material behaviour on redundancy factor for design of structural components in nondeterministic systems. *Structure and Infrastructure Engineering, 11*, 466–485.

After-fracture redundancy analysis of an aged truss bridge in Japan

Weiwei Lin, Heang Lam, Teruhiko Yoda, Haijie Ge, Ying Xu, Hideyuki Kasano, Kuniei Nogami and Jun Murakoshi

ABSTRACT

In the next 20 years, more than half of the bridges in Japan would be in service for more than 50 years. These include truss bridges those are generally considered to be non-redundant and fracture critical. With ageing, these bridges are facing an increasing risk of bridge collapse under extreme conditions. Therefore, it is essential to evaluate alternate load paths on existing bridge structures to avoid bridge collapse or replacement. In this paper, after-fracture redundancy of truss bridges is investigated through a case study for a five-span continuous truss bridge. Field tests were performed, and the displacements on key sections were measured in the test. According to the experimental observations and test results, a numerical model capable of simulating the present truss bridge was built and was used to investigate the after-fracture redundancy of the target bridge. Fracture of truss members in typical locations was considered in the numerical analyses, and the 'R' values of both intact and damaged trusses were compared. On the basis of the numerical results, a critical member in this bridge was also determined.

1. Introduction

Among a large amount of infrastructure systems in Japan, aged bridges are facing an increasing risk of failures, due to the deterioration of structural members caused by corrosion, fatigue and the increasing of traffic load (occasionally overloading), etc. In addition, in spite of these inevitable service reasons, other extreme events resulting from accidents or natural disasters such as ship collision, flood, hurricane and earthquake also threaten bridges' safety. Occurrence probabilities of extreme conditions also positively correlate with service time. In fact, Japan experienced a period of strong economic growth after the World War II. A number of bridges were built during 1960s and 1970s and are reaching an advanced age, as shown in Figure 1. In the next 20 years, more than half of the bridges in Japan would be in service more than their design life (50 years in Japan). Thus, with the intention of increasing reliability of aged bridges, engineers are taking into consideration how to settle the issues of aged bridges.

Most countries and regions in the world including Japan adopt component-oriented design and evaluation techniques to verify the safety of structures. Systematic effects of the structures are ignored in the conventional design codes. Namely, safety of every member in the structures is required to ensure overall safety of structures. Some requirements of current design codes, such as serviceability limit state requirements, demand overall performance of bridges. However, it does not ensure collapse-proof safety of the structures. As for an aged bridge,

damage of members to some extent is inevitable but acceptable as long as it does not cause collapse of the bridge or lose of its serviceability. Although this component-oriented design method has been successfully used for decades, a quantitative conclusion on overall safety of aged bridges cannot be given. Thus, the purpose of this study is to propose a feasible method to address bridge redundancy and to give features of collapse of a steel truss bridge through a case study of an aged steel truss bridge in Japan (Figures 2 and 3).

Frangopol (1985), Frangopol and Curley (1987), Frangopol and Nakib (1987) and Frangopol, Iizuka, and Yoshida (1992) performed fundamental studies about structural redundancy, structural reliability, probability-based design, infrastructures under uncertainty, risk-based assessment and resilience to disasters, etc. Recent and ongoing studies have focused on determining the failure mechanisms of typical structural systems and making recommendations to improve the overall system safety of civil structures, such as the studies performed on girder bridges by NCST (2005), Idriss, White, Woodward, and Jauregui (1995), Tachibana, Tusjikado, Echigo, Takahashi, and Miki (2000), Park, Joe, Park, Hwang, and Choi (2012), Lin, Yoda, Kumagai, and Saigyo (2013), Cha, Lyrenmann, Connor, and Varma (2014), and Lam, Lin, and Yoda (2014).

In recent years, a series studies were performed by Ghosn and Moses (1998), Ghosn and Frangopol (2007), Ghosn, Moses, and Frangopol (2010), and Ghosn and Yang (2014) to investigate and quantify structural redundancy in highway bridges, which can be used for different types of bridge structures including both bridge superstructures and substructures. Both experimental

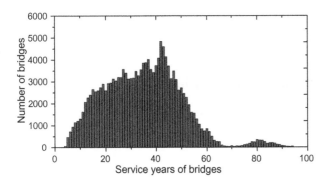

Figure 1. Service years of bridges in Japan (Tamakoshi, Okubo, & Yokoi, 2014).

Figure 3. Image of the former Choshi Bridge.

and numerical analyses were also performed by Nagatani et al. (2009), Yamaguchi, Okamoto, and Yamada (2011) to investigate the redundancy of truss bridges. However, in those studies, member yielding was often used for determining the member failure so as to the structural failure, and post yield capacity of the truss members was rarely considered, resulting in over conservative evaluation results. In addition, progressive failure analyses reflecting the real collapse process of truss bridges are not considered. In 2012, Miyachi, Nakamura, and Manda performed a study on progressive collapse analysis of steel truss bridges, but with complicated non-linear analyses. On this background, the linear analyses considering the progressive analyses to reflect the post fracture redundancy behaviour of truss bridges can benefit the redundancy evaluation of such bridges.

2. Optimised 'R value' method

The optimised 'R value' method consists of the following two equations and four steps. The R value of each truss member is calculated from:

$$R_t = \left(P/P_{ta} \right) + \left(M/M_{pa} \right)_{ip} + \left(M/M_{pa} \right)_{op} \tag{1}$$

$$R_c = \left(P/P_{ua} \right) + \frac{1}{1 - \left(P/P_E \right)_{ip}} \left(M_{eq}/M_{pa} \right)_{ip} + \frac{1}{1 - \left(P/P_E \right)_{op}} \left(M_{eq}/M_{pa} \right)_{op} \tag{2}$$

in which Equation (1) is for tension members; Equation (2) is for compression members; Subscript *ip* denotes in-plane; *op* denotes out of plane; *P*, *M* are the axial force and moment, respectively; P_{ta}, M_{pa}, the allowable values of fully plastic axial force and fully plastic moment; P_{ua}, the allowable load-carrying capacity in axial

compression as a column; P_E, the Euler buckling strength; M_{eq}, the equivalent bending moment of the member.

Definitions of terminology used in this paper are given as follows:

- *Possible critical member (CM)* – a member that has higher probability of failure or whose failure may cause severe consequences.
- *Potential CM* – a member whose failure causes other members' failure. The potential CM must be a possible CM.
- *Critical member (CM)* – a member whose failure also causes collapse of bridge. The critical member must be a potential CM.

The objective of the optimised 'R value' method is to determine the redundancy of truss bridges. Brief evaluation procedure is described as follows: possible CMs will be determined firstly according to numerical analyses results under dead load; then further simulation will be performed on possible CMs to check whether they are potential CMs; then simulation will be performed on potential CMs to check whether they are critical members. If any member is determined as a critical member, the bridge is defined as non-redundant. In order to determine whether a member is CM or not, progressive failure analysis needs to be performed. The evaluation procedure can be described as shown below:

- Step-1: *R* values of members are computed under dead load. The tension members with relatively large *R* values in each span, the members whose failure possibly cause severe consequences and other important members are

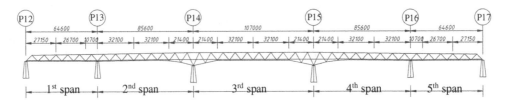

Figure 2. Size dimensions of the former Choshi Bridge.

determined as possible CMs. Damage of the truss member could result from several factors. In the classic sense for fracture critical bridges, it would initiate as a crack at a fatigue-prone detail, such as a plate-to-flange weld connection. Damage could also result from an overload condition, or deterioration mainly due to poor drainage details, vehicle impact, ship collision and other extreme conditions such as earthquakes, flood and hurricane and so on. Linear dead load analysis results and engineering judgment are utilised in this step.

- Step-2: Dead load and most unfavourable lane loads are applied on the intact structure model, and the same loading condition is applied on a damaged bridge model in which one possible CM from step-1 is removed from the intact structure. If R value of any member in the damaged structure exceeds 1.0, the member is determined as failed; thus the selected possible CM is determined as a potential CM, according to the definition. Otherwise, the member is not a potential CM and further analysis is not necessary.
- Step-3: A potential CM's failure causes other members' failure. Then, those failed members will be removed from the structure and numerical analysis will be re-ran under the same loading condition. In such a way, analysis continues until bridge collapses can be confirmed. If it is the case, the potential CM can be determined as a critical member. Otherwise, it is not a critical member.
- Step-4: All the potential CMs need to be analysed in the same way as step-3. If there is no critical member, this bridge is redundant. Otherwise, it is non-redundant.

Design loads and allowable stresses of structural material are used in steps 2, 3 and 4, according to the Japan Road Association (JRA) specification (2012). From step-2 to step-4, members with the R value exceeding 1.0 are defined as either yielding failure or buckling failure. If the R value of a compression member exceeds 1.7 (the safety factor, denoting the ratio of the material's mean yield strength to its allowable strength) according to JRA (2012), then it will be determined as buckling failure for the sake of structural safety. Otherwise, it is defined as yielding failure. The member forces in a buckled member are assumed to be zero, while the member forces of yielded member are assumed to sustain a constant force (yielding load). Therefore, in the numerical analysis, if the failed member buckles, it is removed from the structural model. While if the failed member yields, it is removed from the structural model, but reaction forces when it reaches yielding strength are applied back onto corresponding joints. The redundancy evaluation method considering force redistribution proposed above will be performed using simple linear analysis.

3. The former Choshi Bridge and the FE model

3.1. Description of the target bridge

The former Choshi Bridge was located at Choshi-shi, Chiba Prefecture, close to the river mouth of Tone River connecting Choshi-shi (Chiba Prefecture) and Kamisu-shi (Ibaraki Prefecture). It was a five-span truss highway bridge with a total length of 407 m and a width (main truss) of 8.2 m. The longest span was 107 m and the deck was 7 m in width. The former Choshi Bridge was opened to public in 1962 and was demolished in 2011 due to severe corrosion of the truss members. Thereafter, series experimental and numerical studies were performed by Public Works Research Institute, Tokyo Metropolitan University and Waseda University focusing on the mechanical performance of truss connections (or joints) after severe corrosion of the truss members as well as the gusset plate.

3.2. Numerical model of the truss bridge

The bridge FE model was built using the finite element software *DIANA*, and the whole model consists of two parts. The bridge deck (concrete) was modelled by shell elements (four nodes, with five degrees at each node). The steel truss of the bridge model (structural steel) was modelled by beam elements (two nodes, with six degrees at each node). The shell elements were built at the neutral axis location of the actual concrete deck, and the connection between the concrete deck and the floor system was simulated by short beam elements with relative high stiffness (to fill the void between the shell and beam element, as shown in Figure 4).

Spring connections were used to simulate the cantilever supports in 1st and 5th span of the actual truss bridge, as shown in Figure 4. The stiffness of spring was determined by adjusting the numerical results with the field test results. Although the joints have great importance in the truss bridge, the modelling of the joints will greatly increase the number of the elements and thus greatly increase the computation time as well as cost. For simplicity, details of the joint were not considered in the numerical modelling. In addition, as the corrosion data for all the members are not available, the thickness reduction and spatial variability associated with the steel corrosion were not considered in the numerical analysis.

After the removal of the aged truss bridge, material tests for the old structural steel were conducted. The yield strength of truss members was determined as 253.5 MPa with the Young's modulus of 21GPa. In addition, as a design-based redundancy evaluation method, the allowable strength (140 MPa) of the structural steel according to the Japanese design specifications (JRA, 2012) was used in determining the R values for the conservative purpose and also for considering the time-dependent material deterioration due to corrosion, fatigue and other possible damages. In this study, linear analyses were employed to compare with the field test results as well as the following redundancy analyses.

3.3. Field test and validity of the numerical model

Field test was performed before the removal of the truss bridge. Seven loading cases in total were applied in the field test, in which Case 1 and Case 7 are the cases without truck loads, and other cases are under two heavy trucks on the mid-span section of each span, as shown in Figure 5. The truck used in the field test is shown in Figure 6.

Linear analyses were performed with the same loading conditions as used in the field test. The vertical displacement was measured on key sections in the field test, and its distribution along the bridge axis direction was given in Figure 7 and compared with the numerical results. Taking Case 1 as an

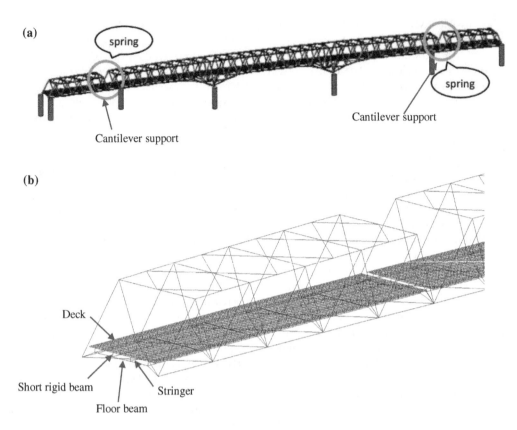

Figure 4. Numerical model of Choshi Bridge. (a) The whole bridge model. (b) Detailed numerical model.

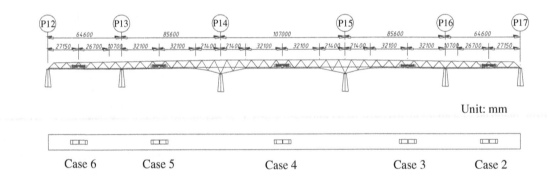

Figure 5. Loading conditions in the field tests (Case 2–Case 6).

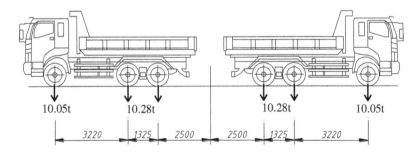

Figure 6. Trucks used in the field test (Unit: mm).

example, when two trucks were placed in the middle of 5th span, the displacement distribution of the truss bridge from the numerical analysis agreed well with the test results. The 5th span had most remarkable downward deformation while the adjacent span had obvious upward deformation due to the continuity of the truss structure. Similar distributions were

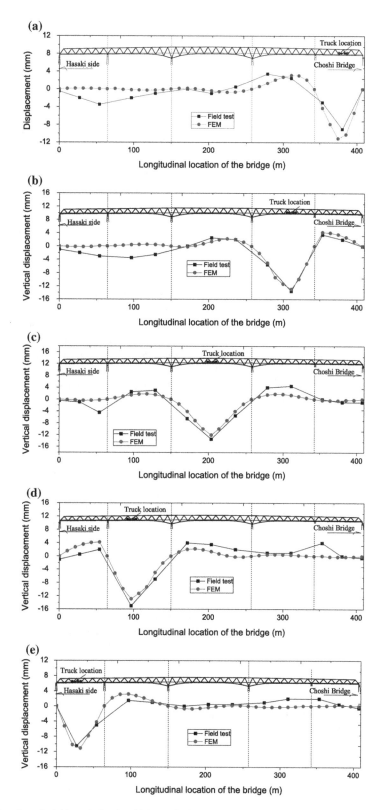

Figure 7. Displacement distribution along the bridge axis direction. (a) Case 2. (b) Case 3. (c) Case 4. (d) Case 5. (e) Case 6.

also confirmed between the numerical test results and the numerical results in other cases, as shown in Figure 7.

In addition, the maximum displacements of both test and numerical simulation results of all the cases were summarised in Table 1, which clearly indicated that difference in the maximum vertical displacements between the numerical results and field tests was relatively small and acceptable. In addition, the residual displacement after the field test (2.5 mm, Case 7) might be caused

111

Table 1. Maximum displacement of field tests and numerical results.

Case #	Field tests (mm)	FEM (mm)
Case 1*	0	0
Case 2	9.0	11.6
Case 3	13.5	13.5
Case 4	13.5	12.1
Case 5	15.0	13.5
Case 6	10.5	11.6
Case 7*	2.5	0

*Dead weight only.

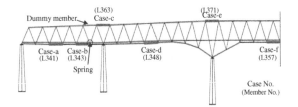

Figure 8. Locations of selected possible CMs.

by deterioration (corrosion, etc.) of the aged structural members, due to the fact that the Choshi Bridge had been in service for almost 50 years. Thus, it is reasonable to conclude that the present FE model can be used to simulate the mechanical behaviours of this truss bridge in the redundancy analyses.

4. Redundancy analysis

4.1. Possible CMs

In the optimised 'R value' method, a member can be determined as a possible CM if it matches either of the following conditions:

(1) Analytical point of view: if a tension member's R value is relatively large, according to the simulation results under dead load.

(2) Structural point of view: if a member is located at a significant location in the structure, according to engineers' expertise.

(3) Practical point of view: if a member is reported badly corroded or damaged in reality.

Possible CMs of the former Choshi Bridge are selected, according to the criteria (1) and (2). The principle (3) was not specially considered in this study because almost all the truss members in this bridge were in similar corrosion conditions on the basis of the field inspection results. The truss bridge consists of two trusses for supporting the deck system. Due to the symmetry, potential critical members were only assumed in 'damaged truss' in this study, and the other truss was denoted as 'intact truss'. Six members in the left half structure on the 'damaged truss' are selected as possible CMs as shown in Figure 8.

- Case-a: left 3rd lower chord in 1st span.
- Case-b: left 5th lower chord in 1st span.
- Case-c: upper chord in between 1st and 2nd span.
- Case-d: left 4th lower chord in 2nd span.
- Case-e: upper chord in between 2nd and 3rd span.
- Case-f: lower chord in the middle of 3rd span.

4.2. Critical members

After possible CMs are determined, analyses on each member should be performed to determine whether they are potential CMs in linear analysis. Damaged structure is created for each case by removing possible CM from intact structure. Since the failure of lower chords, upper chords and diagonal members have higher possibility of causing severe consequences on the overall safety of the bridge, and R values of these members are investigated. If R value of any other members except the possible CM exceeds 1.0, it can be concluded that this possible CM is a potential CM, which means its failure can lead to the failures of others members.

The optimised R value method was used for structural redundancy evaluation of the target truss bridge. Linear analysis was performed to investigate the progressive failure process and determine the CMs in this truss bridge, and the material properties were described in previous section. In the following redundancy analyses, the possible CM corresponding to each case is removed and the design load, including both dead load and design live load, is applied on the bridge structure. The live loads (P1 and P2) were determined according to the Japanese bridge design specifications (JRA, 2012).

4.2.1. Case-a: member L341

In this case, member L341 was selected as a possible CM. R values of truss members in both intact truss and damaged truss of damage bridge models were given and compared. For this model, both Young's modulus and density of the failed member were assumed to be near-zero values. For loading condition, influence line was used to determine the most critical loading condition as shown in Figure 9(a) and (b). Internal forces including axial force, in-plane bending moment and out-plane bending moment were then used to determine the R values using Equations (1) and (2), with the results as shown in Figure 10.

The results shown in Figure 10 indicate that the R value of lower chords in the first span of the damaged bridge changed remarkably in comparison with those of the intact truss. However, for truss members in other spans (far side from the damaged member), no notable difference is observed between intact and damaged trusses. In addition, all R values were smaller than 1.0. Thus, the remaining structure can continue to sustain the load without resulting in new member failures or structural collapse (as shown in Figure 11), indicating that the failure of

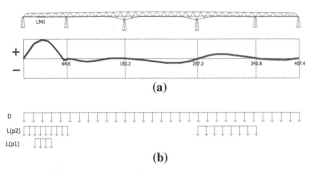

Figure 9. Influence line and loading condition for Case-a. (a) Influence line of Member L341. (b) Loading condition.

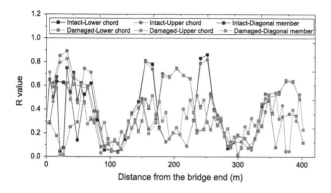

Figure 10. *R* values of Case-a (L341).

Figure 11. Damage member locations in the final stage of progressive analyses (L341).

this member will not affect the overall safety of truss structure. Therefore, the member L341 is not a critical member.

4.2.2. Case-b: member L343

In this case, the member L343 connects springs in the 1st span. Its failure may cause severe consequences due to its important location in structure and thus selected as possible CM. The most critical loading condition is determined according to the influence line as shown in Figure 12(a) and (b). The same as Case-a, internal forces including axial force, in-plane bending moment and out-plane bending moment were then used to determine the *R* values using Equations (1) and (2). According to the *R* value distributions shown in Figure 13, no significant differences were observed between the damaged truss and intact truss except a few members near the damaged member. The *R* value of all members are smaller than 1.0, which means the fracture of member L343 does not lead to severe damage because no other members will fail after the fracture of this member (as shown in Figure 14). Therefore, the member L343 is also not a critical member.

4.2.3. Case-c: member L363

In this case, member L363 was selected as possible CM according to the analytical result under dead load. Figure 15(a) shows the influence line of the assumed damaged member, and Figure 15(b) illustrates the most critical loading condition for this case. According to the numerical result, the member L363 fractures, resulting in significant change of *R* values in 1st and 2nd span, especially diagonal members in the two ends of 1st span, as shown in Figure 16(a). *R* values of 7 members (upper chord, diagonal member and lower chord) exceed 1.0 but are smaller than 1.7, denoting yielding of those members. Therefore, the member L363 can be determined as a potential CM and further analyses are needed.

According to the *R* values in stage-1, 7 members are determined as yielded. In stage-2, potential critical member L363 as well as other members found failed in stage-1 are removed from FE model and the axial forces of those members are added in FE model and linear analysis was performed on the updated

FE model. The internal forces of truss members were generated and used for determining the *R* values as shown Figure 16(b). According to the results, another 4 truss members were determined as yield failure because their *R* values were larger than 1.0 but still smaller than 1.7 for compression members and tensile strain was no larger than .2 for tension members. Thus, the linear progressive failure analysis was continued to stage-3, and the results were shown in Figure 16(c).

In stage-3, more truss member fails due to their *R* values being larger than 1.0, and the damage member locations were summarised in Figure 17. As all truss members in same section fail (such as section A or B in Figure 17), the remaining structure will collapse due to the instable structural system. Therefore, the progressive failure analyses can be stopped and this member can be determined as a critical member.

4.2.4. Case-d: member L348

In this case, the member L348 is located in the middle of the 2nd span and is determined as possible CM due to the dead load analysis results. Same as previous cases, influence line was employed to determine the most critical loading condition, as shown in Figure 18(a) and (b). The numerical result shows that

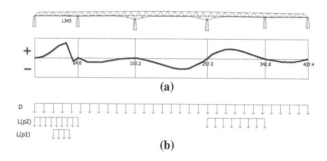

Figure 12. Influence line and loading condition for Case-b. (a) Influence line of Member L343. (b) Loading condition.

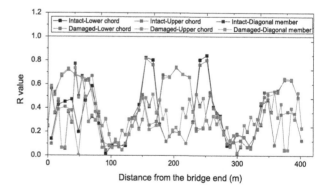

Figure 13. *R* values of Case-b (L343).

Figure 14. Damage member locations in the final stage of progressive analyses (L343).

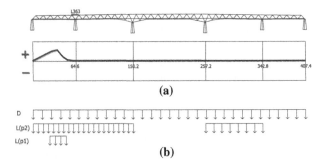

(a)

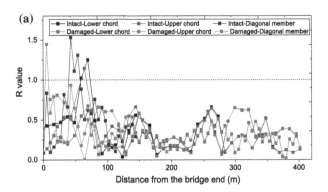

(b)

Figure 15. Influence line and loading condition for Case-c. (a) Influence line of Member L363. (b) Loading condition.

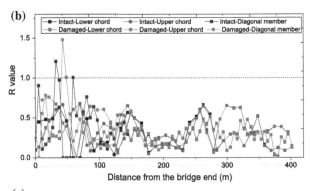

Figure 16. *R* values of Case-c (L363). (a) Stage-1. (b) Stage-2. (c) Stage-3.

Figure 17. Damage member locations in the final stage of progressive analyses (L363).

the fracture of the member L348 causes remarkable effects on the *R* value distribution in the 2nd span but these effects are negligible in the other spans, as shown in Figure 19. Moreover, all of *R* values remain less than 1.0 in both intact and damaged trusses. Thus, this member is not a potential CM and further analyses are not necessary (Figure 20).

4.2.5. Case-e: member L371

In this case, the possible CM L371 is located between 2nd and 3rd span. Like previous cases, the most critical loading condition is determined according to influence line, as shown in Figure 21(a) and (b), respectively. According to the numerical result, after failure of member L371, difference in *R* values of upper chords in 2nd and 3rd spans is considerable. *R* values of upper chords under compression in these two spans increase remarkably, as shown in Figure 22(a). In addition, *R* values of some lower chords and diagonal members increase remarkable but still remain within the safety level. Two upper chords in the 2nd span have *R* values more than 1.0 but smaller than 1.7, indicating that these members are yielding but not buckling. Thus, the member L371 is defined as a potential CM and further analyses are needed to determine whether its failure will cause the collapse of the structure.

According to the *R* value in stage-1, 2 upper chords are determined as yielded. In stage-2, potential critical member L371 as well as other failed members indicating stage-1 are removed from FE model and the axial forces of those members are added in FE model and linear analysis was performed. According to results as shown in Figure 22(b), all the *R* values of truss member are smaller than 1.0 in both intact and damaged trusses, indicating that no more yielded or buckled member were observed in stage-2. The results in Figures 22 and 23 demonstrate that the target truss bridge can continue to carry load after the failure of L371 and other 2 upper chords. Therefore, potential CM L371 is nor a CM.

4.2.6. Case-f: member L357

In this case, the possible CM L357 is a lower chord located in the middle of the 3rd span. The same as previous cases, influence line was employed to determine the most critical loading condition, as presented in Figure 24(a) and (b). The *R* value distribution in Figure 25(a) shows that the fracture of member L357 will cause remarkable change in *R* values of members in the 3rd span. Failure of 2 lower chords was confirmed in the target truss bridge. Since no *R* value of compression member exceeds 1.7, all of the failed members are yielded rather than buckled. Therefore, the failure of potential CM L357 in Case-f causes other members' yielding failure, and progressive failure analysis in stage-2 is necessary.

The *R* value distribution in stage-2 is shown in Figure 25(b). The results show that all the *R* values of truss member are smaller than 1.0, indicating that no more member will fail in stage-2 (Figure 26). Therefore, potential CM L357 is finally determined as nor a CM.

4.3. Critical members

On the basis of the above discussion, failure of truss members of L341, L343 and L348 will not result in the failure of other

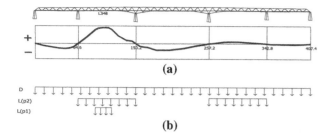

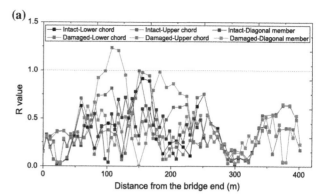

Figure 18. Influence line and loading condition for Case-d. (a) Influence line of Member L348. (b) Loading condition.

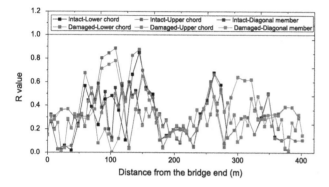

Figure 19. *R* values of Case-d (L348).

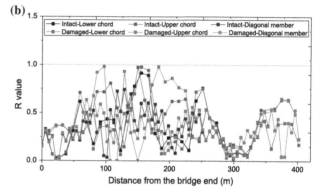

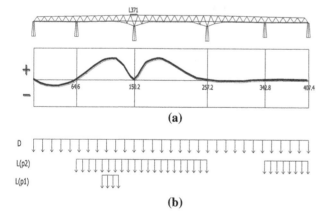

Figure 20. Damage member locations in the final stage of progressive analyses (L348).

Figure 21. Influence line and loading condition for Case-e. (a) Influence line of Member L371. (b) Loading condition for Case-e.

Figure 22. *R* value of Case-e (L371). (a) Stage-1. (b) Stage-2.

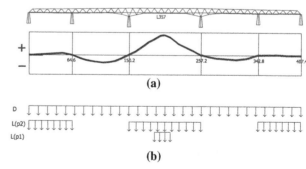

Figure 23. Damage member locations in the final stage of progressive analyses (L371).

Figure 24. Influence line and loading condition for Case-f.

truss members. On the other hand, the failure of truss members in L371 and L357 can cause the failure of other truss members. However, as the remaining structural system is still structurally effective and can continue to sustain the load without collapse,

thus all the members were not critical members. For the member L363, as its failure will cause failure of all lower chord and upper chord in a section and results in instable of the remaining structural system, this member is finally determined as a critical member for the target truss bridge.

115

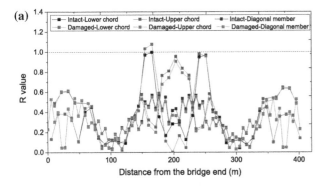

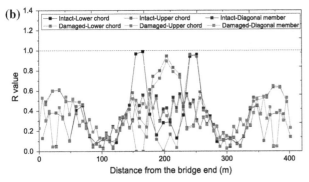

Figure 25. *R* values of Case-f (L357). (a) Stage-1. (b) Stage-2.

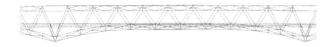

Figure 26. Damage member locations in the final stage of progressive analyses (L357).

5. Conclusions

On the basis of the results presented in this study, the following conclusions deserving priority are made:

(1) A new redundancy evaluation method for truss bridges and detailed analysis steps were proposed in this paper. Also, selection principles for possible CMs in truss bridges considering analytical, structural and practical viewpoints were proposed and used for determining the possible CMs in the present bridge.

(2) A three-dimensional finite element model of the target truss bridge was established. Field tests on the target truss bridge were also performed, and displacements on key sections were used to verify the validity of the present numerical model. The numerical results agree well with those obtained from the field tests, which demonstrates the accuracy and efficiency of the proposed FE model.

(3) *R* values of truss members near a possible CM increase markedly in the damaged cases. While for truss members far away from the possible CM, the *R* values are close to those of the intact truss bridge.

(4) For the present through truss bridge, failure of lower chords will generally not lead to severe consequences due to the fact that the bridge deck and stringers connecting to lower chords would continue to carry load and play as back-up resistance. On the contrary, upper chords are more likely to be potential CMs.

However, as a primary redundancy analysis, only limited number of truss members were considered. Further analysis about more truss members and the verification of the proposed redundancy evaluation method will be beneficial for the engineering practice.

Acknowledgement

The numerical analyses performed in this study (partly) was supported by the Sasakawa Scientific Research Grant from The Japan Science Society [grant number: 28-201]. The financial support is gratefully acknowledged.

Disclosure statement

No potential conflict of interest was reported by the authors.

References

Cha, H., Lyrenmann, L., Connor, R., & Varma, A. (2014). Experimental and numerical evaluation of the postfracture redundancy of a simple span truss bridge. *Journal of Bridge Engineering, 19*, 04014048. doi: http://dx.doi.org/10.1061/(ASCE)BE.1943-5592.0000622

Frangopol, D. (1985). Structural optimization using reliability concepts. *Journal of Structural Engineering, 111*, 2288–2301. doi: http://dx.doi.org/10.1061/(ASCE)0733-9445(1985)111:11(2288)

Frangopol, D., & Curley, J. (1987). Effects of damage and redundancy on structural reliability. *Journal of Structural Engineering, 113*, 1533–1549. doi: http://dx.doi.org/10.1061/(ASCE)0733-9445(1987)113:7(1533)

Frangopol, D., Iizuka, M., & Yoshida, K. (1992). Redundancy measures for design and evaluation of structural systems. *Journal of Offshore Mechanics and Arctic Engineering, 114*, 285–290. doi: http://dx.doi.org/10.1115/1.29199828

Frangopol, D., & Nakib, R. (1987). Accuracy of methods for structural system reliability evaluation. *Engineering Computations, 4*, 90–103. doi: http://dx.doi.org/10.1108/eb023687

Ghosn, M., & Frangopol, D. M. (2007). Redundancy and robustness of structures: A retrospective. *Reliability and optimization of structural systems*. London: Taylor & Francis Group.

Ghosn, M., & Moses, F. (1998). *Redundancy in highway bridge superstructures* (NCHRP Rep. 406). Washington, DC: National Cooperative Highway Research Program.

Ghosn, M., Moses, F., & Frangopol, D. M. (2010). Redundancy and robustness of highway bridge superstructures and substructures. *Structure and Infrastructure Engineering, 6*, 257–278.

Ghosn, M., & Yang, J. (2014). *Bridge system safety and redundancy* (NCHRP Rep. 776). Washington, DC: National Cooperative Highway Research Program.

Idriss, R. L., White, K. R., Woodward, C. B., & Jauregui, D. V. (1995). *After-fracture redundancy of two-girder bridge: Testing I-40 bridges over Rio Grande* (pp. 316–326). Proceeding of 4th International Bridge Engineering Conference, TRB, San Fransisco, CA.

Japan Road Association. (2012). *Specification for highway bridges*, Tokyo: Author.

Lam, H., Lin, W., & Yoda, T. (2014). Effects of bracing systems on redundancy of three-span composite twin I-girder bridge. *Journal of Structural Engineering-JSCE, 60A*, 59–69.

Lin, W., Yoda, T., Kumagai, Y., & Saigyo, T. (2013). Numerical study on post-fracture redundancy of the two-girder steel–concrete composite highway bridges. *International Journal of Steel Structures, 13*, 671–681. doi: http://dx.doi.org/10.1007/s13296-013-4008-8

Miyachi, K., Nakamura, S., & Manda, A. (2012). Progressive collapse analysis of steel truss bridges and evaluation of ductility. *Journal of Constructional Steel Research, 78*, 192–200. doi: http://dx.doi.org/10.1016/j.jcsr.2012.06.015

Nagatani, H., Akashi, N., Matsuda, T., Yasuda, M., Ishii, H., Miyamori, M., … Okui, Y. (2009). Structural redundancy analysis for steel truss bridges in Japan. *Journal of Japan Society of Civil Engineers, 65*, 410–425. doi: http://dx.doi.org/10.2208/jsceja.65.410

National Construction Safety Team (NCST) on the Collapse of the World Trade Center Towers. (2005). *Federal building and fire safety investigation of the World trade Center disaster*. Washington, DC: US Department of Commerce, Draft Report, NIST NCSTAR1, USA.

Park, Y., Joe, W., Park, J., Hwang, M., & Choi, B. H. (2012). An experimental study on after-fracture redundancy of continuous span two-girder bridges. *International Journal of Steel Structures, 12*(1), 1–13.

Tachibana, Y., Tusjikado, M., Echigo, S., Takahashi, S., & Miki, C. (2000). A study of after-fracture redundancy for two-girder bridges. *Journal of Japan Society of Civil Engineers, 2000*, 241–251.

Tamakoshi, T., Okubo, M., & Yokoi, Y. (2014). *Technical note of national institute for land and infrastructure management, No. 776*. Tokyo: Ministry of Land, Infrastructure, Transport and Tourism.

Yamaguchi, E., Okamoto, R., & Yamada, K. (2011). Post-member-failure analysis method of steel truss bridge. *Procedia Engineering, 14*, 656–661. doi: http://dx.doi.org/10.1016/j.proeng.2011.07.083

Experimental investigation of the spatial variability of the steel weight loss and corrosion cracking of reinforced concrete members: novel X-ray and digital image processing techniques

Sopokhem Lim, Mitsuyoshi Akiyama, Dan M. Frangopol and Haitao Jiang

ABSTRACT

The material properties of concrete structures and their structural dimensions are known to be random due to the spatial variability associated with workmanship and various other factors. This randomness produces spatially variable corrosion damages, such as steel weight loss and corrosion cracks. The structural capacity of reinforced concrete (RC) members strongly depends on the local conditions of their reinforcements. Modelling the spatial variability of steel corrosion is important, but steel corrosion in RC members can only be observed after severely damaging the concrete members. To understand the steel corrosion growth process and the change in the spatial variability of steel corrosion with time, continuous monitoring is necessary. In this study, X-ray photography is applied to observe steel corrosion in RC beams. The steel weight loss is estimated by the digital image processing of the X-ray photograms. The non-uniform distribution of steel weight loss along rebars inside RC beams determined using X-ray radiography and its correlation with longitudinal crack widths are experimentally investigated.

1. Introduction

The corrosion of embedded rebars due to chloride attack is a major cause of reductions in the service life of reinforced concrete (RC) structures in marine environments. The rebar inside the RC element is protected from corrosion by the passive alkaline nature of the surrounding concrete cover. However, this protected barrier can be damaged or destroyed as a result of gradual chloride attack. The active corrosion of reinforcement is initiated when the chloride reaches the reinforcement surface (Ann & Song, 2007; Trejo & Pillai, 2004). The built-up corrosion product occupies a greater volume than the original rebar, leading to cracking in the concrete cover, followed by surface cracking.

Previous studies (Andrade, Alonso, & Molina, 1993; Otsuki, Miyazato, Diola, & Suzuki, 2000; Vu, Stewart, & Mullard, 2005) have reported that the corrosion-induced cracking time is far shorter than the propagation and widening time. In addition, the residual rebar cross section loss, which has been associated with the crack width during propagation, does not significantly affect the ultimate limit state; however, it does affect the serviceability state and the long-term performance of the structure. According to Palsson and Mirza (2002), at the time the Dickson Bridge was decommissioned in 1994, the bridge was structurally deficient but did not collapse even its deck was severely delaminated and damaged by chloride contamination, showing exposed corroded rebars, concrete cracking and spalling, leeching of the cement paste from the exposed piers, etc. Consequently, deteriorated RC structures are associated with a considerable economic loss in terms of maintenance and repair costs. Therefore, the development of a reliability model for predicting the long-term performance of deteriorated RC structures is important for minimising repair and maintenance during their life-cycle.

Recently, prediction models based on probabilistic concepts have been widely used to estimate the long-term structural performance of corrosion-affected RC structures (Marsh & Frangopol, 2008; Mori & Ellingwood, 1993; Stewart & Mullard, 2007). In assessing the probabilistic structural performance, Stewart (2004) emphasises the need to consider the spatial variability of steel corrosion. Ignoring the distribution of local steel corrosion can lead to the underestimation of the failure probability. Hence, the localised corrosion damage of rebar is an important input parameter for estimating the remaining service life of corroding RC structures.

However, the limited experimental data on the relationship between the spatial variations in steel weight loss and the width of the surface cracks have been reported to hinder the improvement of the accuracy of the prediction models (Akiyama & Frangopol, 2014; Akiyama, Frangopol, & Yoshida, 2010). The scarcity of experimental data is due to the difficulty in continuously observing the non-uniform spatial corrosion of steel weight loss during various stages of corrosion. Although it is possible to use

Table 1. Details of test specimens.

Notation	Cross section (mm)	Span length (mm)	Bar diameter (mm)	Cover (mm)	W/C (%)	Stirrup
I-1	140 × 80	1460	13	20	50	DB6@100[a]
II-1	140 × 80	1460	13	20	40	DB6@100
II-2	140 × 80	1460	13	20	65	DB6@100
III-1	140 × 80	1460	13	20	50	–
III-2	140 × 80	1460	13	20	50	DB6@165

[a]Deformed bars with a diameter of 6 mm arranged in intervals of 100 mm.

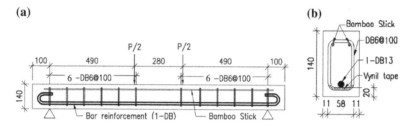

Figure 1. (a) Front view of the specimen for a 100-mm stirrup spacing and (b) typical cross section for all the specimens (all dimensions are in mm).

destructive methods (e.g. Alonso, Andrade, Rodriguez, & Diez, 1998; Vidal, Castel, & François, 2004; Zhang, Castel, & François, 2010) to study the relationship between the two parameters by repeatedly breaking specimens to weigh the rebar at various stages of corrosion, this method can suffer from uncertainties and errors associated with the different conditions encountered when remaking the specimens and the different growth patterns and measurements of the corrosion crack widths.

The use of X-ray technology as a non-destructive method is a promising means of avoiding these problems. This technique enables a continuous investigation of steel corrosion throughout the corrosion process. In concrete engineering, X-ray technology has been used as a non-destructive method to visualise and investigate steel corrosion in concrete structures. Beck, Goebbels, Burkert, Isecke, and Babler (2010) used X-ray computer tomography to examine the surface of a very small steel cylinder (9 mm in diameter and 10 mm long) at various stages of corrosion inside a mortar specimen. The reported difference between the mass loss, which was determined using the constructed 3D X-ray tomography image, and the actual mass loss after breaking the specimen was 40–60%. Akiyama and Frangopol (2013) demonstrated that the X-ray apparatus was a suitable tool for continuously investigating the weight loss of a corroded rebar that was embedded in a cylinder (100 mm × 200 mm) and a prism (100 mm × 100 mm × 400 mm). A digital image analysis based on the X-ray radiography of the shape of a corroded rebar from different viewing angles was used to determine the steel weight loss. The difference between the calculated steel weight loss based on the digital image analysis of the X-ray photogram and the measured value after damaging the specimen is only about 10%.

This paper presents an experimental study aiming to comprehensively investigate the continuous increase in the spatial variability of the steel weight loss along corroded rebars and the associated longitudinal crack widths of RC beams at various stages of corrosion, using novel X-ray and digital image processing techniques. The effects of various water-to-cement (W/C) ratios and stirrups on these two main parameters are discussed. Although the method of investigation is similar to that of Akiyama and Frangopol (2013), larger specimens with longer

rebars are studied. Moreover, using a new, upgraded X-ray apparatus and image intensifier, higher resolution X-ray photograms can be obtained, allowing the detection of the corrosion product. The more advanced digital image analyser used also provides more accurate estimates of the steel weight loss. The experimental outcomes are expected to provide a fundamental understanding of the non-uniform spatial growth of the steel weight loss and the corresponding corrosion cracking. These experimental results can help provide the stochastic fields associated with steel weight loss in RC components and estimate the steel weight loss based on the measured corrosion cracking widths.

2. Experimental programme

2.1. Overview of experimental plan

An experimental plan is established to study the effects of W/C ratio and stirrups on the spatial steel weight loss and corrosion cracking. To achieve these goals, five RC beams were fabricated and divided into two groups. The details of the specimens are shown in Table 1 and Figure 1. For the first group, specimens I-1, II-1 and II-2 were produced with W/C ratios of 50, 40 and 65%, respectively. The second group consists of specimens I-1, III-1 and III-2. Specimens I-1 and III-2 have stirrup spacings of 100 and 165 mm, respectively, whereas specimen III-1 has no stirrups. For the compressive strength test, six cylinders (100 mm × 200 mm) were also produced for each concrete mixture.

The experimental procedure is as follows. The corrosion of embedded longitudinal rebar was accelerated via the electro-chemical technique. At specific time intervals prior to performing the X-ray radiography, the crack widths were recorded by obtaining images of surface cracking on the bottom of the beams. X-ray radiography was performed once before the steel corrosion initiated and several times during the corrosion process to capture photograms of the developing morphology of the non-corroded and corroded rebars from different viewing angles. These photograms were used in the digital image processing to estimate the steel weight loss.

Table 2. Mixing proportions.

Notation	G_{max} (mm)	W/C (%)	s/a[a] (%)	Water (kg/m³)	Cement (kg/m³)	FA[c] (kg/m³)	CA[d] (kg/m³)	AE[b] (ml/m³)
I-1	20	50	44.3	181	362	754	961	2715
II-1	20	40	42.3	178	445	694	961	3338
II-2	20	65	47.3	185	285	829	940	2138
III-1 & III-2	20	50	44.3	181	362	754	961	2715

[a]Fine aggregate ratio.
[b]Air entranced agent.
[c]Fine aggregate.
[d]Coarse aggregate.

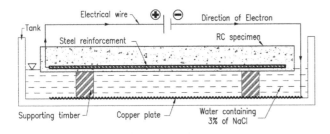

Figure 2. Electrolytic experimental test set-up.

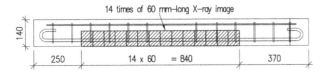

Figure 3. Total steel bar length captured by the X-ray apparatus (all dimensions are in mm).

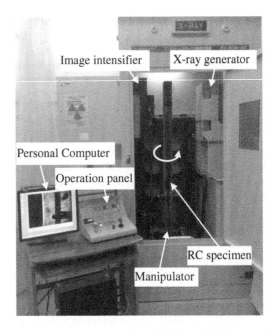

Figure 4. X-ray imaging set-up at Material Engineering Laboratory, Waseda University.

2.2. Materials and concrete mix proportion

All specimens were fabricated using identical material constituents. Ordinary Portland cement with a specific density of 3.16 g/cm³ was used. The fine aggregate has a fineness modulus of 2.64 and a specific density of 2.60 g/cm³. The coarse aggregate has a maximum size of 20 mm ($G_{max} \leq 20$ mm) and a specific density of 2.64 g/cm³. A deformed rebar with a diameter of 13 mm (DB13) was used as the longitudinal rebar, and deformed rebars with a diameter of 6 mm (DB6) were used as stirrups. All rebars were of the same steel quality grade, SD345. The details of the concrete mixing proportions are shown in Table 2.

2.3. Specimen fabrication procedure

The same fabrication procedure was performed for all the specimens. When used, the stirrups were wrapped with vinyl tape to prevent direct contact with the longitudinal rebar. The stirrups were arranged in specified intervals at the shear span to prevent abrupt shear failure of the concrete during the corrosion process. Before pouring the concrete, electrical wire was tied to one end of the steel reinforcement. Two days after fabrication, the mould was stripped off from the specimens, and the specimens were cured in water in a 23–25 °C room for 28 days.

2.4. Electrolytic experiment

After the specimens were cured, the steel corrosion process was initiated using the electrolytic technique. The detailed assembly of the electrolytic experimental test is shown in Figure 2. The RC specimen was placed on two pieces of supporting timber and partially immersed in a 3% sodium chloride (NaCl) solution in a tank in a controlled environment at 23–25 °C. The external copper plate, which is placed under the supporting timbers below the specimen, served as the cathode; the embedded rebar inside the specimen served as the anode. To ensure that the tests could be completed within a reasonable timeframe, the total impressed current was adjusted for each specimen to maintain the same current density (i.e. 1000 µA/cm²) to pass over the surface of the rebar. The accelerated corrosion process proceeded until the accumulated current time reached approximately 620 h (i.e. about 25 days).

2.5. Surface crack width measurement

The external surface cracks that occurred along the bottom of the specimens at various steel weight losses were imaged

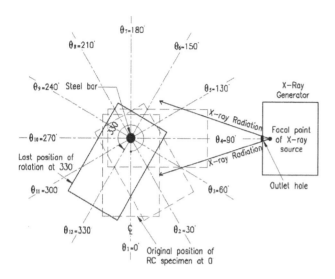

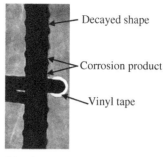

(a) Before enhancement   (b) After enhancement

Figure 7. X-ray images of a corroded steel bar at 0° with a mean steel weight loss of 8.79% (a) before and (b) after enhancement.

Figure 5. Top view of the specimen set-up and views of the steel bar at different angles associated with the rotation of the RC specimen.

2.6. X-ray photogram acquisition procedure

by a digital camera before X-ray radiography was performed. The location of the captured images corresponds to that of the captured X-ray image, i.e. 250–1090 mm from the left side of the specimens, as shown in Figure 3. This required seventeen 50-mm-long images to be continuously obtained along the bottom of the specimens. Note that in this experimental study, the visual longitudinal corrosion cracking occurred only at the bottom of the corroded beams. Crack width measurements on the photographs were continuously performed every 5 mm until a distance of 1090 mm was reached using an advanced image analysis programme.

Using the X-ray configuration in Figure 4, images of the non-corroded and corroded areas of the rebar inside the specimen were captured from different viewing angles once before the initiation of corrosion and several times during the corrosion process at various steel weight losses. Images of the rebar from 12 viewing angles (i.e. 0°, 30°, 60°, 90°, 120°, 150°, 180°, 210°, 240°, 270°, 300° and 330°) were recorded. At each angle, the total length of the rebar imaged by the X-ray apparatus was 840 mm. Fourteen 60-mm-long images were consecutively captured (see Figure 3). The X-ray radiography procedure used to acquire the photograms consists of two main steps, as described below.

The first step is setting the specimen in an appropriate position. The specimen was placed on a manipulator located between the X-ray intensifier and generator in the X-ray chamber. This

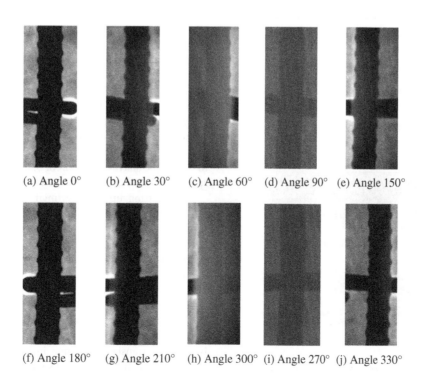

Figure 6. Ten X-ray images of the original steel bar obtained at different viewing angles.

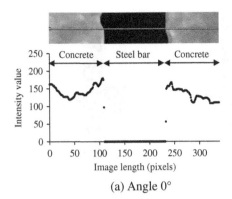

(a) Angle 0°

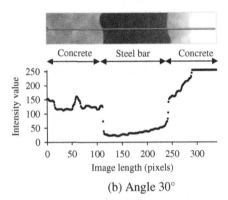

(b) Angle 30°

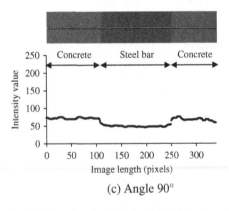

(c) Angle 90°

Figure 8. Profiles of the intensity values for a row of pixels in the X-ray photograms at three different viewing angles.

(a) Non-corroded steel bar at MRw = 0.00%

(b) Corroded steel bar at MRw = 6.05%

Figure 9. Sliced 5-mm-high X-ray photograms in (a) non-corroded steel bar at MRw = .00% and (b) corroded steel bar at MRw = 6.05%.

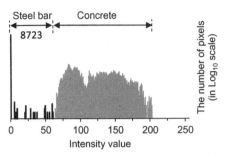

(a) Non-corroded steel bar at MRw = 0.0%

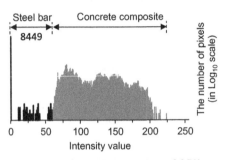

(b) Corroded steel bar at MRw = 6.05%

Figure 10. Histograms of the accumulated number of pixels classified by intensity values for the steel bars in Figures 9(a) and (b).

Table 3. Estimated weight loss vs. actual measured weight loss.

Specimens	Estimated weight loss (%)	Actual weight loss (%)	Absolute difference (%)
I-1	19.65	16.63	3.02
II-1	16.97	13.71	3.26
II-2	23.18	20.62	2.56
III-1	27.23	24.48	2.75
III-2	25.54	22.51	3.03

manipulator was used to translate and rotate the specimen into the desired positions via the operation panel. The specimen was first rotated to $\theta_1 = 0°$ and horizontally adjusted into the position where the centre of the embedded rebar was aligned with the middle point of the X-ray source outlet. This setting fixes the centre of the rebar as the centre point of rotation of the specimen, as indicated in Figure 5. In the vertical direction, the specimen was then shifted up or down relative to the X-ray radiation source to obtain the starting position of 370 mm from the base of the specimen on the computer screen.

After setting up the specimen, radiography images were acquired by attenuating the primary X-ray beam with materials of different densities and thicknesses. The chosen power and current settings for the X-ray radiography are functions of the source-to-specimen distances and concrete thicknesses when the specimen is rotated to different viewing angles: 120 kV, 1.2 mA for viewing angles of 0°, 30°, 60°, 120°, 150°, 180°, 210°, 240°, 300°, and 330° and 145 kV, 1.2 mA for viewing angles of 90° and 270°. Starting from a known position at 370 mm from the base of the specimen and a viewing angle of $\theta_1 = 0°$, the RC specimen is vertically translated 14 times in increments of 60 mm relative to the radiation cone beam supplied by the X-ray generator. The radiation penetrates the RC specimen for visualisation, and

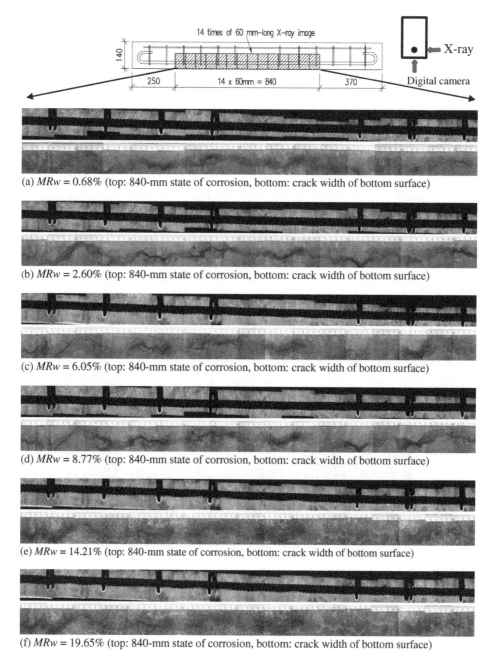

(a) *MRw* = 0.68% (top: 840-mm state of corrosion, bottom: crack width of bottom surface)

(b) *MRw* = 2.60% (top: 840-mm state of corrosion, bottom: crack width of bottom surface)

(c) *MRw* = 6.05% (top: 840-mm state of corrosion, bottom: crack width of bottom surface)

(d) *MRw* = 8.77% (top: 840-mm state of corrosion, bottom: crack width of bottom surface)

(e) *MRw* = 14.21% (top: 840-mm state of corrosion, bottom: crack width of bottom surface)

(f) *MRw* = 19.65% (top: 840-mm state of corrosion, bottom: crack width of bottom surface)

Figure 11. Spatial growth of the steel weight loss and corrosion cracking of specimen I-1 for six different values of *MRw*.

the attenuated X-ray radiation detected by the image intensifier reveals the composition details of various densities. After the attenuated X-rays are converted into visible light on a fluorescent screen, an equipped charge-coupled device camera unit, whose capture command is linked to and controlled by a software program, is used to capture and store the light intensity as digital values. These digital values comprise a 1024 × 768-pixel greyscale image.

The same process was performed repeatedly to capture images from the remaining viewing angles. The viewing angles of 0°, 30°, 60°, 90°,120°, 150°, 180°, 210°, 240°, 270°, 300° and 330° refer to those rebar views at which the specimen was rotated to angles θ_1, θ_2, θ_3, θ_4, θ_5, θ_6, θ_7, θ_8, θ_9, θ_{10}, θ_{11} and θ_{12}, respectively (see Figure 5).

3. Procedure for estimating the steel weight loss

3.1. Image enhancement before analysis

Various 60-mm-long images of the original rebar that were captured using the X-ray apparatus before corrosion are shown in Figure 6. In general, the original images at 0° and 180° provide the clearest views, followed by images at 30°, 150°, 210° and 330° and then by images at 90° and 270°, respectively. The worst images are those at 60° and 330°. This ordering is a result of the differences in the thickness of the concrete composite penetrated by the X-ray radiation during the image capturing for a particular specimen rotation angle. For example, at 0° and 180°, the specimen is in a favourable position, as the concrete thickness

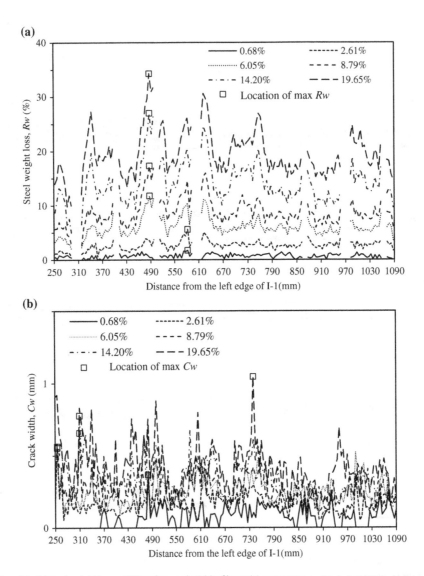

Figure 12. Spatial distribution of the (a) steel weight loss and (b) surface crack width of beam I-1.

encountered by the X-ray radiation is only 80 mm, providing a notably clear image. In contrast, at 90° and 270°, the angle is unfavourable, as 140 mm of concrete is penetrated by the X-ray radiation.

Furthermore, the post-corrosion image at the viewing angle of 0° in Figure 7(a) illustrates the shape of the corroded rebar at a mean steel weight loss of 8.79%, which corresponds to that of non-corroded rebar in Figure 6(a). However, although the image at this angle provides the clearest view, it remains difficult to carefully examine the corrosion products or decayed shape of the rebar. In Figure 7(b), for the image after enhancement, the corrosion products and decayed shape of the corroded rebar can be more easily identified.

Therefore, it is necessary to enhance the image before the analysis to readily obtain detailed information from the image. In the enhancement process, the fine details of the image were revealed or the blurred regions were reduced using Image-Pro Plus software version 7.0 of Media Cybernetics, Inc. (2012) to accentuate the intensity changes and make the high-contrast edges visible. Visualising the high-contrast edges between the rebar and concrete composite allows the area shapes of the

concrete composite and rebar to be easily distinguished, which is important for determining the area of the corroded rebar to estimate the steel weight loss.

3.2. Steel weight loss estimated by digital image analysis

To estimate the steel weight loss using the X-ray photograms, the area of the original rebar before corrosion and that of the corroded rebar at a given time during the corrosion process need to be determined. At each of the eight viewing angles, the two types of rebar areas are determined through the digital image analysis of X-ray photograms (i.e. a manipulation of the stored digital data of the image in terms of numerical representations of pixels). Note that only X-ray images from 8 viewing angles (i.e. $\theta_1 = 0°, \theta_2 = 30°, \theta_4 = 90°, \theta_6 = 150°, \theta_7 = 180°, \theta_8 = 210°, \theta_{10} = 270°$ and $\theta_{12} = 330°$) were used for estimating the steel weight loss along the corroded bars because: (1) the time-consuming and laborious works involved in digital image analysis of considerably large amount of data and (2) a good accuracy of estimated steel weight loss could be obtained which will be presented in the next section.

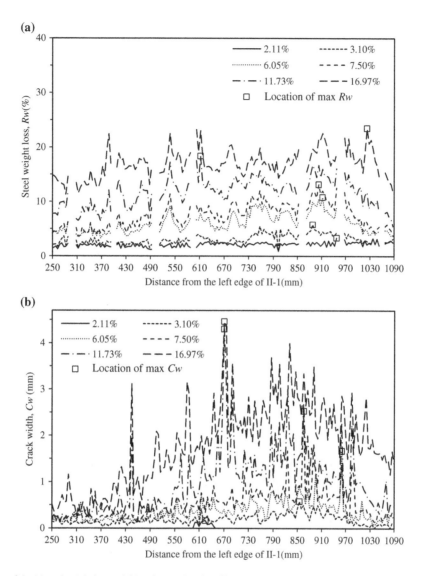

Figure 13. Spatial distribution of the (a) steel weight loss and (b) surface crack width of beam II-1.

In this paper, the acquired X-ray photograms are 8-bit grey-scale images with 1024 × 768 pixels. The greyness levels of the pixels are numerically represented by 256 intensity values ranging from 0 for completely black to 255 for completely white. This numeric representation enables the image software to distinguish pixels of different colours. Because the reinforcing rebar is denser than the concrete composite or corrosion product, it absorbs the X-ray radiation most efficiently. Thus, the rebar always produces the darkest pixels with the lowest intensity range compared to the concrete composite and corrosion product in the images. Figure 8 illustrates the profiles of the intensity values for a row of pixels that correspond to a line drawn on each of the above X-ray photograms. All the profile lines show that the intensity values of the rebar at 0°, 30° and 90° in the middle part of each graph are always lower than those of the corrosion products located close to the corroded rebar and the concrete in the left- and right-hand sides of the graphs. This finding also holds for the images of other viewing angles, as the X-ray images captured at 0° and 180°; 30°, 150°, 210° and 330°; and 90° and 270° are very similar (see Figure 6).

Therefore, by manipulating the intensity values of the pixels of the rebar, the total number of pixels below a minimum threshold of intensity values that represent the area of rebar alone can be selected, counted and classified using the image processing software. To facilitate the analysis of the digital data of the image, the 60-mm-high image of all viewing angles is sliced into twelve 5-mm-high images (see Figures 9(a) and (b)). Figures 10(a) and (b) present histograms of the accumulated numbers of pixels classified by intensity values for the 5-mm images of the non-corroded and corroded rebars in Figures 9(a) and (b), respectively. The histograms explicitly show that the total number of pixels of the selected intensity threshold of 0–62 is 8723 for the non-corroded rebar area, greater than the corresponding value of 8449 pixels of the selected intensity threshold of 0–59 for the corroded rebar area.

After the number of pixels of non-corroded and corroded rebar was determined, the total area was obtained by multiplying the number of pixels by the unit area per pixel as follows:

$$A_{\theta_n} = P_{\theta_n} \times A_p \tag{1}$$

$$A'_{\theta_n} = P'_{\theta_n} \times A_p \tag{2}$$

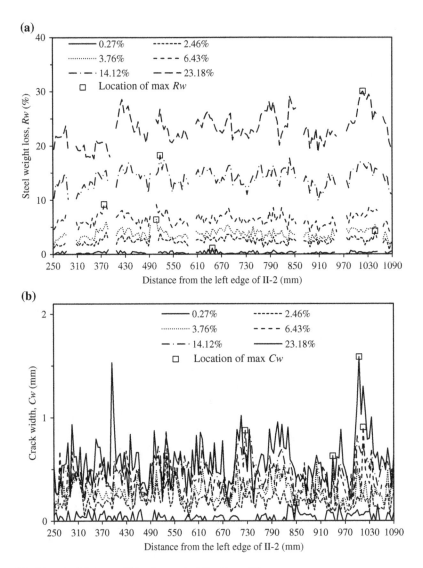

Figure 14. Spatial distribution of the (a) steel weight loss and (b) surface crack width of beam II-2.

where A_{θ_n} and A'_{θ_n} are the areas of the original rebar before corrosion and the corroded rebar at each viewing angles, respectively; θ_n denotes the viewing angles in which the subscript $n = \{1, 2, 4, 6, 7, 8, 10, 12\}$; P_{θ_n} and P'_{θ_n} are the number of pixels of the area of the original rebar and corroded rebar, respectively, at any of n viewing angles; and A_p is the unit area per pixel in the image.

The volume of the rebar before and after corrosion can then be calculated as follows:

$$V_{\theta_n} = \frac{\pi \left(A_{\theta_n}\right)^2}{4L} \qquad (3)$$

$$V'_{\theta_n} = \frac{\pi \left(A'_{\theta_n}\right)^2}{4L} \qquad (4)$$

where V_{θ_n} and V'_{θ_n} are the volumes of the original and corroded rebars, respectively, at a viewing angle θ_n; and L is the length of the rebar, which is 5 mm herein.

The steel weight loss per length L (mm) of the rebar is determined by taking the average of each value of the steel weight loss for each viewing angle as follows:

$$Rw = \frac{1}{k} \sum \frac{\left(V_{\theta_n} - V'_{\theta_n}\right)}{V_{\theta_n}} \times 100 \qquad (5)$$

where Rw is the steel weight loss in percentage (%) per length L (mm) of rebar, θ_n denotes the viewing angles in which $n = \{1, 2, 4, 6, 7, 8, 10, 12\}$, and k is the number of viewing angles ($k = 8$ for the eight different viewing angles considered).

4. Results and discussion

4.1. Accuracy of the estimation method

After the completion of all the tests, the concrete specimens were demolished to remove the embedded rebars from them. All the rebars were immersed in a water tank containing 10% diammonium hydrogen citrate solution for 24 h to remove the corrosion products. Next, the weights of the corroded rebars

126

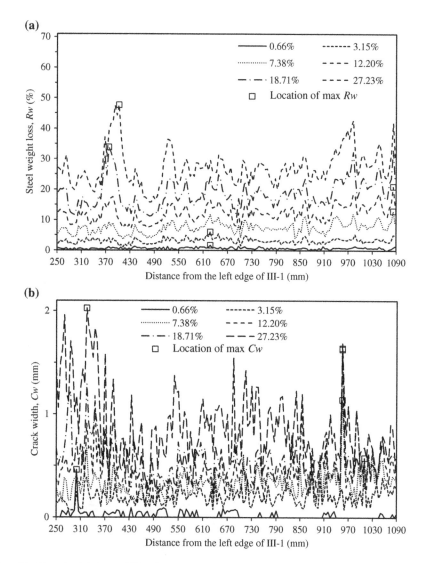

Figure 15. Spatial distribution of the (a) steel weight loss and (b) surface crack width of beam III-1.

were measured using a laboratory digital scale. The measured steel weights and the estimated weight loss of rebars calculated using digital image processing are compared in Table 3.

Note that the differences between the steel weight losses estimated using the digital image analysis and the actual measured steel weight losses are approximately 3%. This result demonstrates the good accuracy of the estimation method used for steel weight loss. The weight losses quantified via the X-ray images appear to be higher than the actual measured amounts, indicating that the employed X-ray method marginally overestimates the actual measured weight loss of embedded rebars in RC members. One possible cause of the overestimation is the inability of the projected 2D X-ray images to provide information about pit corrosion on the rebar surface. Consequently, the estimated cross section areas based on the projected areas of the rebar using the X-ray image are slightly smaller than those based on the actual areas of rebars embedded in the RC beams.

4.2. Spatial variability of the steel weight loss and crack widths

Figure 11 illustrates the continuous spatial growth of the steel corrosion visualised using the X-ray technique and the propagation of longitudinal corrosion cracking at the bottom of the beams at different stages of corrosion. Comparing the X-ray images and photographs, it is found that the steel weight loss Rw occurs gradually and at locations near the areas exacerbated by the longitudinal surface crack widths Cw. These weight loss locations are noticeable at some of the locations of the arranged stirrups and in the centre of the span.

Figures 12–16 show the spatial distribution of the steel weight loss Rw and surface crack width Cw at their corresponding locations along beams I-1, II-1, II-2, III-1, and III-2, at various mean steel weight losses MRw. Note that because Rw could not be obtained at the locations of the arranged stirrups, there are regular gaps in the graphs at both of the shear spans, from 250

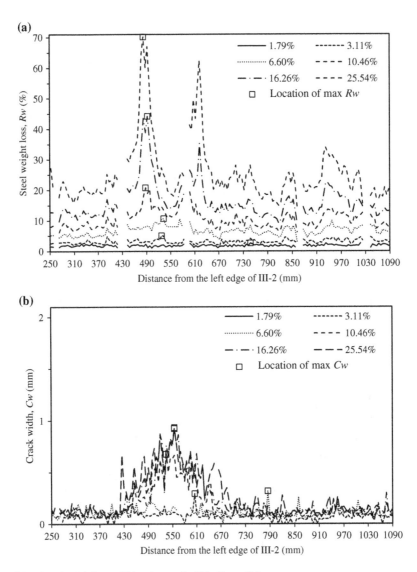

Figure 16. Spatial distribution of the (a) steel weight loss and (b) surface crack width of beam III-2.

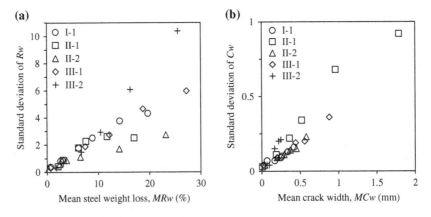

Figure 17. Relationship between the standard deviations and mean values of Rw and Cw.

to 610 mm and from 850 to 1090 mm, except for specimen III-1, which had no stirrups. In general, it can be observed that Cw increases with Rw. The distribution of Rw and Cw at various stages of steel corrosion is spatially non-uniform because Rw and Cw fluctuate erratically along the specimens. The non-uniformity of Rw and Cw becomes increasingly prominent with increasing MRw and MCw. In Figure 17, the standard deviations of Rw and Cw indicate that the distributions of Rw and Cw increasingly

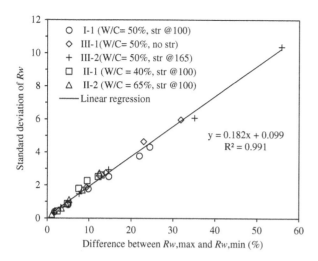

Figure 18. Relationship between the standard deviation and the difference between Rw, max and Rw, min.

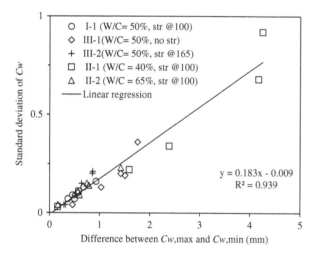

Figure 19. Relationship between the standard deviation and the difference between Cw, max and Cw, min.

diverge from their means as MRw and MCw increase. Therefore, the spatial variability in Rw and Cw is becoming larger as the steel weight loss and crack width increase. Yamamoto, Oyado, Mikata,

Kobayashi, and Shimomura (2011) studied the distribution of steel weight loss along the corroded bar using a great number of corroded RC beams. They also reported that the non-uniform degree of steel cross section losses of the corroded reinforcement increases with the corrosion loss. The cause of this larger spatial variability might result from the greater contribution of the larger surface cracks to the increase of Rw. When chloride penetrated from outside, corrosion cracking allowed much higher concentration of chloride to reach the rebar surface more easily, according to Andrade et al. (1993).

On the other hand, the maximum steel weight loss Rw, max and crack width Cw, max, which are denoted by square symbols in the graphs, do not usually occur at the corresponding locations. Several maxima of Rw, denoted by square symbols in the graphs, and other peaks (not maximum points) of Rw occur close to the locations of the stirrups (i.e. approximately at 300, 400, 500, 600, 860, 960 and 1060 mm) for specimens I-1, II-1 and II-2 in Figures 12(a), 13(a) and 14(a). This is probably due to the fact that the thinner concrete covers at the locations of the installed stirrups expose the longitudinal rebar to chloride and thus cause it to corrode more quickly than the rebar at other locations. The locations of the peaks often shift depending on the mean values. For example, the Rw, max of specimen I-1 is located at approximately 580 mm at MRw of .68% and 2.60% but shifts to approximately 490 mm at larger values of MRw. However, Figure 12(a) and (b) shows that some peaks of Rw and Cw also occur at approximately the same locations, namely 310, 490, 610 and 730 mm. This finding emphasises the influence of steel weight loss on the increase in crack widths at corresponding locations along the specimen.

4.3. Trend of steel weight loss and crack widths

In addition to the previously mentioned non-uniform spatial distributions of Rw and Cw, a trend is consistently observed among the graphs in Figures 12–16. Although the spatial variabilities of Rw and Cw increase as their mean values increase, it is worth noting that their erratic shapes tend to have a clear trend as MRw exceeds approximately 5%. For example, in Figure 12(a) for specimen I-1, the spatial distributions of Rw at MRw of 8.79, 14.20 and 19.65% seem to increase, following a very similar fluctuating pattern to the previous distributions of Rw at

(a)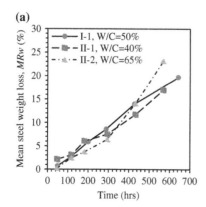

(b)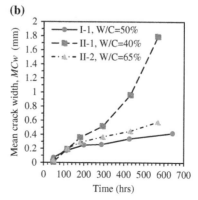

Figure 20. Effects of W/C ratios on the development of steel corrosion and crack width.

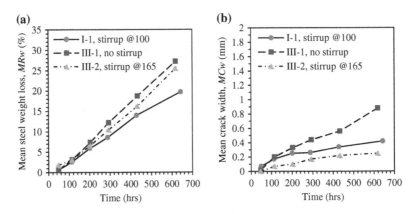

Figure 21. Effects of stirrups on the development of steel corrosion and crack width.

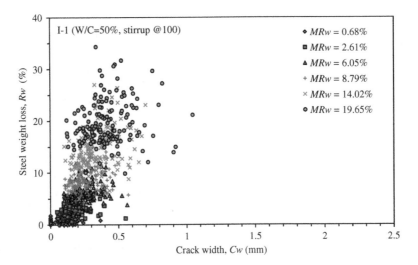

Figure 22. Relationship between the steel weight loss and crack width of specimen I-1.

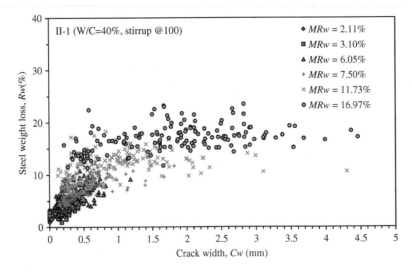

Figure 23. Relationship between the steel weight loss and crack width of specimen II-1.

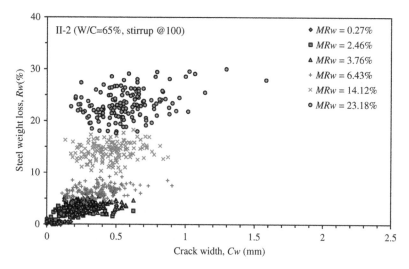

Figure 24. Relationship between the steel weight loss and crack width of specimen II-2.

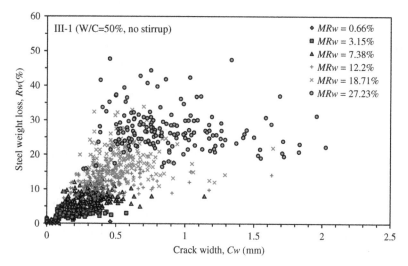

Figure 25. Relationship between the steel weight loss and crack width of specimen III-1.

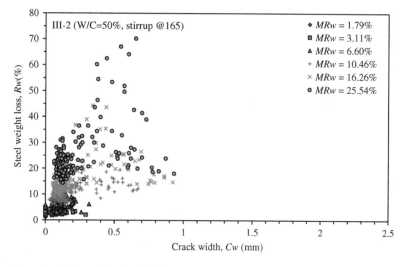

Figure 26. Relationship between the steel weight loss and crack width of specimen III-2.

$MRw = 6.05\%$. The spatial distributions of Rw in Figures 13(a), 14(a), 15(a) and 16(a) also exhibit a similar spatial growth behaviour as MRw increases beyond 5%. Similarly, the fluctuating pattern of the distribution of crack widths in Figures 12(b), 13(b), 14(b), 15(b) and 16(b) seems to follow the same trend as MRw exceeds 5%; however, the trend of the spatial distribution of crack widths appears to be weaker than that of Rw.

With respect to this behaviour, a strong relationship can be found between the standard deviation and differences between the maximum and minimum values of steel corrosion (Rw, max − Rw, min) and crack widths (Cw, max − Cw, min), as shown in Figures 18 and 19. The significant merit of this relationship is that it might be possible to estimate the spatial variability of Rw and Cw between two different points if the maximum and minimum values of Rw or Cw between these points can be determined via *in situ* inspection. Due to this consistent spatial growth trend, it might be preferable to estimate the distribution of steel corrosion of RC members using the inspection results of steel corrosion in the *in situ* structure when the mean steel weight loss exceeds 5%.

4.4. Effect of water-to-cement ratio

Figure 20 shows the increase in the means of Rw and Cw as a function of corrosion time for specimens with different W/C ratios. From Figure 20(a), because the trend lines of the three specimens exhibit similar behaviour, there appears to be no clear effect of W/C ratios on the growth of steel corrosion. Figure 20(b) shows the effects of different W/C ratios on the surface crack widths. It can be seen that the crack widths of the specimens with W/C ratios of 40% grew more quickly than those with W/C ratios of 50% and 65%. Therefore, the crack widths of specimens with lower W/C ratios tend to grow more quickly than those with higher W/C ratios. These results are in agreement with the findings of Alonso et al. (1998). There is a delay in the increase of the crack widths for specimens with higher W/C ratios, which have greater porosity to accommodate the corrosion product and reduce the internal pressure.

4.5. Effect of stirrups

Figure 21 shows the effects of stirrups on the development of steel corrosion and crack widths for the beams exposed to approximately the same amount of accumulated current. Figures 21(a) and (b) indicates that the steel corrosion and surface crack width of the specimen having no stirrups (i.e. III-1) increase more quickly than those of other specimens having stirrups (i.e. I-1 and III-2). For the steel corrosion of the specimens having stirrups, Figure 21(a) shows that the longitudinal rebar of specimen III-2, with a stirrup spacing of 165 mm, is corroded more quickly than that of specimen I-1, with a stirrup spacing of 100 mm. Thus, decreasing the stirrup spacing or increasing the number of stirrups slows the steel corrosion. In Figure 21(b), because the mean crack width of specimen I-1 increases more quickly than that of specimen III-2, one can conclude that increasing the number of stirrups might accelerate the growth of crack widths.

4.6. Relationship between steel weight loss and crack widths

The graphs in Figures 22–26 show the relationships between the steel weight loss and the surface crack widths at their corresponding locations for all the specimens. Generally, the scattered points are increasingly highly dispersed with increasing MRw. Therefore, the relationship between Rw and Cw weakens significantly at higher values MRw.

However, it is also found that the steel weight loss increases with increasing crack width. In particular, in the initial part of the corrosion process, up to a crack width of about .3 mm, both the steel weight loss and crack width appear to be linearly related for all the specimens. This result is inconsistent with previous findings reported in the literature. For example, Vidal et al. (2004) found a linear approximation between the two parameters for the RC specimens with corrosion crack widths of about 1 mm, which is similar to the finding by Alonso et al. (1998) although these authors reported that the scatter of their relationships increased with the corrosion crack width and became significant as the crack width was over 1 mm. This inconsistency might be caused by the differences in the experimental procedure such as the magnitude of current density to corrode the specimens. Further research is needed to investigate the effects of the current densities, structural details and concrete qualities on the relationship between the steel weight loss and corrosion crack width.

5. Conclusions

In this paper, a procedure using an X-ray technique to continuously investigate the spatial growth of corroded rebars inside RC beams has been illustrated, and a method for estimating steel weight loss via the digital image analysis of X-ray photograms has been presented. The accuracy of the estimation method was also discussed. The following conclusions can be drawn:

(1) The estimated rebar weight loss calculated using the digital image analysis of X-ray photograms was found to be only 3% higher than that of the actual measured steel weight loss. This demonstrated the good accuracy of the application of the X-ray technique for investigating the spatial growth of corroded rebars in RC members.

(2) The distributions of the steel weight loss and crack width are spatially non-uniform, and their degree of non-uniformity significantly increases as the steel weight loss and corrosion cracking increase. The cause of this larger spatial variability might result from the greater contribution of the larger surface cracks to the increase in steel weight loss (i.e. the larger corrosion cracking allows the much higher chloride concentration to penetrate through and reach the reinforcement more easily).

(3) The maximum steel weight loss usually does not occur at the locations corresponding to the maximum crack width and often occurs close to the stirrups. Additionally, these locations vary depending on the mean steel weight loss.

(4) A strong relationship was found between the standard deviations of the steel weight loss or crack width and the differences between their maximum and minimum values when the mean steel weight loss exceeded 5%. This relationship may allow the estimation of the variability between two points after data from the *in situ* inspection of steel corrosion are obtained.

(5) The effect of W/C ratio on the increased steel weight loss is not obvious. Meanwhile, the crack width of the specimens with low W/C ratios increased faster than those with high W/C ratios.

(6) The crack width and steel corrosion of the specimen having no stirrups increased more quickly than those of specimens having stirrups. Increasing the number of stirrups slows the steel corrosion but accelerates the growth of the crack widths.

The parameters used to produce the stochastic field associated with the steel weight loss in RC structures could be determined using the experimental results presented in this paper. This in turn can facilitate the life-cycle reliability assessment of ageing RC structures by incorporating the spatial variability of steel weight loss (as presented in Akiyama & Frangopol, 2014). In addition, a relationship between the steel weight loss and width of the longitudinal surface cracks might be established with corrosion cracks smaller than .3 mm. This relationship might be incorporated into an estimation model of the long-term structural performance of RC elements in a chloride-contaminated environment. At larger crack widths, a weaker relationship was found. This suggests that it is desirable to estimate the steel corrosion based on the inspected data of the steel corrosion levels of the *in situ* structures when the crack width is above .3 mm.

On the other hand, the difficulty in quantifying the non-uniform distribution of corrosion pits along the corroded rebars has been reported as the key issue in assessing the performance and structural reliability of corrosion-affected RC members (e.g. Coronelli & Gambarova, 2004; Shimomura, Saito, Takahashi, & Shiba, 2011; Stewart, 2009; Stewart & Al-Harthy, 2008; Val, 2007) and also in assessing their residual life (e.g. Torres-Acosta, Navarro-Gutierrez, & Terán-Guillén, 2007). Although the derivation of residual cross-sectional loss according to time is not presented in this paper, one can imagine that, by employing a similar estimating method to that reported in this paper, the X-ray images of non-corroded and corroded bars from the 12 different viewing angles can be used to derive the pitting corrosion and residual cross-sectional loss at any specific location along the corroded bars. The research related to the derivation of corrosion pits using X-ray images and assessment of corrosion-affected RC members using a finite element model with the consideration of the spatial variability of steel weight loss is under development and will be presented in the near future.

Disclosure statement

No potential conflict of interest was reported by the authors.

References

Akiyama, M., & Frangopol, D. M. (2013). Estimation of steel weight loss due to corrosion in RC members used on digital image processing of X-ray photogram. In A. Strauss, D. M. Frangopol, & K. Bergmeister (Eds.), *Life-cycle and sustainability of civil infrastructure systems* (pp. 1885–1891). London: CRC Press/Balkema, Taylor & Francis Group plc.

Akiyama, M., & Frangopol, D. M. (2014). Long-term seismic performance of RC structures in an aggressive environment: Emphasis on bridge piers. *Structure and Infrastructure Engineering, 10*, 865–879.

Akiyama, M., Frangopol, D. M., & Yoshida, I. (2010). Time-dependent reliability analysis of existing RC structures in a marine environment using hazard associated with airborne chlorides. *Engineering Structures, 32*, 3768–3779.

Alonso, C., Andrade, C., Rodriguez, J., & Diez, J. M. (1998). Factors controlling cracking of concrete affected by reinforcement corrosion. *Materials and Structures, 31*, 435–441.

Andrade, C., Alonso, C., & Molina, F. J. (1993). Cover cracking as a function of bar corrosion: Part I – Experimental test. *Materials and Structures, 26*, 453–464.

Ann, K. Y., & Song, H.-W. (2007). Chloride threshold level for corrosion of steel in concrete. *Corrosion Science., 49*, 4113–4133.

Beck, M., Goebbels, J., Burkert, A., Isecke, B., & Babler, R. (2010). Monitoring of corrosion processes in chloride contaminated mortar by electrochemical measurement and X-ray tomography. *Materials and Corrosion, 61*, 475–479.

Coronelli, D., & Gambarova, P. (2004). Structural Assessment of corroded reinforced concrete beams: Modeling guidelines. *Journal of Structural Engineering, 130*, 1214–1224.

Marsh, P. S., & Frangopol, D. M. (2008). Reinforced concrete bridge deck reliability model incorporating temporal and spatial variations of probabilistic corrosion rate sensor data. *Reliability Engineering & System Safety, 93*, 394–409.

Media Cybernetics, Inc. (2012). Image-Pro Plus version 7.0 for Windows, start-up guide. Rockville, MD: Media Cybernetics.

Mori, Y., & Ellingwood, B. R. (1993). Reliability-based service-life assessment of aging concrete structures. *Journal of Structural Engineering, 119*, 1600–1621.

Otsuki, N., Miyazato, S.-I., Diola, N., & Suzuki, H. (2000). Influences of bending crack and water–cement ratio on chloride-induced corrosion of main reinforcing bars and stirrups. *ACI Material Journal, 97*, 454–464.

Palsson, R., & Mirza, M. S. (2002). Mechanical response of corroded steel reinforcement of abandoned concrete bridge. *ACI Structural Journal, 99*, 157–162.

Shimomura, T., Saito, S., Takahashi, R., & Shiba, A. (2011). Modelling and nonlinear FE analysis of deteriorated existing concrete structures based on inspection. In C. Andrade & G. Mancini (Eds.), *Modelling of corroding concrete structures*, RILEM Bookseries 5 (pp. 259–272). Dordrecht, Heidelberg, London, New York: Springer.

Stewart, M. G. (2004). Spatial variability of pitting corrosion and its influence on structural fragility and reliability of RC beams in flexure. *Structural Safety, 26*, 453–470.

Stewart, M. G. (2009). Mechanical behaviour of pitting corrosion of flexural and shear reinforcement and its effect on structural reliability of corroding RC beams. *Structural Safety, 31*, 19–30.

Stewart, M. G., & Al-Harthy, A. (2008). Pitting corrosion and structural reliability of corroding RC structures, experimental data and probabilistic analysis. *Reliability Engineering System and Safety, 93*, 273–382.

Stewart, M. G., & Mullard, J. A. (2007). Spatial time-dependent reliability analysis of corrosion damage and the timing of first repair for RC structures. *Engineering Structures, 29*, 1457–1464.

Torres-Acosta, A. A., Navarro-Gutierrez, S., & Terán-Guillén, J. (2007). Residual flexure capacity of corroded reinforced concrete beams. *Engineering Structures, 29*, 1145–1152.

Trejo, D. & Pillai, R. G. (2004). Accelerated chloride threshold testing – Part II: Corrosion resistant reinforcement. *ACI Materials Journal, 101*, 57–64.

Val, D. V. (2007). Deterioration of strength of RC beams due to corrosion and its influence on beam reliability. *Journal of Structural Engineering, 133*, 1297–1306.

Vidal, T., Castel, A., & François, R. (2004). Analyzing crack width to predict corrosion in reinforced concrete. *Cement and Concrete Research, 34*, 165–174.

Vu, K., Stewart, M. G., & Mullard, J. (2005). Corrosion-induced cracking: Experimental data and predictive models. *ACI Structural Journal, 102*, 719–726.

Yamamoto, T., Oyado, M., Mikata, Y., Kobayashi, K., and Shimomura, T. (2011). Systematic laboratory test on structural performance of corroded reinforced concrete and its utilization in practice. In C. Andrade & G. Mancini (Eds.), *Modelling of corroding concrete structures*, RILEM Bookseries 5 (pp. 113–124). Dordrecht, Heidelberg, London, New York: Springer.

Zhang, R., Castel, A., & François, R. (2010). Concrete cover cracking with reinforcement corrosion of RC beam during chloride-induced corrosion process. *Cement and Concrete Research, 40*, 415–425.

Modelling processes related to corrosion of reinforcement in concrete: coupled 3D finite element model

Joško Ožbolt, Filip Oršanić and Gojko Balabanić

ABSTRACT

Aggressive environmental conditions, such as exposure to the sea climate or use of de-icing salts, have a strong influence on durability of reinforced concrete (RC) structures due to reinforcement corrosion-induced damage. In the present paper, a recently developed three-dimensional (3D) chemo-hygro-thermo-mechanical model for concrete is briefly discussed. The model was implemented into a 3D finite element code and its application is illustrated through numerical analysis of a RC beam-end specimen with stirrups, exposed to aggressive environmental conditions. Damage of concrete cover due to expansion of corrosion products and transport of rust through concrete pores and cracks are computed. Subsequently, the influence of corrosion-induced damage of concrete cover on pull-out resistance of deformed reinforcement is investigated. The comparison between numerical results and experimental evidence shows that the complex coupled mathematical model is able to realistically predict the phenomena related to corrosion of steel reinforcement in concrete.

Introduction

Corrosion-induced damage of concrete is a major problem for durability of reinforced concrete (RC) elements exposed to severe climate conditions. The onset of corrosion can be caused by carbonation of the concrete cover or by reaching a critical concentration of free chloride ions in the vicinity of the reinforcement bar. Here, presented work is focused on damage due to the chloride-induced corrosion.

RC structures in maritime environment, such as bridges, oil platforms or structures treated with de-icing salts during winter periods show a great vulnerability to corrosion processes (Cairns, 1998; Tuutti, 1993). Damage is usually exhibited in the form of cracking and spalling of concrete cover, which reduces the mechanical properties of RC elements, leading to a shorter serviceability. Moreover, due to corrosion, the cross-section area of reinforcement bars is reduced, which has negative implications on the bearing capacity of the RC structures.

Additionally, as a consequence of the pitting effect (Apostolopoulos & Papadakis, 2008; Cairns, Plizzari, Du, Law, & Franzoni, 2005), ductility and bond properties of reinforcement can be significantly reduced (Fischer, 2012; Tuutti, 1993). Damage due to the chloride-induced corrosion in RC structures exerts relatively large costs of repair. Approximately 300 million euros were spent in 2004 for the maintenance of highway bridges in Germany (Jacob, 2008). Roughly 40% of the costs were due to repair of concrete bridges, damaged mostly through chloride contamination and the subsequent corrosion-induced damage.

Therefore, it is important to have reliable computational model, which can realistically predict influence of different parameters relevant for the corrosion-induced damage of RC structures. Such relatively complex models cannot be used in daily engineering practice; however, they contribute to better understand the phenomena related to corrosion-induced damage and can be effectively used for the improvement of current design codes.

Generally, computation of corrosion current density depends upon the following physical, electrochemical and mechanical processes: (1) Transport of capillary water, oxygen and chloride through the concrete cover; (2) Immobilisation of chloride in the concrete; (3) Drying and wetting of concrete and related hysteretic property of concrete; (4) Transport of OH^- ions through electrolyte in concrete pores; (5) Cathodic and anodic polarisation; (6) Transport of corrosion products; (7) Creep of concrete around the reinforcement bar and (8) Damage of concrete due to mechanical and non-mechanical actions (Bažant, 1979).

Currently, there are a number of models in the literature that are able to simulate processes before and after depassivation of reinforcement in uncracked concrete (Andrade, Díez, & Alonso, 1997; Balabanić, Bićanić, & Đureković, 1996a, 1996b; Bažant, 1979; Glass & Buenfeld, 2000; Martin-Perez, 1999; Page, Short, & El Tarras, 1981; Thomas & Bamforth, 1999; Tuutti, 1993; Zhang & Gjørv, 1996). For such computational models to be considered as realistic, chemo-hygro-thermal processes have to be coupled with mechanical processes and also vice versa. Presently, there are only a limited number of coupled three-dimensional (3D) chemo-hygro-thermo-mechanical (CHTM) models capable of realistically simulating corrosion processes in cracked concrete (Ishida, Iqbal, & Anh, 2009; Marsavina, Audenaert, De Schutter, Faur, & Marsavina, 2009; Ožbolt, Balabanić, & Kušter, 2011; Ožbolt, Balabanić, Periškić, & Kušter, 2010; Ožbolt, Oršanić, & Balabanić, 2014). Transport of corrosion products in cracked concrete and its effect on the corrosion-induced damage so far has not been addressed by any model, i.e. the current models do not account for the transport of soluble species that are involved in the process of corrosion. The main problem in defining such models is quantifying controlling parameters for processes before and after depassivation of the reinforcement (Fischer, 2012; Wong, Zhao, Karimi, Buenfeld, & Jin, 2010).

The recently developed 3D chemo-hyrgo-thermo mechanical model for concrete is here briefly discussed. The model is able to simulate non-mechanical and mechanical processes and their interaction before and after depassivation of steel reinforcement (Ožbolt, Balabanić et al., 2010; Ožbolt, Balabanić, & Kušter 2011). The model was implemented into a 3D finite element (FE) code and the results showed that it is able to realistically predict relevant processes before (Ožbolt et al., 2010) and after depassivation of reinforcement (Ožbolt, Balabanić et al., 2011; Ožbolt, Oršanić et al., 2014). The application of the model is illustrated on numerical examples in which corrosion of reinforcement of the beam-end specimen with stirrups is experimentally and numerically investigated.

Uncertainty associated with the parameters of the proposed model, which is an important aspect of the problem, is not accounted for because the presented model is deterministic. It is well known that the mechanical and non-mechanical parameters over the volume of a concrete structure are not uniform. However, using the presented deterministic model, the influence of random spatial distribution of each relevant model parameter can be investigated and subsequently stochastic study can be performed.

CHTM model for concrete

A surface layer of ferric oxide, covers and protects the steel in concrete. With this layer being damaged or depassivated the corrosion of reinforcement in concrete can be activated. Depassivation of the protective layer can occur by reaching a critical threshold concentration of free chloride ions near the reinforcement bar surface or as a consequence of carbonation of concrete (Glass & Buenfeld, 1995; Sandberg, Peterson, Sarensen, & Arup, 1995; Tuutti, 1993). Here, only chloride type of the corrosion that is the most severe one is discussed. In the literature (Bažant, 1979; Glasstone, 1964; Page & Treadaway, 1982) corrosion is described as an electrochemical process dependent on electrical conductivity of concrete and steel surfaces, presence of electrolyte in concrete and concentration of dissolved oxygen in the pore water near the reinforcement. The developed 3D CHTM model couples the above-mentioned physical and electrochemical processes with the mechanical behaviour of concrete (damage).

Non-mechanical processes before depassivation of reinforcement

In the model, the transport of capillary water is described in terms of volume fraction of pore water in concrete by Richard's

equation (Bear & Bachmat, 1990), based on the assumption that transport processes take place in aged concrete, i.e. the hydration of cement paste is completed:

$$\frac{\partial \theta_w}{\partial t} = \nabla \cdot \left[D_w(\theta_w) \nabla \theta_w \right] \tag{1}$$

where θ_w is volume fraction of pore water (m^3 of water/m^3 of concrete) and $D_w(\theta_w)$ is capillary water diffusion coefficient (m^2/s) described as a strongly non-linear function of moisture content (Leech, Lockington, & Dux, 2003). Transport of chloride ions through a non-saturated concrete occurs as a result of convection, diffusion and physical and chemical binding by cement hydration product (Bear & Bachmat, 1990):

$$\theta_w \frac{\partial C_c}{\partial t} = \nabla \cdot \left[\theta_w D_c(\theta_w, T) \nabla C_c \right] + D_w(\theta_w) \nabla \theta_w \nabla C_c - \frac{\partial C_{cb}}{\partial t} \tag{2a}$$

$$\frac{\partial C_{cb}}{\partial t} = k_r \left(\alpha C_c - C_{cb} \right) \tag{2b}$$

where C_c is the concentration of free chloride dissolved in pore water (kgCl$^-$/m^3 of pore solution), $D_c(\theta_w, T)$ is the effective chloride diffusion coefficient (m^2/s) expressed as a function of water content θ_w and concrete temperature T, C_{cb} is concentration of bound chloride (kgCl$^-$/m^3 of concrete), k_r is binding rate coefficient, $\alpha = .7$ is constant (Saetta, Scotta, & Vitaliani, 1993).

It is well known that diffusivity of chloride depends on many parameters such as temperature, porosity, pore relative humidity, degree of cement hydration and cement type (Saetta et al., 1993). However, the results of experimental investigations (Tang & Nilsson, 1996) show that chloride diffusivity is not dependent on equivalent hydration age (t_e) if $t_e > 180$ days. Here, in the presented model, it is assumed that $t_e > 180$ days, i.e. diffusivity of chloride is assumed to be independent of the concrete age.

Assuming that oxygen does not participate in any chemical reaction before depassivation of steel, transport of oxygen through concrete is considered as a convective diffusion problem (Bear & Bachmat, 1990):

$$\theta_w \frac{\partial C_o}{\partial t} = \nabla \cdot \left[\theta_w D_o(\theta_w) \nabla C_o \right] + D_w(\theta_w) \nabla \theta_w \nabla C_o \tag{3}$$

where C_o is oxygen concentration in pore solution (kg of oxygen/m^3 of pore solution) and $D_o(\theta_w)$ is the effective oxygen diffusion coefficient (Ožbolt et al., 2010), dependent on concrete porosity p_{con} and water saturation of concrete S_w.

Based on the constitutive law for heat flow and conservation of energy, the equation which describes temperature distribution in continuum reads:

$$\lambda \Delta T + W(T) - c\rho \frac{\partial T}{\partial t} = 0 \tag{4}$$

where λ is thermal conductivity (W/(m K)), c is heat capacity per unit mass of concrete (J/(K kg)), ρ is mass density of concrete (kg/m^3) and W is internal source of heating (W/m^3). More detail related to the strong and weak formulations of the processes up to the depassivation of reinforcement can be found in Ožbolt et al. (2010).

Non-mechanical processes after depassivation of reinforcement

The corrosion of steel is activated with the depassivation of the steel reinforcement in concrete. The non-mechanical processes important for the propagation stage of steel corrosion in concrete are: (1) mass sinks of oxygen at steel surface due to cathodic and anodic reaction, (2) the flow of electric current through pore solution and (3) the cathodic and anodic potential.

The oxygen consumption at the cathodic and anodic surfaces is a result of the following reactions:

- Reaction of dissolved oxygen in the pore water with the electrons on the cathode:

$$2H_2O + O_2 + 4e^- \rightarrow 4OH^- \tag{5}$$

- Transport of hydroxyl ions to the anode, where corrosion products forms:

$$Fe^{2+} + 2OH^- \rightarrow Fe(OH)_2 \tag{6}$$

$$4Fe(OH)_2 + O_2 + 2H_2O \rightarrow 4Fe(OH)_3 \tag{7}$$

and can be calculated as:

$$D_o(\theta_w)\frac{\partial C_o}{\partial n}\bigg|_{cathode} = -k_c i_c \quad k_c = 8.29 \times 10^{-8}\frac{kg}{C} \tag{8a}$$

$$D_o(\theta_w)\frac{\partial C_o}{\partial n}\bigg|_{anode} = -k_a i_a \quad k_a = 4.14 \times 10^{-8}\frac{kg}{C} \tag{8b}$$

where n is the outward normal to the steel bar surface and i_c and i_a are cathodic and anodic current density (A/m²), respectively. The constants k_c and k_a are calculated using the stoichiometry of chemical reactions (5–7) and Faraday's law.

According to Butler–Volmer kinetics, in the present model kinetics of reaction at the cathodic and anodic surface can be estimated from:

$$i_c = i_{0c}\frac{C_o}{C_{ob}}e^{2.3(\Phi_{0c}-\Phi)/\beta_c} \quad i_a = i_{0a}e^{2.3(\Phi-\Phi_{0a})/\beta_a} \tag{9}$$

where C_{ob} is the oxygen concentration at the surface of concrete element exposed to sea-water (kg/m³), Φ is the electric potential in pore solution near the reinforcement surface (V), i_{0c} and i_{0a} are the exchange current density of the cathodic and anodic reaction (A/m²), Φ_{0c} and Φ_{0a} are the cathodic and anodic equilibrium potential (V), β_c and β_a are the Tafel slope for cathodic and anodic reaction (V/dec), respectively.

The electric current through the electrolyte is a result of motion of charged particles and, if the electrical neutrality of the system and the uniform ions concentration are assumed, can be written as:

$$\mathbf{i} = -\sigma \nabla\Phi \tag{10}$$

where σ is electrical conductivity of concrete.

The equation of electrical charge conservation, if the electrical neutrality is accounted for and the electrical conductivity of concrete is assumed as uniformly distributed, reads:

$$\nabla^2\Phi = 0 \tag{11}$$

Rate of rust production J_r (kg/m²s) and mass of hydrated red rust per related surface (A_r) of rebar m_r (kg), respectively, are calculated as:

$$J_r = 5.536 \times 10^{-7} i_a$$
$$m_r = J_r \Delta t A_r \tag{12}$$

where Δt is the time interval in which the corrosion is taking place and A_r is the corresponding surface of the steel reinforcement. The coefficient of proportionality between the anodic current density i_r and rate of rust production J_r is calculated using the stoichiometry of chemical reactions and Faraday's law (Martin-Perez, 1999).

Recent experimental investigations (Fischer, 2012; Wong et al., 2010) have shown that the penetration of corrosion products into the pores and their relatively large ingress through the radial cracks, generated around the bar, has a significant effect on the development of corrosion-induced damage. The influence can be summarised as: (1) the distribution of rust and radial pressure over the anodic surface is not uniform, and (2) damage due to expansion of products is less pronounced. The distribution of the corrosion products (red rust) R (kg/m³ of pore solution) into the pores and through the cracks in concrete has been mathematically formulated as a convective diffusion problem:

$$\theta_w\frac{\partial R}{\partial t} = \nabla \cdot [\theta_w D_r \nabla R] + D_w(\theta_w)\nabla\theta_w\nabla R \tag{13}$$

in which D_r is the diffusion coefficient (m²/s) of corrosion product. It is important to note that Equation (13) does not directly describe the transport of the red rust, but rather the distribution of the rust formed in the concrete pores and cracks as a consequence of soluble species (which can dissolve in concrete pore solution and subsequently migrate in pores and cracks) reacting with oxygen in the pore water (Wong et al., 2010).

Currently, the above-mentioned process of transport of corrosion products is considered more in a qualitative sense. Detailed experimental investigations needed for the quantitative calibration of the model are currently not available. Therefore, to adjust model parameters further development of the model is planned in the context of experimental measuring of the transported amount of corrosion products through the cracks.

In the presented computations in this study, the diffusion coefficient of corrosion product D_r is adopted by the calibration using experimental results obtained by Fischer (2012). Principally, with the increase in D_r, more corrosion products will be transported through pores and cracks of concrete. Consequently, damage of concrete will be less pronounced.

CHTM coupling

The microplane model for concrete with relaxed kinematic constraints (Ožbolt, Li, & Kožar, 2001) is applied in the mechanical part of the model. In the FE analysis cracks are treated in a smeared way, i.e. smeared crack approach is employed. To assure the objectivity of the results with respect to the size of the FEs, the crack band method is used (Bažant & Oh, 1983).

The governing equation for the mechanical behaviour of a continuous body in the case of static loading condition (momentum equation) reads:

$$\nabla \big[D_m \big(u, \theta_w, T \big) \nabla u \big] + \rho b = 0 \qquad (14)$$

In which D_m is material stiffness tensor, ρb is specific volume load and u is displacement field. In the mechanical part of the model the total strain tensor is decomposed into mechanical strain, thermal strain, hygro strain (swelling–shrinking) and strain due to expansion of corrosion products.

One-dimensional (1D) corrosion contact elements are employed in the model to account for the inelastic strains due to the expansion of corrosion products. They are placed radially around the bar surface and their main function is to simulate the contact between reinforcement and surrounding concrete. These contact elements can take up only shear forces in direction parallel to reinforcement axis and compressive forces perpendicular to the adjacent surface of the reinforcement. The inelastic radial expansion due to corrosion Δl_r is calculated as:

$$\Delta l_r = \frac{m_r}{A_r} \left(\frac{1}{\rho_r} - \frac{.523}{\rho_s} \right) \qquad (15)$$

where $\rho_r = 1.96 \times 10^3$ (kg/m^3) and $\rho_s = 7.89 \times 10^3$ (kg/m^3) are densities of rust and steel, respectively, .523 is the ratio between the mass of steel (m_s) and the corresponding mass of rust (m_r) over the related surface of reinforcement A_r that corresponds to the contact element.

Note that the mass of steel m_s consumed to produce m_r is calculated from the ratio of molar mass of metal iron and molar mass of hydrated red rust, $m_s = .523\, m_r$ (Bažant, 1979).

The stiffness of the rust layer is assumed to be $E_r = 100$ MPa. In the model it is represented by the axial stiffness of the corrosion contact elements. The shear resistance of the contact elements, defined by the bond-slip relationship, is used to model the bond between deformed steel reinforcement and concrete. The deformed surface of rebar is not modelled explicitly; however, their contribution to the bond resistance is accounted for through the constitutive bond-slip relationship (mechanical bond). Moreover, there is no degradation of the bond-slip relationship as a function of the corrosion penetration. This effect comes automatically out from the 3D FE model since due to the expansion of the corrosion products the concrete cover becomes damaged and consequently the bond resistance decreases.

Numerical implementation

Using the FEs to solve the partial differential equations implemented in the model, the strong form has to be rewritten into a weak form. The weak form of the system of partial differential equations is carried out by employing the Galerkin weighted residual method (Belytschko, Liu, & Moran, 2001). The model is then implemented into a 3D FE code. The non-mechanical part of the problem is solved using direct integration method of implicit type (Belytschko et al., 2001). To solve the mechanical part, Newton–Rapshon iterative scheme is used. Coupling between mechanical and non-mechanical part of the model is performed by continuous update of governing parameters during the incremental transient FE analysis using staggered solution scheme. For more details see Ožbolt, Balabanić et al. (2010), Ožbolt, Balabanić, and Kušter (2011), Ožbolt, Oršanić, Balabanić, and Kušter (2012), Ožbolt, Oršanić, Kušter, and Balabanić (2012), and Ožbolt, Oršanić, & Balabanić (2014).

Numerical example: beam-end specimen with stirrups

The application of the presented 3D CHTM model is demonstrated through a numerical simulation of the pull-out of the reinforcement bar from the beam-end specimen with stirrups. The presented numerical results are verified with experimental data obtained by Fischer (2012). Detailed results of the numerical study in regard to the application of the 3D CHTM model for the specimens without stirrups can be found in Ožbolt et al. (2014).

Experiments performed by Fischer (2012) were focused on the influence of the corrosion-induced damage on the bond behaviour of the embedded steel bars. Series of beam-end specimens were chosen according to Chana (1990), where the specimen geometry was optimised in order to prevent yielding of reinforcement and to exclude the influence of neighbouring bars on each other while the pull-out test is conducted in each separate corner. The specimens were first exposed to aggressive environmental conditions, which caused the corrosion of the embedded reinforcement bar and subsequently the reinforcement was pulled out from the specimen.

Four specimen types with stirrups, with four bars arranged in corners, were studied by Fischer (2012). The cross-section in all cases was 200 mm^2 × 200 mm^2. The diameter of the main reinforcement bar was either 12 mm with a concrete cover of 20 or 16 mm with a cover of 35 mm (Figure 1). The total embedment length of the rebar in all cases was 180 mm, whereas the rest of the length was isolated with a plastic sleeve (Figure 1). The stirrups with the diameter ϕ6 mm were chosen with a spacing of 90 mm (Figure 1). Two additional series were also studied, in which the direct electric contact with the main rebar was prevented by isolating the stirrup corners with tape. All specimen types with stirrups which were investigated by Fischer (2012) and their main characteristics are summarised in Table 1. Types 1 and

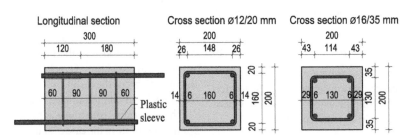

Figure 1. Geometry of the specimen with stirrups used in experiments by Fischer (2012).

Table 1. Investigated specimen types with stirrups from the experiments by Fischer (2012).

Types	Diameter, d (mm)	Concrete cover, c (mm)	c/d	Stirrups (mm)	Stirrup – el. contact
2	12	20	1.67	6/90	Yes
4	16	35	2.19	6/90	Yes
5	12	20	1.67	6/90	No
6	16	35	2.19	6/90	No

3 without stirrups, i.e. φ12/20 mm and φ16/35 mm respectively, are discussed in detail in Ožbolt et al. (2014).

In the following sections only the conditions of the series 5 and 6 are simulated, with the stirrups not being in the direct contact with the main reinforcement. These are chosen to compare the numerical results more easily with the numerical data obtained from the models without stirrups (Ožbolt et al., 2014).

Numerical model

Emphasis of the here presented numerical analysis is on the processes after the depassivation of reinforcement bars. Specifically, the anodic and cathodic regions were chosen upfront and kept unchanged throughout the calculation, which is an assumption made to simplify an already complicated process. The processes before depassivation of reinforcement are not computed, i.e. certain length sections of the rebar are assumed to be activated as anode (depassivated) at the start of the analysis and kept unchanged during the computation. This is not a limitation of the 3D CHTM model, which is able to simultaneously calculate the processes before and after depassivation, but rather an approach chosen due to time and CPU requirements that such calculation would require. Note that the application of the model for computation of processes before depassivation of reinforcement can be found in Ožbolt et al. (2010).

Symmetry conditions are used to model only half of the specimen for the types 5 and 6. The width of the modelled section for the type 6 is taken 10 mm wider than the symmetry plane to accommodate for a bigger bar diameter and a deeper concrete cover while analysing the pull-out resistance (Figure 2(a)). Eight node solid elements are used to model the main reinforcement and the stirrups (Figure 2(b)). To avoid a contact problem between the main rebar and stirrups, the stirrup position is offset 2 mm towards the upper and side concrete surfaces. The .1 mm thick interface layer is placed

around the steel reinforcement and modelled with eight node solid elements. Corresponding 1D corrosion contact elements are placed in the interface layer. Due to the presence of the curved stirrup geometry, the concrete is discretised by using the tetrahedral solid elements (Figure 2(a)).

The results of the numerical study by Ožbolt et al. (2014), which was focused on investigating the beam-end specimens without stirrups from the experiments by Fischer (2012), have shown a significant influence of the anodic–cathodic configuration over the rebar cross-section on the predicted corrosion-induced damage. Types 1B and 2B (Ožbolt et al., 2014) for φ12/20 and φ16/35 mm without stirrups, respectively, with only half of the rebar circumference being depassivated, gave the most realistic crack patterns compared to the ones from the experiment. Hence, the same distribution of the anodic and cathodic regions along the rebar length (Figure 3(a)) and along the rebar cross-section (Figure 3(b)) are chosen for the types 5 and 6 with stirrups.

So far in computations of corrosion rate of steel in concrete cathodic and anodic areas are determined (assumed) in advance. Since the results of the computation show good agreement with the experimental results it is concluded that the assumption agrees well with reality. If the areas of the cathode and anode cannot be determined experimentally, corrosion rate cannot be computed properly without additional conditions. Such additional conditions could be deduced from irreversible thermodynamics using the principle of maximum entropy production (or the principle of maximum dissipation of energy), generated in the dissipative processes such as cathodic and anodic polarisation, the flow of ions through the electrolyte and mass transfer through porous media.

Due to the stirrups being electrically isolated from the rebar in the experiment (Fischer, 2012), it is assumed that the whole stirrup area is acting as a cathode. Type 2 and 4 specimens, where the electrical contact was enabled, show similar results regarding the corrosion-induced damage and the behaviour of the pull-out capacity as compared to type 5 and 6 (Fischer, 2012).

Mechanical properties of concrete, steel and rust used in the numerical analysis are summarised in Table 2. For the numerical simulation of the transport processes, the water saturation is assumed to be constant over the entire volume of the specimen (S_w = 50%). The initial concentration of oxygen in the concrete bulk is the same as the concentration on the exposed concrete surfaces and is equal to .0085 kg of dissolved oxygen / m³ of pore solution. Oxygen diffusivity and electrical conductivity values for concrete are chosen for a good quality

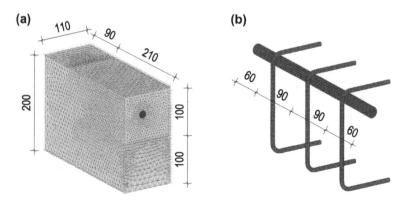

Figure 2. Model geometry (in mm) with the finite element discretisation of the specimen-type 6 (a) and the used main rebar and stirrups (b).

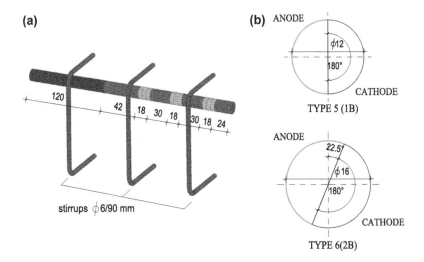

Figure 3. Distribution of the anodic and cathodic regions along the rebar and stirrups (a) and over the rebar circumference for types 5 and 6 (b).

Table 2. Mechanical properties used in the analysis.

Parameter	Value
Modulus of elasticity of concrete, E_c (N/mm^2)	25000.0
Poisson's ratio of concrete, v_c	.18
Tensile strength of concrete, f_t (N/mm^2)	3.0
Uniaxial compressive strength of concrete, f_c (N/mm^2)	41.0
Fracture energy of concrete, G_f (J/m^2)	90.0
Modulus of elasticity of steel, E_s (N/mm^2)	200000.0
Poisson's ratio of steel, v_s	.33
Modulus of elasticity of rust, E_r (N/mm^2)	100.0
Volume expansion factor of rust, $a_r = \rho_s/\rho_r$	4.0

Table 3. Parameters used in the numerical simulations.

Parameter	Value
Faraday constant, F (C/mol)	96486.7
Anodic exchange current density, i_{0a} (A/m^2)	1.875×10^{-4}
Cathodic exchange current density, i_{0c} (A/m^2)	6.25×10^{-6}
Anodic equilibrium potential, Φ_{0a} (V vs. SCE)	−.780
Cathodic equilibrium potential, Φ_{0c} (V vs. SCE)	.160
Tafel slope for anodic reaction, β_a (V/dec)	.06
Tafel slope for cathodic reaction, β_c (V/dec)	.160
Diffusivity coefficient of rust, D_r (m^2/s)	2.2×10^{-16}

concrete with a water–cement ratio, w/c = .4. The parameters used for the calculation of electric potential, consumption of oxygen and rust distribution are summarised in Table 3. They approximately correspond to severe splash environmental condition (Ožbolt et al., 2010, 2011). It is important to note, that in the experiments the corrosion was accelerated and the corrosion rate was approximately five times higher than the most severe rate observed in nature. However, the chosen model parameters result in relatively high corrosion rate, assuming natural corrosion conditions. Therefore, the model parameters (saturation of 50%) are chosen such that the corrosion-induced damage in real time (after 7 years) corresponds to the experimentally obtained damage under accelerated conditions (after 6 months).

The importance of modelling the distribution of corrosion products into the pores and cracks was shown in Ožbolt et al. (2014). Therefore, the transport of rust is accounted for in the following numerical simulation.

In the numerical analysis the specimen type 5 and 6 are exposed to natural corrosion conditions over 7 years. Subsequently, for different levels of corrosion-induced damage, i.e. after 1–5 and 7 years of active corrosion, the pull-out capacity of the rebar is calculated.

Numerically predicted corrosion-induced damage

As mentioned in the previous section, the corrosion-induced damage is simulated for a period of 7 years. The numerically obtained crack patterns for the specimen type 6, after 2, 3, 5 and 7 years of active corrosion processes, are shown in Figure 4. Red colour corresponds to the crack width greater or equal to the critical crack width. For normal strength concrete the critical crack width is approximately equal to .10 mm.

To validate the numerical results, sum of average crack widths on the upper and side concrete surface is plotted as a function of the average corrosion penetration on the anodic–cathodic transition zone, for the specimen type 5 and 6. These results are then compared with the experimental data and presented in Figure 5. It can be seen that for both cases, a good agreement is obtained with the experimental data (Fischer, 2012), showing that the corrosion-induced damage (cracks) are developing realistically with the active corrosion process. Note that the average crack width is here used because along the beam axis the crack width is not constant, i.e. the average crack width along the beam axis is measured.

Numerical analysis of the pull-out capacity for different corrosion levels

For each level of corrosion-induced damage, i.e. 1–5 and 7 years after the depassivation, the corresponding pull-out capacity is calculated. The pull-out failure mode for different levels of corrosion damage for the specimen type 6 is shown in Figure 6. Experimentally obtained failure mode (Fischer, 2012) due to the cracking of the concrete cover is replicated by the numerical results.

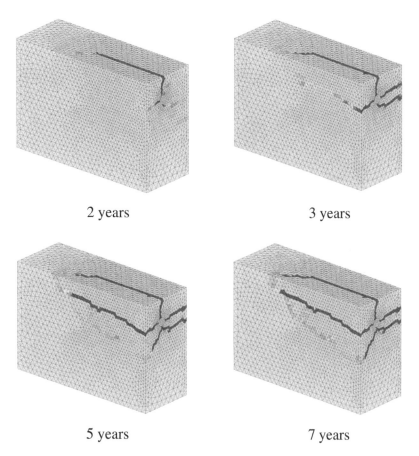

2 years 3 years

5 years 7 years

Figure 4. Numerically predicted crack patterns for the specimen type 6 after 2, 3, 5 and 7 years of active corrosion processes.

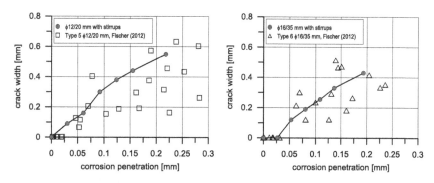

Figure 5. Sum of average crack widths on the concrete surfaces as a function of the average corrosion penetration for the type 5 (left) and type 6 (right).

To validate the influence of the corrosion damage level on the corresponding pull-out capacity of the rebar, the relative values of the bond resistance are plotted in Figure 7 as a function of the average corrosion penetration of the rebar. Comparison with the experimental results shows that in the case of the specimen type 6 (ϕ16/35 mm with stirrups), the same trend, slow decrease of pull-out capacity, is realistically captured by the model (Figure 7(b)).

In the case of the specimen type 5 (ϕ12/20 mm with stirrups), the model is not able to simulate a slight increase of the bond resistance with the ongoing corrosion (Figure 7(a)), although the predicted decrease is relatively low and kept constant after

reaching the relative value of .93 of the reference pull-out resistance (without corrosion).

The relative values of the pull-out capacity are also plotted as a function of the corresponding sum of average crack widths on the upper and side concrete surfaces. The average crack widths are calculated for each level of corrosion-induced damage, i.e. from 1 to 7 years after depassivation. The predicted values are compared with the experimentally obtained data (Fischer, 2012) in Figure 8. It can be seen that in the case of larger concrete cover and larger bar diameter (Type 6) the predicted decrease of bond resistance is realistically following the slight decrease of the

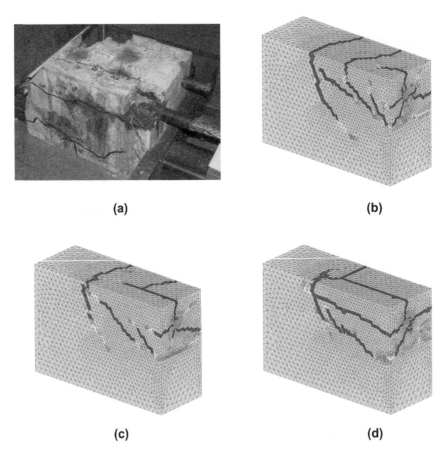

Figure 6. Pull-out failure mode obtained for the specimen type 6: (a) Fischer (2012), corresponds approximately to $t = 7$ years of the analysis (b) analysis (reference) $t = 0$; (c) analysis, $t = 2$ years and (d) analysis, $t = 7$ years.

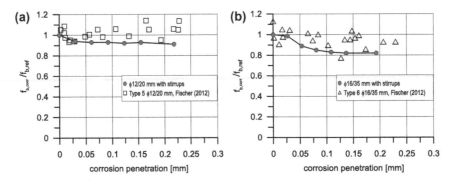

Figure 7. Predicted and measured values of the relative pull-out capacity as a function of the average corrosion penetration of the rebar, for the specimen type 5 (a) and type 6 (b).

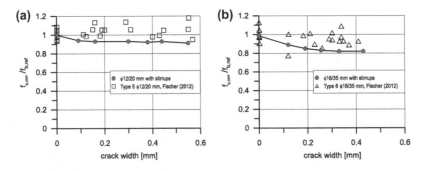

Figure 8. Predicted and measured values of the relative pull-out capacity as a function of the average crack widths for the specimen type 5 (a) and type 6 (b).

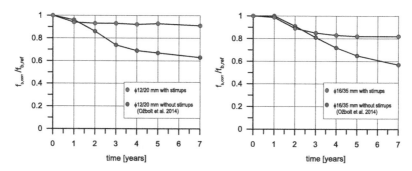

Figure 9. Relative pull-out capacity as a function of time (years) for specimens φ12/20 mm (left) and φ16/35 mm (right), with and without (Ožbolt et al., 2014) stirrups.

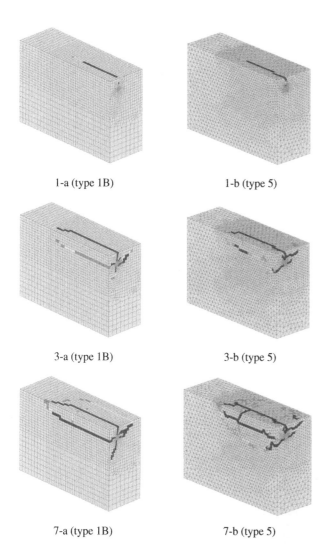

1-a (type 1B)   1-b (type 5)

3-a (type 1B)   3-b (type 5)

7-a (type 1B)   7-b (type 5)

Figure 10. Predicted crack patterns due to corrosion-induced damage, 1, 3 and 7 years after depassivation for: (a) without – type 1B (Ožbolt et al., 2014) and (b) with stirrups – type 5.

pull-out capacity (Figure 8(b)). In the case of a smaller bar diameter and concrete cover (Type 5), the minor relative increase of the pull-out force is not obtained (Figure 8(a)). Nonetheless, the decrease is relatively small at the beginning of the active corrosion process, i.e. with the onset of the visible corrosion-induced

damage (cracks) while afterwards is following a relatively constant trend line, same as in the experiments.

Note that the stirrups influence the crack width and crack pattern. Compared with the beam cross-section without stirrups, for the same corrosion penetration the case with stirrups exhibits different crack pattern and smaller crack width. Moreover, because of the nature of the problem it can be seen that experimental results from Figures 5 and 7 exhibit large scatter.

Comparison of numerical results for the specimens with and without stirrups

Experimental data and numerical analysis show a significant influence of stirrups on the pull-capacity of rebars in the case of beam-end specimen exposed to aggressive corrosion conditions. To investigate this effect, the relative pull-out capacity is plotted as a function of time (1–5 and 7 years) for specimen types φ12/20 and φ16/35 mm, with and without stirrups (Figure 9). The numerical data for the specimens without stirrups are taken from Ožbolt et al. (2014). It can be seen that with time, i.e. with the progressing corrosion damage, the relative bond resistance is decreasing much moderately for specimens with stirrups. After reaching a certain level of corrosion, the rebar is exhibiting approximately constant value of pull-out capacity for specimen type 5 and 6, i.e. φ12/20 and φ16/35 mm with stirrups, respectively.

To demonstrate the influence of stirrups on corrosion-induced crack patterns and crack widths, the damage after 1, 3 and 7 years is shown in Figure 10 for specimen φ12/20 mm, without (Type 1B – left column) and with stirrups (Type 5 – right column). Correspondingly, experimental and numerical average crack widths as a function of average corrosion penetration of the main rebar are given for all investigated specimens in Figure 11. It can be concluded, that for smaller levels of corrosion-induced damage and for the assumed anodic–cathodic positions along the bar, the crack widths and crack patterns for the stirrup cases differ only slightly compared to the specimens without stirrups. Nonetheless, the experimental data by Fischer (2012) showed that the differences become more pronounced for higher levels of corrosion penetration, i.e. the crack widths for cases with stirrups were smaller compared to the cases where the shear reinforcement was not present.

Furthermore, the average values of the principal tensile stresses for the stirrup elements, shown in Figure 12(a), are

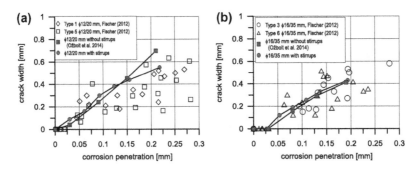

Figure 11. Predicted and measured values of the average crack width as a function of the average corrosion penetration for the specimens φ12/20 mm, with and without stirrups (a) and for the specimens φ16/35 mm, with and without stirrups (b).

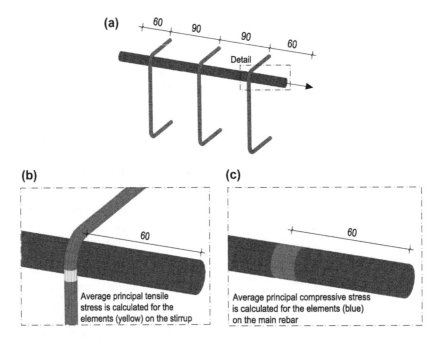

Figure 12. (a) Detail of the stirrups elements (yellow) used to calculate the average values of the principal tensile stresses. (b) Detail of the rebar elements (blue) used to calculate the average values of the principal compressive stresses.

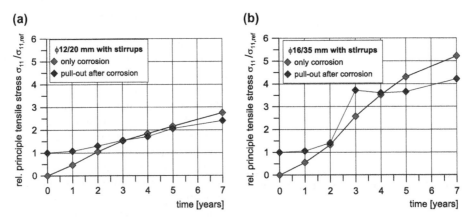

Figure 13. Relative average principal tensile stresses in a stirrup section (during corrosion and after reaching a peak load during the pull-out of the rebar) plotted as a function of time for type 5 (a) and type 6 (b).

monitored for types 5 and 6. Two cases are considered for each specimen: the average principal tensile stress calculated after a certain period (1–5 and 7 years) of active corrosion and the corresponding principal tensile stress obtained when the maximum bond capacity is reached in the subsequent pull-out of the main rebar. The results are shown in Figure 13 as relative values to the average principal tensile stress of the reference pull-out case (no corrosion damage), $\sigma_{11,\,ref}$.

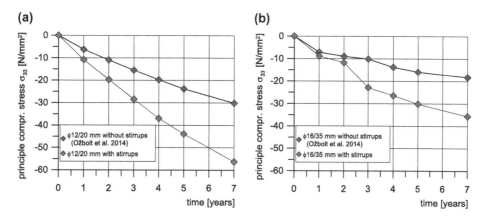

Figure 14. Average values of the principal compressive stresses in the rebar cross-section (with and without stirrups) as a function of time for type φ12/20 mm (a) and φ16/35 mm (b).

It can be seen that for the specimen type 5 (φ12/20 mm), the expansion of the corrosion products around the reinforcement bar activates the tensile stresses in the stirrup, which imposes confined conditions on the main rebar. Therefore, with the pull-out of the rebar the principal tensile stress in the investigated section at the peak load continuously rises and reaches almost 2.5 higher value after 7 years (Figure 13(a)). Due to a smaller bar and a smaller concrete cover, the confined conditions are already reached after one year of active corrosion and the relative bond resistance is afterwards kept approximately constant (Figure 7(a)). On the other hand, in the case of larger bar diameter and concrete cover, i.e. in the case of type 6 (φ16/35 mm) the confined conditions are reached 3 years after the start of corrosion. This is clearly seen from the Figure 13(b) with a sudden jump of the principal tensile stress at the peak load. Furthermore, this also explains a relatively larger decrease of the pull-out capacity at lower levels of corrosion-induced damage for the specimen type 6 (Figure 7(b)) compared to type 5.

To illustrate the effect of the stirrups presence with the ongoing expansion of the corrosion products, the average values of the principal compressive stresses in the rebar cross-section (Figure 12(b)), located under the stirrups for type 5, and in the corresponding position on the rebar for type 1B (without stirrups), are plotted in Figure 14 as a function of time. The values are numerically obtained after 1–7 years of active corrosion, before calculating the bond resistance. It can be seen that in both cases (with and without stirrups), the compressive forces are exerted on the rebar due to the expansion of corrosion products. In the case where the stirrups are present, confined conditions contribute to higher compressive stresses which have a positive effect of minimizing the reduction of the pull-out capacity with the ongoing corrosion damage.

Conclusions

The coupled 3D CHTM model for numerical analysis of non-mechanical and mechanical processes related to the corrosion of steel reinforcement in concrete is presented. The numerical analysis is performed on the beam-end specimen with stirrups, assuming aggressive environmental conditions (severe splash zone) and the results are compared with the experimental data

from the literature. Corrosion-induced damage is predicted for different levels of corrosion and subsequently the influence of the corrosion-induced damage on the pull-out capacity of reinforcement bar is investigated.

The numerical analysis of the beam-end specimens reinforced with stirrups shows that the model is able to realistically predict the corrosion-induced damage. The increase of the average crack width with the ongoing corrosion penetration of the rebar is verified by comparing the results with the experimental data. Furthermore, the model realistically accounts for the influence of stirrups on the bond resistance with the ongoing corrosion damage. It is shown that the expansion of the corrosion products around the rebar activates the tensile forces in the stirrups. Therefore, the additional confinement through the stirrups minimises the reduction of the pull-out capacity.

By comparing the two investigated cases (φ12/20 and φ16/35 mm), it can be concluded that, for a constant stirrups diameter, the effect of the confinement conditions depends on the depth of the concrete cover and on the size of the rebar. For smaller cover and smaller bar diameter, the confinement effect is reached sooner with the ongoing corrosion process and hence a lower decrease of the pull-out capacity is obtained.

Acknowledgement

The authors are grateful for the financial supports of "Deutsche Forschungsemeinschaft" (DFG, Grant Nr. 601984) and "Croatian Science Foundation" (HRZZ, Grant Nr. 9068).

Disclosure statement

No potential conflict of interest was reported by the authors.

References

Andrade, C., Díez, J. M., & Alonso, C. (1997). Mathematical modeling of a concrete surface "Skin Effect" on diffusion in chloride contaminated media. *Advanced Cement Based Materials, 6*, 39–44.

Apostolopoulos, C. A., & Papadakis, V. G. (2008). Consequences of steel corrosion on the ductility properties of reinforcement bar. *Construction and Building Materials, 22*, 2316–2324.

Balabanić, G., Bićanić, N., & Đureković, A. (1996a). The influence of w/c ratio, concrete cover thickness and degree of water saturation on the corrosion rate of reinforcing steel in concrete. *Cement and Concrete Research, 26*, 761–769.

Balabanić, G., Bićanić, N., & Đureković, A. (1996b). Mathematical modeling of electrochemical steel corrosion in concrete. *Journal of Engineering Mechanics, 122*, 1113–1122.

Bažant, Z. P. (1979). Physical model for steel corrosion in concrete sea structures – Theory. *Journal of the Structural Division, 105*, 1137–1153.

Bažant, Z. P., & Oh, B. H. (1983). Crack band theory for fracture of concrete. *Materials and Structures, 93*, 155–177.

Bear, J., & Bachmat, Y. (1990). *Introduction to modeling of transport phenomena in porous media* (Vol. 4). Dordrecht: Springer.

Belytschko, T., Liu, W. K., & Moran, B. (2001). *Nonlinear finite elements for continua and structures*. Chistester: Wiley.

Cairns, J. (1998). *State of the art report on bond of corroded reinforcement* (Tech Rep No. CEB-TG-2/5). Lausanne: CEB.

Cairns, J., Plizzari, G. A., Du, Y., Law, D. W. & Franzoni, C. (2005). Mechanical properties of corrosion-damaged reinforcement. *ACI Materials Journal, 102*, 256–264.

Chana, P. S. (1990). A test method to establish realistic bond stresses. *Magazine of Concrete Research, 42*, 83–90.

Fischer, C. (2012). *Auswirkungen der Bewehrungskorrosion auf den Verbund zwischen Stahl und Beton* [The influence of the corrosion of steel reinforcement on the bond between steel and concrete]. Stuttgart: Universität Stuttgart.

Glass, G. K., & Buenfeld, N. R. (1995). Chloride threshold levels for corrosion induced deterioration of steel in concrete. *Chloride penetration into concrete: Proceedings of the RILEM international workshop*, Saint-Remy-Les-Chevreuse, France. (pp. 429–440).

Glass, G. K., & Buenfeld, N. R. (2000). The influence of chloride binding on the chloride induced corrosion risk in reinforced concrete. *Corrosion Science, 42*, 329–344.

Glasstone, S. (1964). An introduction to electrochemical behaviour of steel in concrete. *ACI Journal, 61*, 177–188.

Ishida, T., Iqbal, P. O., & Anh, H. T. L. (2009). Modeling of chloride diffusivity coupled with non-linear binding capacity in sound and cracked concrete. *Cement and Concrete Research, 39*, 913–923.

Jacob, A. (2008). Composite bridge decks cut life cycle costs. *Reinforced Plastics, 52*, 30–32.

Leech, C., Lockington, D., & Dux, P. (2003). Unsaturated diffusivity functions for concrete derived from NMR images. *Materials and Structures, 36*, 413–418.

Marsavina, L., Audenaert, K., De Schutter, G., Faur, N., & Marsavina, D. (2009). Experimental and numerical determination of the chloride penetration in cracked concrete. *Construction and Building Materials, 23*, 264–274.

Martin-Perez, B. (1999). *Service life modelling of RC highway structures exposed to chlorides* (PhD thesis). University of Toronto, Canada.

Ožbolt, J., Balabanić, G., & Kušter, M. (2011). 3D Numerical modelling of steel corrosion in concrete structures. *Corrosion Science, 53*, 4166–4177.

Ožbolt, J., Balabanić, G., Periškić, G., & Kušter, M. (2010). Modelling the effect of damage on transport processes in concrete. *Construction and Building Materials, 24*, 1638–1648.

Ožbolt, J., Li, Y., & Kožar, I. (2001). Microplane model for concrete with relaxed kinematic constraint. *International Journal of Solids and Structures, 38*, 2683–2711.

Ožbolt, J., Oršanić, F., Balabanić, G., & Kušter, M. (2012). Modeling damage in concrete caused by corrosion of reinforcement: coupled 3D FE model. *International Journal of Fracture, 178*, 233–244.

Ožbolt, J., Oršanić, F., Kušter, M., & Balabanić, G. (2012). Modelling bond resistance of corroded reinforcement. *Proceedings of the International Conference Bond in Concrete, 2012*, 437–444.

Ožbolt, J., Oršanić, F., & Balabanić, G. (2014). Modeling pull-out resistance of corroded reinforcement in concrete: Coupled three-dimensional finite element model. *Cement and Concrete Composites, 46*, 41–55.

Page, C. L., Short, N. R., & El Tarras, A. (1981). Diffusion of chloride ions in hardened cement pastes. *Cement and Concrete Research, 11*, 395–406.

Page, C. L., & Treadaway, K. W. J. (1982). Aspects of the electrochemistry of steel in concrete. *Nature Publishing Group, 297*, 109–115.

Saetta, A. V., Scotta, R. V., & Vitaliani, R. V. (1993). Analysis of chloride diffusion into partially saturated concrete. *ACI Materials Journal, 90*, 441–451.

Sandberg, P., Peterson, K., Sarensen, H., & Arup, H. (1995). Critical chloride concentrations for the onset of active reinforcement corrosion. *Chloride penetration into concrete: Proceedings of the RILEM international workshop*, Saint-Remy-Les-Chevreuse, France. (pp. 435–439).

Tang, L., & Nilsson, L.-O. (1996). Service life prediction for concrete structures under seawater by a numerical approach. In C. Sjostrom (Ed.), *Durability of building materials and components 7* (Vol. 1, pp. 97–106). London: E & FN Spon.

Thomas, M. D., & Bamforth, P. B. (1999). Modelling chloride diffusion in concrete: effect of fly ash and slag. *Cement and Concrete Research, 29*, 487–495.

Tuutti, K. (1993). The effect of individual parameter on chloride induced corrosion. *Chloride Penetration into Concrete Structures*, 18–25.

Wong, H. S., Zhao, Y. X., Karimi, A. R., Buenfeld, N. R., & Jin, W. L. (2010). On the penetration of corrosion products from reinforcing steel into concrete due to chloride-induced corrosion. *Corrosion Science, 52*, 2469–2480.

Zhang, T., & Gjørv, O. E. (1996). Diffusion behavior of chloride ions in concrete. *Cement and Concrete Research, 26*, 907–917.

Efficient multi-objective optimisation of probabilistic service life management

Sunyong Kim and Dan M. Frangopol

ABSTRACT

The inspection and maintenance plans to ensure the structural safety and extend the service life of deteriorating structures can be established effectively through an optimisation process. When several objectives are required for inspection and maintenance strategies, a multi-objective optimisation process needs to be used in order to consider all objectives simultaneously and to rationally select a well-balanced solution. However, as the number of objectives increases, additional computational efforts are required to obtain the Pareto solutions, for decision-making to select well-balanced solutions, and for visualisation of the solutions. This paper presents a novel approach to multi-objective optimisation process of probabilistic service life management with four objectives: minimising the damage detection delay, minimising the probability of failure, maximising the extended service life and minimising the expected total life-cycle cost. With these four objectives, the single, bi-, tri- and quad-objective optimisation processes are investigated using the weighted sum method and genetic algorithms. The objective reduction approach with the Pareto optimal solutions is applied to estimate the degree of conflict among the objectives, and to identify the redundant objectives and minimum essential objective set. As a result, the efficiency in decision-making and visualisation for service life management can be improved by removing the redundant objectives.

Introduction

The structural performance of a deteriorating structure can be improved by the application of appropriate inspection and maintenance actions (IAEA, 2002; Liang, Lin, & Liang, 2002; NCHRP, 2006; Stewart & Rosowsky, 1998). Efficient and effective inspection and maintenance planning can be made based on an optimisation process (Onoufriou & Frangopol, 2002; Kong & Frangopol, 2004; Luki & Cremona, 2001; Miyamoto, Kawamura, & Nakamura, 2000). In the last few decades, significant efforts have been made in the field of the optimum inspection and maintenance planning of deteriorating structures (Frangopol, 2011; Frangopol & Soliman, 2016). The formulation of optimisation for this planning requires damage occurrence and propagation prediction, and the estimation of the effects of inspection and maintenance on the service life of a deteriorating structure under uncertainty (Ellingwood & Mori, 1997; Madsen, Torhaug, & Cramer, 1991; Soliman & Frangopol, 2014). The objectives of minimising the total life-cycle cost, maximising the service life, maximising the structural performance indicators such as reliability index, availability and redundancy during the service life, and minimising the damage detection delay have been applied to determine the optimum inspection and maintenance planning (Frangopol, Lin, & Estes, 1997; Furuta, Kameda, Nakahara, Takahashi, & Frangopol, 2006; Kim & Frangopol, 2011a, 2011c, Kim, Frangopol, & Zhu, 2011; Liu, Li, Huang, & Yuanhui, 2009; Okasha & Frangopol, 2010).

Inspection and maintenance planning are functions of the optimisation problem. For example, the inspection and maintenance planning associated with the minimisation of the total life-cycle cost will not be the same as the plan associated with the minimisation of the probability of failure during the service life or the maximisation of the expected extended service life of a structure. Practically, structural managers will need to make a decision on which objectives need to be used, or how all the possible objectives will be considered in a rational way because practical structural management might require only one plan for inspection and maintenance of a single structural system (Brockhoff & Zitzler, 2009). The multi-objective optimisation provides multiple trade-off solutions for the decision-makers, and one of the obtained solutions can be selected using higher level information to be considered (Deb, 2001; Fonseca & Fleming, 1998). Most existing studies on optimum inspection and maintenance planning have been based only on a few (i.e. one, two or three) objectives (Frangopol & Kim, 2014; Frangopol & Liu, 2007). The development of new concepts and approaches for optimum inspection and maintenance planning can increase the number of objectives to be optimised simultaneously. In general, optimisation with a large number of objectives involves difficulties with respect to computation, decision-making and visualisation (Saxena, Duro, Tiwari, Deb, & Zhang, 2013; Verel, Liefooghe, Jourdan, & Dhaenens, 2011). Recently, objective reduction approaches to address these difficulties have been developed

Table 1. Objectives, design variables, required estimations and given conditions in optimum service life management.

Objective	Design variables	Required estimations	Given conditions
Minimisation of expected damage detection delay $E(t_{\text{delay}})$	Optimum inspection times	E① and E②	C① and C②
Minimisation of probability of failure P_{fail}	Optimum inspection times	E① and E②	C① and C②
Maximisation of expected extended service life $E(t_{\text{life}})$	Optimum inspection and maintenance times	E①, E② and E③	C①, C② and C③
Minimisation of expected total life-cycle cost C_{lcc}	Optimum inspection and maintenance times	E①, E②, E③ and E④	C①, C② and C③

Notes: E① = Damage occurrence/propagation under uncertainty.
E② = Relation between degree of damage and probability of damage detection of inspection method.
E③ = Effect of available maintenance on service life extension.
E④ = Life-cycle cost over time.
C① = Number and quality of inspections.
C② = Available inspection types.
C③ = Available maintenance types and damage criteria to determine the types of maintenance.

by Deb and Saxena (2006), Brockhoff and Zitzler (2009), and Singh, Isaacs, and Ray (2011), among others.

This paper deals with the efficient multi-objective optimisation of probabilistic service life management associated with four objectives: minimising the damage detection delay, minimising the probability of failure, maximising the extended service life and minimising the expected total life-cycle cost. With these objectives, the single, bi-, tri- and quad-objective optimisation processes are investigated based on both the preference-based and Pareto front-based approaches represented by the weighted sum method and genetic algorithms (GA), respectively. In the Pareto front-based approach, the objective reduction approach is applied to estimate the degree of conflict among the objectives, and to identify both the redundant objectives and the minimum essential objective set. As a result, the efficiency in decision-making and visualisation for service life management is significantly improved by removing the redundant objectives. In this paper, the multi-objective optimisation of service life management is applied to an existing reinforced concrete bridge under corrosion.

Objectives for optimum service life management

The optimum service life management of a deteriorating structure is based on objectives related to structural performance, total life-cycle cost and service life (Fu & Frangopol, 1990; Frangopol & Kim, 2014; Frangopol & Liu, 2007). In this study, minimisation of the expected damage detection delay, minimisation of the probability of failure, maximisation of extended service life and minimisation of expected total life-cycle cost are used as the objectives for optimum service life management. Table 1 shows the required estimations and conditions for optimisation formulations with these four objectives. As indicated, if only the optimum inspection plan needs to be found, and the probabilistic models are available (which represent damage occurrence and propagation over time and quantify the relation between the degree of damage and probability of damage detection of an inspection method), then the optimisation can be performed with the objective of minimising the expected damage detection delay or minimising the probability of failure. Furthermore, in the case where optimum inspection and maintenance planning needs to be addressed, considering the effect of maintenance on service life or life-cycle cost of a deteriorating structure, managers can use the objective of maximising the expected service life or minimising the expected total life-cycle cost. Detailed formulations of these four objectives are summarised in this section.

Expected damage detection delay

The damage detection delay, defined as the time-lapse from damage occurrence to the time for the damage to be detected, can be used to establish the optimum inspection plan. Maintenance actions generally follow inspection, and the type of maintenance/repair depends on the inspection results. Therefore, a reduced damage detection delay can lead to effective and timely maintenance actions. If the damage occurrence time is known, and inspection methods are perfect, there will be no damage detection delay. However, due to the uncertainties associated with damage occurrence/propagation and inspection methods, damage detection delay cannot be avoided. Considering these uncertainties, the expected damage detection delay was formulated and applied to deteriorating structures subjected to corrosion and fatigue (Kim & Frangopol, 2011a, 2011b).

When N_{insp} inspections are applied to detect damage, the expected damage detection delay $E(t_{\text{delay}})$ is formulated as:

$$E(t_{\text{delay}}) = \sum_{i=1}^{N_{\text{insp}}+1} \left(\int_{t_{\text{insp}, i-1}}^{t_{\text{insp}, i}} \left[t_{\text{delay}, i} \cdot f_T(t) \right] dt \right) \quad (1)$$

where $t_{\text{delay}, i}$ = damage detection delay for the damage to occur in the time interval $t_{\text{insp}, i-1} \leq t < t_{\text{insp}, i}$; $t_{\text{insp}, i}$ = ith inspection time; and $f_T(t)$ = probability density function (PDF) of the damage occurrence time t. The damage detection delay $t_{\text{delay}, i}$ for a given damage occurrence time of $t_{\text{insp}, i-1} \leq t < t_{\text{insp}, i}$ is:

$$t_{\text{delay}, i} = \sum_{k=i}^{N_{\text{insp}}+1} \left(\prod_{j=1}^{k} \left(1 - P_{\text{insp}, j-1} \right) \right) \cdot P_{\text{insp}, k} \cdot \left(t_{\text{insp}, k} - t \right) (2)$$

where $P_{\text{insp}, i}$ = probability of damage detection of the ith inspection, $P_{\text{insp}, 0} = 0$ for $j = 1$, and $P_{\text{insp}, N_{\text{insp}}+1} = 1.0$ for $k = N_{\text{insp}+1}$. This is the conditional probability that damage is detected at the jth inspection when the damage intensity at time t is $\delta(t)$.

The probability of damage detection P_{insp} can be expressed based on the normal cumulative distribution function (CDF) as (Frangopol et al., 1997):

$$P_{\text{insp}} = \Phi \left(\frac{\delta(t) - \delta_{.5}}{\sigma_\delta} \right) \quad (3)$$

where $\Phi(\cdot)$ is the standard normal CDF, and σ_δ is the standard deviation (SD) of the damage intensity δ. The damage intensity δ ranges from 0 (i.e. no damage) to 1.0 (i.e. full

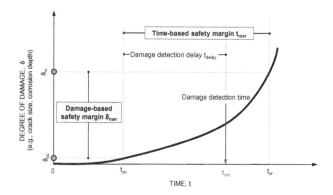

Figure 1. Damage-based and time-based safety margins.

damage), and can be estimated in terms of crack size or corrosion depth. σ_δ is assumed to be $.1\delta_{.5}$ herein, where $\delta_{.5}$ is the damage intensity at which the inspection method has a probability of detection of .5. The quality of an inspection method is represented by $\delta_{.5}$. The times $t_{\text{insp, }0}$ and $t_{\text{insp, }Ninsp + 1}$ in Equation (1) are defined as:

$$t_{\text{insp, }0} = F_T^{-1}(\Phi(-u)) \tag{4a}$$

$$t_{\text{insp, }N_{\text{insp}}+1} = F_T^{-1}(\Phi(u)) \tag{4b}$$

where $F_T^{-1}(\cdot)$ = inverse CDF of the damage occurrence time t, and $u > 0$, and $t_{\text{insp, }0}$ and $t_{\text{insp, }Ninsp + 1}$ serve as the lower and upper bounds of damage occurrence time t, respectively. These bounds can be determined, assuming that the probabilities of damage occurrence and damage detection are both very small before the lower bound $t_{\text{insp, }0}$, and that these probabilities are very close to one near the upper bound $t_{\text{insp, }Ninsp + 1}$. For example, if the damage occurrence time t is represented by the normal distribution with the mean of 5 years and SD of 1 year, and u in Equations (4a) and (4b) is 3.0, $t_{\text{insp, }0}$ and $t_{\text{insp, }Ninsp + 1}$ are 2 and 8 years, respectively (i.e. $F_T^{-1}(\Phi(-3)) = 2$, $F_T^{-1}(\Phi(3)) = 8$). The probabilities that the damage will occur before 2 and 8 years are .0013 and .9987, respectively (i.e. $F_T(2) = .0013$, $F_T(8) = .9987$). In this study, u is assumed to be 3.0 (Kim & Frangopol, 2011a). The expected damage detection delay $E(t_{\text{delay}})$ can be reduced by increasing the number of inspections and/or inspection quality. Detailed application and results regarding minimising the expected damage detection delay for the optimum inspection planning can be found in Kim and Frangopol (2011a, 2011b).

Time-based safety margin and probability of failure

During the evolution of the damage process, the safety margin of deteriorated structures can be expressed by both damage intensity and time. The safety margin based on the damage intensity δ_{mar} is expressed as:

$$\delta_{\text{mar}} = \delta_{cr} - \delta_{oc} \tag{5}$$

where δ_{cr} = critical damage intensity that can lead to structural failure; and δ_{oc} = damage intensity when the damage can be detected. A time-based safety margin t_{mar} is defined as:

$$t_{\text{mar}} = t_{cr} - t_{oc} \tag{6}$$

where t_{cr} = time when the damage intensity reaches the critical damage (i.e. damage threshold) δ_{cr}; and t_{oc} = damage occurrence time.

As shown in Figure 1, the damage intensities δ_{cr} and δ_{oc} in Equation (5) are associated with t_{cr} and t_{oc}, respectively. The damage has to be detected before reaching δ_{cr}. This means that the damage detection delay $t_{\text{delay}} = t_{\text{det}} - t_{oc}$ should be less than t_{mar} of Equation (6), where t_{det} is the time for the damage to be detected. Accordingly, assuming that the appropriate and immediate maintenances are applied after damage is detected, the time-based state function g can be expressed as:

$$g = t_{cr} - t_{\text{det}} = t_{\text{mar}} - t_{\text{delay}} \tag{7}$$

The relations among the critical time t_{cr}, damage detection time t_{det}, time-based safety margin t_{mar} and damage detection delay t_{delay} are illustrated in Figure 1. Considering the uncertainties associated with damage detection and damage initiation/propagation, t_{cr}, t_{det} t_{mar} and t_{delay} in Equation (7) can be treated as random variables. The time-based probability of failure is defined as:

$$P_{\text{fail}} = P(t_{cr} - t_{\text{det}} \leq 0) = P(t_{\text{mar}} - t_{\text{delay}} \leq 0) \tag{8}$$

The inspection plan can be established based on the optimisation problem with the objective of minimising the time-based probability of failure P_{fail}. Increasing the number of inspections and/or inspection quality can lead to a reduction of P_{fail}.

Extended service life under uncertainty

The service life of a deteriorating structure is limited; it can be extended through appropriate and timely maintenance actions. The extended service life with inspection and maintenance is formulated using the decision tree model shown in Figure 2. Considering the damage occurrence time, inspection time, damage detection and maintenance types, the extended service life after the ith scheduled inspection $t_{\text{life, }i}$ is formulated as:

$$t_{\text{life, }i} = (1 - P_{\text{insp}}) \cdot t_{\text{life, }i-1} \\ + P_{\text{insp}} \cdot (t_{life, i-1} + t_{ex}) \quad \text{for } t_{\text{insp, }i} \leq t_{\text{life, }i-1} \tag{9a}$$

$$t_{\text{life, }i} = t_{\text{life, }i-1} \quad \text{for } t_{\text{insp, }i} > t_{\text{life, }i-1} \tag{9b}$$

where t_{ex} = service life extension after damage detection. If the damage is detected by the scheduled inspection at time $t_{\text{insp, }i}$, the damage intensity is measured by the in-depth inspection following the scheduled inspection, and a maintenance type is determined according to the damage intensity δ. Furthermore, if the damage is not detected or the inspection application time is later than the service life, the service life will not be extended (i.e. $t_{ex} = 0$) as shown in Equation (9b) and Figure 2. When two maintenance types are available and the service life extensions associated with these two maintenance types are $t_{ex, \text{I}}$ and $t_{ex, \text{II}}$, respectively, the service life extension t_{ex} is estimated as:

$$t_{ex} = 0 \quad \text{for } \delta(t_{\text{insp, }i}) \leq \delta_\text{I} \tag{10a}$$

$$t_{ex} = t_{ex, \text{I}} \text{ for } \delta_\text{I} < \delta(t_{insp,i}) \leq \delta_\text{II} \tag{10b}$$

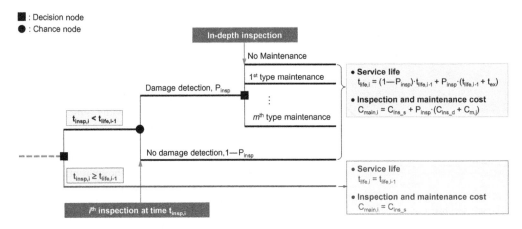

Figure 2. Decision tree for extended service life and the associated cost formulation.

$$t_{ex} = t_{ex,\,\mathrm{II}} \quad \text{for } \delta(t_{\mathrm{insp},\,i}) > \delta_{\mathrm{II}} \qquad (10c)$$

where δ_{I} and δ_{II} are the damage thresholds. In the case of $\delta(t_{\mathrm{insp},\,i}) \leq \delta_{\mathrm{I}}$, no maintenance is applied, and t_{ex} becomes zero.

Life-cycle cost analysis

The life-cycle cost analysis has been applied in various fields of engineering for effective and efficient inspection and maintenance management considering structural performance and service life under uncertainty (Frangopol, Lin et al., 1997; Frangopol, Saydam, & Kim, 2012). Decision-makers can allocate the optimum financial resources and establish the optimum inspection and maintenance actions while minimising the life-cycle cost during the service life of a deteriorating structure (Kong & Frangopol, 2003). The life-cycle cost analysis requires evaluating and integrating the initial construction cost C_{ini}, total inspection and maintenance cost C_{main} and the expected failure cost C_{fail} of deteriorating structures. The expected total life-cycle cost C_{lcc} during the target lifetime t^{*}_{life} can be expressed as (Chang & Shinozuka, 1996; Enright & Frangopol, 1999; Frangopol et al., 1997; Thoft-Christensen & Sørensen, 1987; Wallbank, Tailor, & Vassie, 1999):

$$C_{\mathrm{lcc}} = C_{\mathrm{ini}} + C_{\mathrm{main}} + C_{\mathrm{fail}} \qquad (11)$$

Using the decision tree shown in Figure 2, the inspection and maintenance cost $C_{\mathrm{main},\,i}$ associated with ith inspection time $t_{\mathrm{insp},\,i}$ is:

$$C_{\mathrm{main},\,i} = C_{\mathrm{ins\_s}} + P_{\mathrm{insp}}(C_{\mathrm{ins\_d}} + C_{m,\,i}) \quad \text{for } t_{\mathrm{insp},\,i} \leq t_{\mathrm{life},\,i-\mathrm{I}} \quad (12a)$$

$$C_{\mathrm{main},\,i} = C_{\mathrm{ins\_s}} \quad \text{for } t_{\mathrm{insp},\,i} > t_{\mathrm{life},\,i-\mathrm{I}} \qquad (12b)$$

where $C_{\mathrm{ins\_s}}$ = cost of the scheduled inspection; $C_{\mathrm{ins\_d}}$ = cost of the in-depth inspection; and $C_{m,\,i}$ = maintenance cost. As mentioned previously, a maintenance type is determined according to the damage intensity δ measured by the in-depth inspection following the scheduled inspection, and the maintenance cost $C_{m,\,i}$ in Equation (12a) is expressed in a similar way when the service life extension t_{ex} is estimated as indicated in Equations (10a)–(10c).

When the number of scheduled inspections N_{insp} is available, the total inspection and maintenance cost C_{main} in Equation (11) is:

$$C_{\mathrm{main}} = \sum_{i=1}^{N_{\mathrm{insp}}} C_{\mathrm{main},\,i} \qquad (13)$$

Furthermore, the expected failure cost C_{fail} in Equation (11) is computed as:

$$C_{\mathrm{fail}} = C_f P_{\mathrm{fail}} \qquad (14)$$

where C_f is the expected monetary loss associated with structural failure. P_{fail} is the probability that the extended service life $t_{\mathrm{life},\,i}$ after the ith inspection is less than the target lifetime t^{*}_{life}, computed as:

$$P_{\mathrm{fail}} = P(t_{\mathrm{life},\,i} \leq t^{*}_{\mathrm{life}}) \qquad (15)$$

Efficient multi-objective optimisation

Depending on the objective of the optimisation process, the inspection and maintenance planning can be varied. Only one inspection and maintenance strategy is able to be applied for a single structural system. For this reason, the structural managers should: (a) select the most appropriate objective among the possible objectives to establish the best inspection and maintenance plan through the single objective optimisation process, or (b) consider all the possible objectives simultaneously using the multi-objective optimisation approach so that the structural managers select the most well-balanced solution among the multiple trade-off solutions. In general, the methods used to solve the multi-objective optimisation problem are categorised into preference-based and Pareto front-based approaches, which are represented by the weighted sum method and evolution algorithm, respectively (Saxena et al., 2013; Verel et al., 2011).

Preference-based approach

The most common preference-based approach is the weight sum method, in which a relative preference vector is used to convert multiple objectives into a single objective. The weight sum method with M objective functions to be minimised can be formulated as (Arora, 2012):

$$\text{Minimise } f_w(\mathbf{x}) = \sum_{i=1}^{M} w_i f_i^{\text{norm}}(\mathbf{x}) \qquad (16)$$

where $\mathbf{x}$ = vector of design variables; f_i^{norm} = ith normalised objective function; and w_i = weight factor reflecting the relative preference among the objective set $\Phi_M = \{f_1, f_2, ..., f_M\}$ such that $\sum_{i=1}^{M} w_i = 1$ and $w_i \geq 0$. In order to estimate in a rational way the weight factor, fuzzy preference approach can used. More detailed information on this approach can be found in Parmee, Cvetkovic, Watson, and Bonham (2000). The normalised objective function $f_i^{\text{norm}}(\mathbf{x})$ is:

$$f_i^{\text{norm}}(\mathbf{x}) = \frac{f_i(\mathbf{x}) - f_i^{\min}}{f_i^{\max} - f_i^{\min}} \qquad (17)$$

where $f_i(\mathbf{x})$ is the ith objective function; and $f_i^{\max}$ and $f_i^{\min}$ are the maximum and minimum of all the $f_i(\mathbf{x})$ values, respectively. It is known that the weight sum method is easy to use even though the number of objectives is large. However, the complete Pareto optimal set cannot be obtained, even though the continuous weight factor w_i is applied (Arora, 2012; Deb, 2001).

Pareto front-based approach

In a Pareto front-based approach, a number of Pareto optimal solutions (i.e. Pareto front) are obtained (Jin & Sendhoff, 2008). Multi-objective genetic algorithms (MOGA) has been known as the most common and attractive Pareto front-based approach (Singh et al., 2011). The main reasons for using MOGA are: (a) it is not necessary to consider the convexity, concavity and/or continuity of the objective function and (b) the representative Pareto front can be obtained after a sufficient number of generations (Arora, 2012). More detailed information on MOGA is summarised in Deb (2001).

Recent studies on MOGA (Deb & Saxena, 2006; Purshouse & Fleming, 2007; Teytaud, 2007) show that increasing the number of objectives leads to high computational cost, low search efficiency and difficulty in visualisation and decision-making. With an increase in the number of objectives, the dimensionality of the Pareto front increases, the required number of points needed to represent the Pareto optimal front increases and the Pareto front searching ability of MOGA deteriorates. Even though the data used to represent the Pareto front are computed, visualising more than a three-dimensional (3D) Pareto front will be difficult for proper decision-making.

Objective reduction approach

In order to address the aforementioned difficulties, several objective reduction approaches have been developed recently. The objective reduction approach is the process to identify the essential and redundant objectives among the original objective set (Brockhoff & Zitzler, 2009; Deb & Saxena, 2006; Singh et al., 2011). The Pareto front is only affected by the essential objectives. The redundant objectives can be removed without changing the Pareto front obtained with the original objectives. In general, the objective reduction approach is used during the Pareto front search process (Deb & Saxena, 2006; Singh et al., 2011) or during the decision-making process as a posteriori (Brockhoff & Zitzler,

2009; Saxena et al., 2013). Two recent approaches for objective reduction are summarised in this section.

Correlation-based objective reduction approach

The multi-objective optimisation provides the Pareto front, only when the objectives of multi-criteria decision analysis conflict with each other (Deb, 2001). According to Carlsson and Fullér (1995), the relations between the two objectives of minimising both f_1 and f_2 are defined as:

(i) f_1 is in conflict with f_2 on $\mathbf{X}$ if $f_1(x_1) \geq f_1(x_2)$ implies $f_2(x_1) \leq f_2(x_2)$ for all $x_1, x_2 \in \mathbf{X}$.

(ii) f_1 supports f_2 on $\mathbf{X}$ if $f_1(x_1) \geq f_1(x_2)$ implies $f_2(x_1) \geq f_2(x_2)$ for all $x_1, x_2 \in \mathbf{X}$.

(iii) f_1 and f_2 are independent on $\mathbf{X}$ otherwise,

where $\mathbf{X}$ is a feasible solution set.

The correlation coefficient can be used to quantify the conflict among the objectives (Jaimes, Aguirre, Tanaka, & Coello, 2010). If the correlation coefficient between the objective functions f_1 and f_2 is negative, f_1 decreases while f_2 increases and vice versa. This means that f_1 and f_2 conflict with each other. On the other hand, a positive correlation coefficient indicates that f_1 and f_2 support each other because f_1 and f_2 decrease or increase simultaneously. The objective space of f_1 and f_2, and the possible Pareto front associated with the above cases of (i), (ii) and (iii) are illustrated in Figure 3. When f_1 supports f_2 with the correlation coefficient of 1.0 (see Figures 3(b)), a single optimum solution is available instead of the Pareto front. Therefore, in this case, f_1 or f_2 is a redundant objective, because the optimum solution is not changed, even though one of these objectives is removed. Based on this concept, the correlation-based objective reduction approaches, such as the principal component analysis (Deb & Saxena, 2006; Saxena et al., 2013) and objective partitioning method (Jaimes et al., 2010), have been developed and applied to search the Pareto front using MOGA.

Dominance relation-based objective reduction approach

The objective reduction approach developed by Brockhoff and Zitzler (2006, 2009) is based on the dominance relation among the objective values. In this approach, a solution $\mathbf{x}_1 \in \mathbf{X}$ is said to dominate another solution $\mathbf{x}_2 \in \mathbf{X}$ if $\mathbf{x}_1 \preceq_{\Phi^*} \mathbf{x}_2$ and $\mathbf{x}_2 \npreceq_{\Phi^*} \mathbf{x}_1$ for a particular objective function set $\Phi^* \subseteq \Phi_M^* := \{f_1, f_2, ..., f_M^*\}$, in which f_i is the ith objective function among M objective functions, $\mathbf{X}$ is the design space and $\subseteq$ denotes the subset. All the objective functions of Φ_M have to be minimised. The notation $\mathbf{x}_1 \preceq_{\Phi^*} \mathbf{x}_2$ is defined as:

$$\mathbf{x}_1 \preceq_{\Phi^*} \mathbf{x}_2 : \Leftrightarrow \forall f_i \in \Phi^* : f_i(\mathbf{x}_1) \leq f_i(\mathbf{x}_2) \qquad (18)$$

If no solution can be found that dominates $\mathbf{x}^* \in \mathbf{X}$, the solution $\mathbf{x}^*$ is called the Pareto optimal. An objective of minimising f_i is considered to be redundant and non-conflicting with the other objectives in a given objective set Φ_M, when the objective function f_i can be removed without changing the dominance relation among the objective values. Since the dominance relation is represented by the Pareto front, it can be said that the Pareto front remains unchanged, when the redundant objectives are removed (Brockhoff & Zitzler, 2006, 2009).

(a)

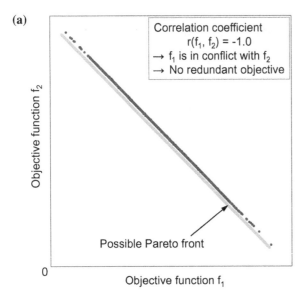

(b)

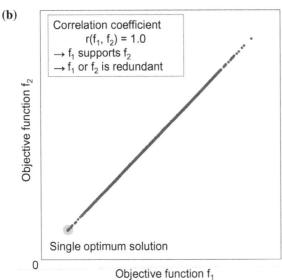

(c)

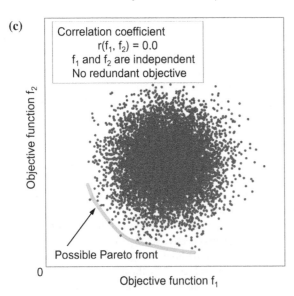

Figure 3. Objective space and Pareto front for minimising f_1 and f_2: (a) conflict; (b) support; (c) independence.

The dominance relation-based objective reduction approach is useful for investigating the effect of adding or removing a particular objective set $\Phi^* \subseteq \Phi_M$ on the Pareto front, to estimate the degree of conflict between Φ^* and Φ_M, and to find the redundant objectives and minimum essential objective set. The degree of conflict between Φ^* and Φ_M is expressed as the extent Δ to which the Pareto front changes when Φ^* is removed. Δ needs to be normalised with objectives that have various units and orders of magnitudes. The normalised degree of conflict is denoted as Δ_{norm}. The value of Δ_{norm} ranges from 0 (i.e. non-conflict) to 1.0 (i.e. full-conflict). In this paper, the algorithms associated with the dominance relation-based reduction approach, which are provided by Brockhoff and Zitzler (2006, 2009), are applied for an efficient decision-making process with the Pareto optimal solution set obtained from MOGA.

Application to an existing highway bridge

The approach for efficient service life management using multi-objective optimisation is applied to the I-39 Northbound Bridge over the Wisconsin River. More information on this bridge is given in Mahmoud, Connor, and Bowman (2005) and Kim et al. (2011). This application focuses on the corrosion of the top transverse reinforcement bars of the interface between spans, where the maximum negative moment can occur as indicated in Kim et al. (2011).

Initial service life estimation

The corrosion initiation and propagation under uncertainty are estimated based on both Fick's second law (Crank, 1975) and the pitting corrosion model (Val & Melchers, 1997). For a given corrosion rate r_{corr} (mm/year), the maximum pit depth $PT(t)$ at time t is expressed as (Val & Melchers, 1997):

$$PT(t) = r_{corr}R(t - t_{oc}) \quad \text{for } t > t_{oc} \qquad (19)$$

where R = ratio of maximum pit depth to average pit depth; t_{oc} = corrosion initiation time (mm/year). Moreover, the damage intensity $\delta(t)$ based on the pitting corrosion model is (Kim et al., 2011):

$$\delta(t) = \frac{PT(t)}{d_0} \qquad (20)$$

where d_0 (mm) = initial diameter of a reinforcement bar. In this paper, the initial service life is considered to be the time when the maximum pit depth PT reaches the allowable maximum pit depth PT_{allow} = 4.43 mm. The random variables to estimate the initial service life are provided in Table 2. Using Equation (19) and Monte Carlo simulation with 10^6 samples, the initial service life $t_{life, 0}$ when the PT reaches PT_{allow} is computed with a mean and SD of 22.14 and 4.91 years, respectively (denoted as LN (22.14; 4.91 years), where LN denotes lognormal distribution.

Formulation of objective functions

In this study, four objectives are used: (a) minimising the expected damage detection delay $f_1 = E(t_{delay})$; (b) minimising the time-based probability of failure $f_2 = P_{fail}$; (c) maximising the

Table 2. Variables for estimating initial service life.

Random variables	Notation and units	Mean	Coefficient of variation	Type of distribution
Corrosion initiation time	t_{oc} (years)	8.59	.27	Lognormal
Rate of corrosion	r_{corr} (mm/year)	.06	.3	Lognormal
Coefficient representing ratio between maximum and average corrosion penetrations	R	6	.1	Normal
Initial diameter of reinforcement	d_0	19.05	.02	Lognormal
Allowable maximum pit depth	PT_{allow} (mm)	4.43	–	Deterministic

Notes: Based on information provided in González, Andrade, Alonso, and Feliu (1995), Torres-Acosta and Martinez-Madrid (2003), and Kim, Frangopol, and Zhu (2011).

expected extended service life $f_3 = E(t_{life})$; and (d) minimising the expected total life-cycle cost $f_4 = C_{lcc}$. In this application, corrosion initiation is considered as the damage occurrence time. As defined in Equation (1), $E(t_{delay})$ can be formulated using the PDF $f_T(t)$ of corrosion initiation time in Table 2, probability of corrosion damage detection P_{insp} (see Equation (3)) and damage intensity (see Equation (20)). The objective function P_{fail} is defined in Equation (8), where the time-based safety margin t_{mar} is the difference between the time associated with the critical degree of corrosion damage t_{cr} and corrosion initiation time t_{oc} as shown in Figure 1. The initial service life $t_{life, 0} = t_{cr}$ is considered lognormal with a mean and SD of 22.14 and 4.91 years, respectively. The descriptors of random variable t_{oc} are provided in Table 2.

The formulation of $E(t_{life})$ is based on Equations (9a) and (9b) considering the uncertainty associated with the corrosion damage occurrence and propagation. In this application, it is assumed that when the damage intensity δ is between damage intensity bounds $\delta_I = .03$ and $\delta_{II} = .05$, corrosion protection is used, and the associated service life extension t_{ex} is treated as a lognormal random variable: LN (6; .5 years). If the damage intensity δ is larger than $\delta_{II} = .05$, the deck repair is applied, and the original structural performance is expected (i.e. $t_{ex} = t_{life, 0}$) (Kim, Frangopol, & Soliman, 2013).

Furthermore, in order to formulate the objective function of the expected total life-cycle cost C_{lcc} (see Equation (11)), it is assumed that the initial cost C_{ini}, in-depth inspection C_{ins} and expected monetary loss C_f are 1000, 20 and 10,000, respectively. The scheduled inspection cost C_{ins_s} in Equations (12a) and (12b) was estimated by Mori and Ellingwood (1994):

$$C_{ins\_s} = C_{ins}(1 - .7\delta_{.5})^{10} \qquad (21)$$

where $C_{ins} = 30$. As mentioned previously, the damage intensity $\delta_{.5}$ at which the given inspection method has 50% probability of damage detection represents the quality of inspection in this paper. The maintenance costs $C_{m, i}$ for corrosion protection and deck repair are assumed 60 and 300, respectively (Kim et al., 2013).

Correlation among objective functions

As mentioned previously, the coefficient of correlation can be used to quantify the conflict among objective functions. The coefficient of correlation between the objective functions f_i and f_j $r(f_i, f_j)$ is defined as:

$$r(f_i, f_j) = \frac{E\left[\{f_i(\mathbf{x}) - E(f_i(\mathbf{x}))\} \cdot \{f_j(\mathbf{x}) - E(f_j(\mathbf{x}))\}\right]}{\sigma_i \cdot \sigma_j} \qquad (22)$$

where $E(\cdot)$ denotes the expected value, $\mathbf{x}$ is the variable in the design space, and σ_i and σ_j are the SDs of $f_i(\mathbf{x})$ and $f_j(\mathbf{x})$, respectively. When $\delta_{.5}$, representing the quality of inspections, is .03, and the design space is associated with the time interval between the inspection times (i.e. 1 year $\leq t_{insp, i} - t_{insp, i-1} \leq 20$ years), the correlation coefficients among the four objective functions of $f_1 = E(t_{delay}), f_2 = P_{fail}, f_3 = E(t_{life})$ and $f_4 = C_{lcc}$ are computed as listed in Table 3. Considering the two objective functions of f_1 and f_2, the correlation coefficients for one, two or three inspections, $N_{insp} = 1, 2$ and 3, are .939, .926 and .929, respectively.

Single-objective optimisation

The single-objective optimisation problem with f_1, f_2, f_3 or f_4 is formulated as:

$$\text{Find } t_{insp} = \{t_{insp, 1}, t_{insp, 2}, \cdots, t_{insp, N_{insp}}\} \qquad (23a)$$

$$\text{for minimising } f_1 = E(t_{delay}) \qquad (23b)$$

$$\text{or for minimising } f_2 = P_{fail} \qquad (23c)$$

$$\text{or for maximising } f_3 = E(t_{life}) \qquad (23d)$$

$$\text{or for minimising } f_4 = C_{lcc} \qquad (23e)$$

$$\text{such that } 1 \text{ year} \leq t_{insp, i} - t_{insp, i-1} \leq 20 \text{ years} \qquad (23f)$$

$$\text{given } N_{insp} = 2 \text{ and } \delta_{.5} = .03 \qquad (23g)$$

The design variables of this problem are the inspection times t_{insp} (years). The time interval between inspections has to be between 1 year and 20 years (see Equation (23f)). The number of inspections $N_{insp} = 2$ and $\delta_{.5} = .03$ representing the quality of inspections are given as indicated in Equation (23g). The optimisation problem was solved using the global optimisation toolbox (i.e. GA) of MATLAB® version R2015b (MathWorks, 2015) with 200 generations.

Figure 4 depicts the optimum inspection times with objectives of Equations (23b)–(23e). If two inspections with $\delta_{.5} = .03$ are applied to minimise $E(t_{delay})$, the inspection must be performed at 10.13 and 13.12 years, and the corresponding $E(t_{delay})$ will be 3.65 years. Two inspections applied at 17.02 and 34.04 years

Table 3. Coefficients of correlation among objective functions.

		f_1	f_2	f_3	f_4
Number of inspections	f_1	1.0	.939	−.338	.931
$N_{insp} = 1$	f_2	.939	1.0	−.469	.954
	f_3	−.338	−.469	1.0	−.650
	f_4	.931	.954	−.650	1.0
Number of inspections	f_1	1.0	.926	−.074	.300
$N_{insp} = 2$	f_2	.926	1.0	−.172	.421
	f_3	−.074	−.172	1.0	−.794
	f_4	.300	.421	−.794	1.0
Number of inspections	f_1	1.0	.929	−.190	−.055
$N_{insp} = 3$	f_2	.929	1.0	−.323	.061
	f_3	−.190	−.323	1.0	−.786
	f_4	.055	.061	−.786	1.0

Notes: f_1 = expected damage detection delay $E(t_{delay})$.
f_2 = probability of failure P_{fail}.
f_3 = expected extended service life $E(t_{life})$.
f_4 = expected total life-cycle cost C_{lcc}.

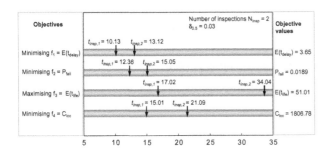

Figure 4. Inspection times based on single-objective optimisations.

result in the maximum expected extended service life $E(t_{life})$ of 51.01 years.

Bi-objective optimisation

The inspection planning can be formulated as the bi-objective optimisation consisting of three combinations of two objectives among the four objective functions f_1, f_2, f_3 and f_4 (i.e. $\{f_1, f_2\}$, $\{f_1, f_3\}$ and $\{f_1, f_4\}$). Such bi-objective optimisation for optimum inspection planning is:

$$\text{Find } t_{insp} = \{t_{insp,1}, t_{insp,2}, \dots, t_{insp,N_{insp}}\} \qquad (24a)$$

$$\text{for } \Phi_2 := \{f_1, f_2\} = \text{minimising both } E(t_{delay}) \text{ and } P_{fail} \qquad (24b)$$

$$\text{or for } \Phi_2 := \{f_1, f_3\} = \text{minimising } E(t_{delay}) \text{ and maximising } E(t_{life}) \qquad (24c)$$

$$\text{or for } \Phi_2 := \{f_1, f_4\} = \text{minimising both } E(t_{delay}) \text{ and } C_{lcc} \qquad (24d)$$

The design variables are the inspection times t_{insp}. The constraints and given conditions of this problem are the same as those in Equations (23f) and (23g).

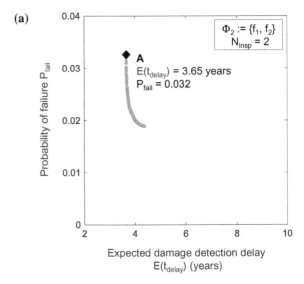

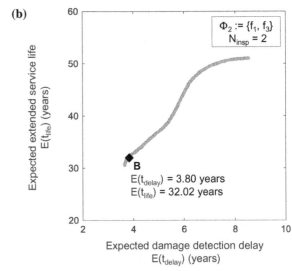

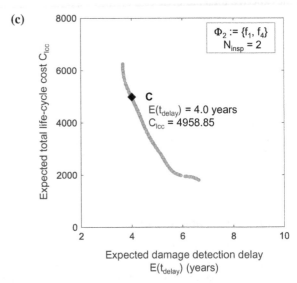

Figure 5. Pareto solution sets of bi-objective optimisation: (a) $\Phi_2 := \{f_1, f_2\}$; (b) $\Phi_2 := \{f_1, f_3\}$; (c) $\Phi_2 := \{f_1, f_4\}$.

Table 4. Design variable and objective function values associated with Pareto optimum solutions in Figures 5–8.

| | Objective function values | | | | Design variables | | | Given conditions | |
| | | | | | Optimum inspection times (years) | | | | |
Pareto optimum solution	Expected damage detection delay $E(t_{delay})$	Probability of failure P_{fail}	Expected extended service life (years) $E(t_{life})$	Expected total life-cycle cost C_{lcc}	$t_{insp,1}$	$t_{insp,2}$	$t_{insp,3}$	Number of inspections N_{insp}	Number of objectives to be considered
A	3.65	.032	–	–	10.13	13.12	–	2	2
B	3.80	–	32.02	–	10.26	14.30	–	2	2
C	4.00	–	–	4958.85	11.46	14.93	–	2	2
D	3.65	.032	30.55	–	10.13	13.12	–	2	3
E	4.82	.056	30.17	1497.66	11.84	–	–	1	4
F	3.65	.032	30.55	6251.06	10.13	13.12	–	2	4
G	3.13	.023	30.29	10,624.2	9.32	11.36	13.91	3	4

In this application, MOGA provided in MATLAB® version R2015b (MathWorks, 2015) is used to solve the multi-objective optimisation. After 300 generations with 100 populations, the Pareto optimal set are found as shown in Figure 5. Table 4 provides the values of objective functions and design variables of three representative solutions (A, B and C) in Figure 5. If the bi-objective optimisation with minimising both $E(t_{delay})$ and P_{fail} is considered for inspection planning, the Pareto optimal solutions are shown in Figure 5(a). Solution A leads to $E(t_{delay})$ = 3.65 years and P_{fail} = .032. The inspection plan for solution A requires two inspections with $\delta_{.5}$ = .03 at 10.13 and 13.12 years.

As indicated in Table 3, the correlation coefficient between these two objectives is .926. Consequently, these two objectives support each other, but are not perfectly correlated. Therefore, the Pareto set exists as shown in Figure 5(a), and f_1 or f_2 cannot be redundant. Furthermore, when the bi-objective optimisation is formulated with both f_1 and f_3, the Pareto optimal solutions in Figure 5(b) can be obtained. For Pareto point B in Figure 5(b), the associated mean values $E(t_{delay})$ and $E(t_{life})$ are 3.80 and 32.02 years, respectively. The Pareto optimal solutions including solution C for the bi-objective optimisation formulated with Equation (24d) can be found in Figure 5(c).

Tri-objective optimisation

By considering three objectives f_1, f_2 and f_3, the tri-objective optimisation is formulated as:

$$\text{Find } t_{insp} = \{t_{insp,1}, t_{insp,2}, \cdots, t_{insp,N_{insp}}\} \tag{25a}$$

for Φ_3: $= \{f_1, f_2, f_3\}$
$= $ minimising both $E(t_{delay})$ and P_{fail}, and maximising $E(t_{life})$
(25b)

By considering N_{insp} = 2 and $\delta_{.5}$ = .03 and the constraint associated with Equation (23f), the Pareto optimal set of 300 populations is obtained through the MOGA with 1000 generations as shown in Figure 6. Figure 6(a) illustrates this Pareto solution set in the 3D Cartesian coordinate system. Also, this solution set is visualised in the parallel coordinate system as shown in Figure 6(b). This parallel coordinate plot consists of three vertical axes indicating the values of $E(t_{delay})$, P_{fail} and $E(t_{life})$ of Figure 6(a).

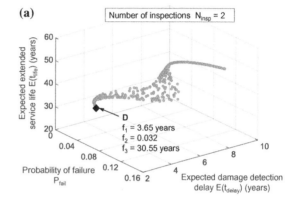

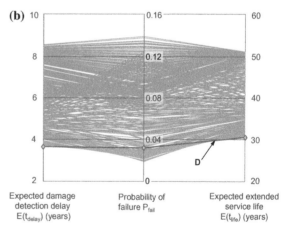

Figure 6. Pareto solution sets of tri-objective optimisation with Φ_3: $= \{f_1, f_2, f_3\}$ for N_{insp} = 2: (a) 3D Cartesian coordinate system; (b) parallel coordinate system.

Each point in Figure 6(a) is represented by a polyline connecting the corresponding value on the vertical axis in Figure 6(b).

For example, if solution D in Figure 6(a) is selected for optimum inspection planning, the values of $E(t_{delay})$, P_{fail} and $E(t_{life})$ are 3.65 years, .032 and 30.55 years, respectively, and the required inspection times can be found in Table 4. In Figure 6(b), the solution D is visualised as the polyline connecting these three values (i.e. $E(t_{delay})$ = 3.65 years, P_{fail} = .032 and $E(t_{life})$ = 30.55 years) on vertical axes. In the parallel coordinate plot, the number of objectives able to be visualised is not limited (Liebscher, Witowski, & Goel, 2009).

Table 5. Optimum solutions obtained using the weighted sum method for N_{insp} = 2.

Normalised weighting factor				Design variables (years)		Objective values			
w_1	w_2	w_3	w_4	$t_{insp, 1}$	$t_{insp, 2}$	$E(t_{delay})$ (years)	C_{fail}	$E(t_{life})$ (years)	C_{lcc}
1	0	0	0	10.134	13.119	3.654	.032	30.550	6251.056
.4	.2	.2	.2	12.732	16.010	4.685	.021	33.824	3693.087
0	1	0	0	12.358	15.051	4.391	.019	33.031	4262.287
.2	.4	.2	.2	13.164	16.117	4.967	.022	34.173	3298.787
0	0	1	0	17.020	34.039	8.612	.144	51.012	3025.736
.2	.2	.4	.2	14.850	29.700	6.925	.080	49.224	2523.269
0	0	0	1	15.009	21.093	6.643	.060	36.669	1806.777
.2	.2	.2	.4	14.134	25.334	6.314	.074	43.135	2070.887

(a)

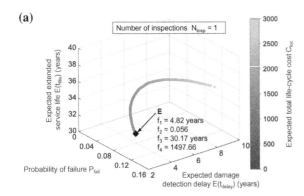

(b)

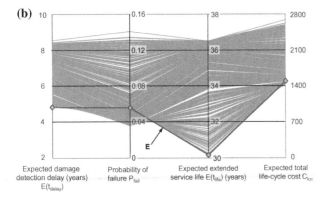

Figure 7. Pareto solution sets of quad-objective optimisation with $\Phi_4 := \{f_1, f_2, f_3, f_4\}$ for N_{insp} = 1: (a) 3D Cartesian coordinate system; (b) parallel coordinate system.

(a)

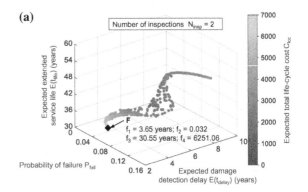

(b)

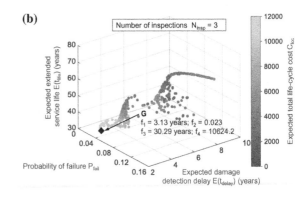

Figure 8. Pareto solution sets of quad-objective optimisation with $\Phi_4 := \{f_1, f_2, f_3, f_4\}$ in 3D Cartesian coordinate system: (a) N_{insp} = 2; (b) N_{insp} = 3.

Quad-objective optimisation

The four objectives of f_1, f_2, f_3 and f_4 are considered simultaneously as a quad-objective optimisation. This quad-objective optimisation is formulated as:

$$\text{Find } t_{insp} = \{t_{insp, 1}, t_{insp, 2}, \cdots, t_{insp, N_{insp}}\} \quad (26a)$$

$$\begin{aligned} \text{for } \Phi_4 := \{f_1, f_2, f_3, f_4\} \\ = \text{minimising both } E(t_{delay}) \text{ and } P_{fail}, \text{ maximising} \\ E(t_{life}), \text{ and minimising } C_{lcc} \end{aligned} \quad (26b)$$

$$\text{given } N_{insp} = 1, 2 \text{ or } 3 \text{ and } \delta_{.5} = .03 \quad (26c)$$

The constraint of this problem is provided by Equation (23f). The number of inspections and value of $\delta_{.5}$ are given in Equation (26c). In order to solve the quad-objective optimisation problem, both the weight sum method and MOGA are used.

Using the weight sum method with various normalised weighting factors, eight Pareto optimal solutions for the number of inspections N_{insp} = 2 are obtained as indicated in Table 5. When the normalised factor w_1 is equal to 1, only the objective function f_1 is taken into account for the quad-objective optimisation problem, and the associated application times of the two inspections are 10.13 and 13.12 years (see Table 5 and Figure 4).

MOGA, with the number of generations fixed at 1000 with 300 populations, is applied in order to find the Pareto optimal solution set. Because the dimensionality of the obtained Pareto solutions is four, the solutions are visualised using both the 3D Cartesian coordinate and parallel coordinate systems as shown in Figures 7 and 8. In the 3D Cartesian coordinate system, the x, y and z axes correspond to $E(t_{delay})$, P_{fail} and $E(t_{life})$, respectively,

Table 6. Normalised degree of conflict between the quad-objective set $\Phi_4 = \{f_1, f_2, f_3, f_4\}$ and the reduced objective set Φ_R.

Number of inspections $N_{insp} = 1$		Number of inspections $N_{insp} = 2$		Number of inspections $N_{insp} = 3$	
Reduced objective set Φ_R	Normalised degree of conflict Δ_{norm}	Reduced objective set Φ_R	Normalised degree of conflict Δ_{norm}	Reduced objective set Φ_R	Normalised degree of conflict Δ_{norm}
$\{f_1, f_2\}$	1.0	$\{f_1, f_2\}$	1.0	$\{f_1, f_2\}$	1.0
$\{f_1, f_3\}$	.0	$\{f_1, f_3\}$	.57	$\{f_1, f_3\}$	.64
$\{f_1, f_4\}$	1.0	$\{f_1, f_4\}$	.78	$\{f_1, f_4\}$	.64
$\{f_2, f_3\}$	.51	$\{f_2, f_3\}$	.51	$\{f_2, f_3\}$	.51
$\{f_2, f_4\}$	1.0	$\{f_2, f_4\}$	.78	$\{f_2, f_4\}$	.57
$\{f_3, f_4\}$	.21	$\{f_3, f_4\}$	1.0	$\{f_3, f_4\}$	1.0
$\{f_1, f_2, f_3\}$	.0	$\{f_1, f_2, f_3\}$	.19	$\{f_1, f_2, f_3\}$	.50
$\{f_1, f_2, f_4\}$	1.0	$\{f_1, f_2, f_4\}$	.78	$\{f_1, f_2, f_4\}$	.57
$\{f_1, f_3, f_4\}$	.0	$\{f_1, f_3, f_4\}$	.53	$\{f_1, f_3, f_4\}$	.64
$\{f_2, f_3, f_4\}$	.21	$\{f_2, f_3, f_4\}$	.50	$\{f_2, f_3, f_4\}$	.49

(a)

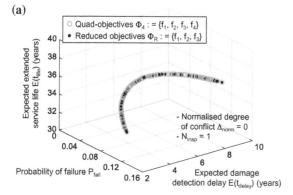

(b)

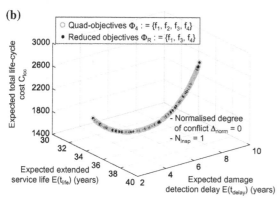

(c)

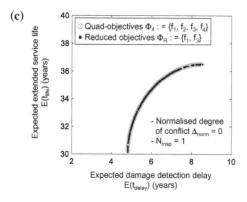

Figure 9. Pareto solution sets of the quad-objective and reduced objective optimisation problems with the normalised degree of conflict $\Delta_{norm} = 0$: (a) $\Phi_{R'} := \{f_1, f_2, f_3\}$; (b) $\Phi_{R'} := \{f_1, f_3, f_4\}$; (c) $\Phi_{R'} := \{f_1, f_3\}$.

and the value of C_{lcc} is also provided. Table 4 provides the values of design variables and objective functions associated with three representative solutions for $N_{insp} = 1$, 2 and 3 (i.e. E in Figure 7(a),

F in Figure 8(a), G in Figure 8(b)). It should be noted that the inspection application times of the three representative solutions A, D and F are identical, as indicated in Table 4, and these solutions result in the minimum $E(t_{delay})$ of Figure 4 for $N_{insp} = 2$ and $\delta_{.5} = .03$.

Table 6 provides the normalised degrees of conflict among the four objectives Δ_{norm}, which are estimated through the dominance relation-based approach. As shown, the normalised degree of conflict between quad-objective set $\Phi_4 := \{f_1, f_2, f_3, f_4\}$ and the reduced objective set $\Phi_R := \{f_1, f_2, f_3\}$ is zero (i.e. $\Delta_{norm} = 0$) for $N_{insp} = 1$. Figure 9(a) illustrates a comparison between the Pareto fronts of $\Phi_4 := \{f_1, f_2, f_3, f_4\}$ and $\Phi_R := \{f_1, f_2, f_3\}$. From this result, it can be seen that there will be no change in the Pareto front when the objective f_4 is removed from the quad-objective set Φ_4. The normalised degree of conflict Δ_{norm} between $\Phi_4 := \{f_1, f_2, f_3, f_4\}$ and $\Phi_R := \{f_1, f_3, f_4\}$ is zero, and the Pareto fronts associated with $\Phi_4 := \{f_1, f_2, f_3, f_4\}$ and $\Phi_R := \{f_1, f_3, f_4\}$ are compared in Figure 9(b). This means that f_2 has no effect on the Pareto front. Furthermore, Table 6 and Figure 9(c) show that, considering the relation between $\Phi_4 := \{f_1, f_2, f_3, f_4\}$ and $\Phi_R := \{f_1, f_3\}$, the normalised degree of conflict Δ_{norm} is zero, and there will be no change if the objectives f_2 and f_4 are omitted. It can be concluded that the objectives f_2 and f_4 are redundant, the minimum essential objective set $\Phi_R := \{f_1, f_3\}$ for $N_{insp} = 1$, and in the decision-making process and visualisation for service life management, only the minimum essential objectives of f_1 and f_3 can therefore be considered instead of the four objectives of f_1, f_2, f_3 and f_4.

For $N_{insp} = 2$ and 3, any normalised degree of conflict Δ_{norm} is not zero, as shown in Table 6. However, if the allowable normalised degree of conflict $\Delta_{norm, all}$ for $N_{insp} = 2$ is .2, the minimum essential objective set can be $\Phi_R := \{f_1, f_2, f_3\}$, and the Pareto solutions computed from the tri-objective optimisation problem associated with $\Phi_R := \{f_1, f_2, f_3\}$ can be used instead of the solutions from the quad-objective optimisation.

Conclusions

In this paper, an efficient multi-objective probabilistic optimisation of service life management of deteriorating structures is presented. Through the objective reduction approach, the degree of conflict among objectives is estimated, and the minimum essential objectives are identified to improve efficiency in decision-making and visualisation. Multi-objective optimisation of service life management is investigated using four objectives: minimising the expected damage detection delay, minimising the probability of failure, maximising the extended service life and minimising the expected total life-cycle cost. The optimum inspection plans resulting from single and multi-objective optimisation processes with these four objectives are computed using the weighted sum method and GAs, respectively.

Based on the presented study, the following conclusions are drawn:

(1) The correlation among objective functions can be used as an indicator to show how the objectives are interrelated. It should be noted that the correlation coefficient does not provide sufficient information to identify the redundant objectives or the minimum essential objectives. From the correlation results provided in this paper, it can be seen that the expected damage detection delay and the probability of failure are strongly interrelated. This is because the probability of failure is based on the state function, which is expressed by the difference between the time-based safety margin and damage detection delay, and the increase in damage detection delay leads to the increase in the probability of failure. For this reason, even though the damage propagation is slow, the expected damage detection delay and probability of failure can be strongly interrelated.

(2) In the case of a single inspection, minimising the probability of failure and minimising the expected total life-cycle cost are redundant objectives. Therefore, decision-makers can use the Pareto solution set associated with minimising the expected damage detection delay and maximising the extended service life. Therefore, the required dimensions of the Pareto front decreases from four to two.

(3) As the number of objectives for service life management of deteriorating structures increases, more rational and well-balanced solutions can be expected, but more effort is required for computation to obtain the Pareto solutions, for decision-making to select well-balanced solutions, and for the visualisation of the solutions. The objective reduction approach used in this paper focuses on improving the efficiency in decision-making and visualisation. Further research is needed to integrate the algorithms for improving both the ability to search the Pareto front and the efficiency in decision-making and visualisation.

(4) This paper is limited to the four objectives associated with damage detection, probability of failure, service life and life-cycle cost. However, additional objectives for optimum inspection and maintenance planning can be considered using other structural performance indicators for deteriorating structures such as redundancy, robustness, risk, resilience and sustainability.

Acknowledgements

The opinions presented in this paper are those of the authors and do not necessarily reflect the views of the sponsoring organisations.

Disclosure statement

No potential conflict of interest was reported by the authors.

Funding

This work was supported by the National Science Foundation (NSF) [Award CMMI-1537926]; the Commonwealth of Pennsylvania, Department of Community and Economic Development, through the Pennsylvania Infrastructure Technology Alliance (PITA); the U.S. Federal Highway Administration (FHWA) [Cooperative Agreement Award DTFH61-07-H-00040]; the U.S. Office of Naval Research (ONR) [Awards N00014-08-1-0188, N00014-12-1-0023, and N00014-16-1-2299]; the National Aeronautics and Space Administration (NASA) [Award NNX10AJ20G]; and the Wonkwang University.

References

Arora, J. S. (2012). *Introduction to optimum design* (3rd ed.). London: Elsevier.

Brockhoff, D., & Zitzler, E. (2006). *Dimensionality reduction in multiobjective optimization with (partial) dominance structure preservation: Generalized minimum objective subset problems* (TIK Report 247). Zurich: ETH Zurich.

Brockhoff, D., & Zitzler, E. (2009). Objective reduction in evolutionary multiobjective optimization: Theory and applications. *Evolutionary Computation, 17*, 135–166.

Carlsson, C., & Fullér, R. (1995). Multiple criteria decision making: The case for interdependence. *Computers & Operations Research, 22*, 251–260.

Chang, S. E., & Shinozuka, M. (1996). Life-cycle cost analysis with natural hazard risk. *Journal of Infrastructure Systems, 2*, 118–126.

Crank, J. (1975). *The mathematics of diffusion* (2nd ed.). Oxford: Oxford University Press.

Deb, K. (2001). *Multi-objective optimization using evolutionary algorithms*. New York, NY: Wiley.

Deb, K., & Saxena, D. (2006, July 16–21). Searching for Pareto-optimal solutions through dimensionality reduction for certain large-dimensional multi-objective optimization problems. *Proceedings of the IEEE congress on evolutionary computation (CEC2006)*, Vancouver, Canada.

Ellingwood, B. R., & Mori, Y. (1997). Reliability-based service life assessment of concrete structures in nuclear power plants: optimum inspection and repair. *Nuclear Engineering and Design, 175*, 247–258.

Enright, M. P., & Frangopol, D. M. (1999). Maintenance planning for deteriorating concrete bridges. *Journal of Structural Engineering, 125*, 1407–1414.

Fonseca, C. M., & Fleming, P. J. (1998). Multiobjective optimization and multiple constraint handling with evolutionary algorithms. I. A unified formulation, *IEEE Transactions on Systems, Man, and Cybernetics – Part A: Systems and Humans, 28*, 26–37.

Frangopol, D. M. (2011). Life-cycle performance, management, and optimization of structural systems under uncertainty: Accomplishments and challenges. *Structure and Infrastructure Engineering, 7*, 389–413.

Frangopol, D. M., & Kim, S. (2014). Life-cycle performance analysis and optimization, chapter 18. In Wai-Fah Chen & Lian Duan (Eds.), *Bridge engineering handbook* (2nd Ed., Vol. 5, pp. 537–566). London: CRC Press-Taylor & Francis Group.

Frangopol, D. M., Lin, K. Y., & Estes, A. C. (1997). Life-cycle cost design of deteriorating structures. *Journal of Structural Engineering, 123*, 1390–1401.

Frangopol, D. M., & Liu, M. (2007). Maintenance and management of civil infrastructure based on condition, safety, optimization, and life-cycle cost. *Structure and Infrastructure Engineering, 3*, 29–41.

Frangopol, D. M., Saydam, D., & Kim, S. (2012). Maintenance, management, life-cycle design and performance of structures and infrastructures: A brief review. *Structure and Infrastructure Engineering, 8*(1), 1–25.

Frangopol, D. M., & Soliman, M. (2016). Life-cycle of structural systems: Recent achievements and future directions. *Structure and Infrastructure Engineering, 12*(1), 1–20.

Fu, G., & Frangopol, D. M. (1990). Reliability-based vector optimization of structural systems. *Journal of Structural Engineering, 116*, 2143–2161.

Furuta, H., Kameda, T., Nakahara, K., Takahashi, Y., & Frangopol, D. M. (2006). Optimal bridge maintenance planning using improved multi-objective genetic algorithm. *Structure and Infrastructure Engineering, 2*, 33–41.

González, J. A., Andrade, C., Alonso, C., & Feliu, S. (1995). Comparison of rates of general corrosion and maximum pitting penetration on concrete embedded steel reinforcement. *Cement and Concrete Research, 25*, 257–264.

IAEA. (2002). *Safety and effective nuclear power plant life cycle management towards decommissioning* (IAEA-TECDOC-1305). Vienna: International Atomic Energy Agency.

Jaimes, A., Aguirre, H., Tanaka, K., & Coello, C. A. C. (2010). Objective space partitioning using conflict information for many-objective optimization. In Robert Schaefer, Carlos Cotta, Joanna Kołodziej, & Günter Rudolph (Eds.), *Parallel problem solving from nature, PPSN XI* (pp. 657–666). Berlin: Springer.

Jin, Y., & Sendhoff, B. (2008). Pareto-based multiobjective machine learning: An overview and case studies. *Systems, Man, and Cybernetics, Part C: Applications and Reviews, 38*, 397–415.

Kim, S., & Frangopol, D. M. (2011a). Optimum inspection planning for minimizing fatigue damage detection delay of ship hull structures. *International Journal of Fatigue, 33*, 448–459.

Kim, S., & Frangopol, D. M. (2011b). Inspection and monitoring planning for RC structures based on minimization of expected damage detection delay. *Probabilistic Engineering Mechanics, 26*, 308–320.

Kim, S., & Frangopol, D. M. (2011c). Cost-based optimum scheduling of inspection and monitoring for fatigue-sensitive structures under uncertainty. *Journal of Structural Engineering, 137*, 1319–1331.

Kim, S., Frangopol, D. M., & Soliman, M. (2013). Generalized probabilistic framework for optimum inspection and maintenance planning. *Journal of Structural Engineering, 139*, 435–447.

Kim, S., Frangopol, D. M., & Zhu, B. (2011). Probabilistic optimum inspection/repair planning to extend lifetime of deteriorating RC structures. *Journal of Performance of Constructed Facilities, 25*, 534–544.

Kong, J., & Frangopol, D. M. (2003). Evaluation of expected life-cycle maintenance cost of deteriorating structures. *Journal of Structural Engineering, 129*, 682–691.

Kong, J., & Frangopol, D. M. (2004). Cost–reliability interaction in life-cycle cost optimization of deteriorating structures. *Journal of Structural Engineering, 130*, 1704–1712.

Liang, M.-T., Lin, L.-H., & Liang, C.-H. (2002). Service life prediction of existing reinforced concrete bridges exposed to chloride environment. *Journal of Infrastructure Systems, 8*, 76–85.

Liebscher, M., Witowski, K., & Goel, T. (2009, June 1–5). Decision making in multiobjective optimization for industrial applications – Data mining and visualization of Pareto data. *Proceedings of the 8th world congress on structural and multidisciplinary optimization*, Lisbon, Portugal.

Liu, Y., Li, Y., Huang, H.-Z., & Yuanhui, K. (2009). An optimal sequential preventive maintenance policy under stochastic maintenance quality. *Structure and Infrastructure Engineering, 7*, 315–322.

Luki, M., & Cremona, C. (2001). Probabilistic optimization of welded joints maintenance versus fatigue and fracture. *Reliability Engineering & System Safety, 72*, 253–264.

Madsen, H. O., Torhaug, R., & Cramer E. H. (1991). Probability-based cost benefit analysis of fatigue design, inspection and maintenance.

Proceedings of the marine structural inspection, maintenance and monitoring symposium (pp. 1–12). Arlington, VA: SSC/SNAME.

Mahmoud, H. N., Connor, R. J., & Bowman, C. A. (2005). *Results of the fatigue evaluation and field monitoring of the I-39 Northbound Bridge over the Wisconsin River* (ATLSS Report No. 05-04). Bethlehem, PA: Lehigh University.

MathWorks. (2015). *Optimization Toolbox™ user's guide.* Natick, MA: Author.

Miyamoto, A., Kawamura, K., & Nakamura, H. (2000). Bridge management system and maintenance optimization for existing bridges. *Computer-Aided Civil and Infrastructure Engineering, 15*, 45–55.

Mori, Y., & Ellingwood, B. R. (1994). Maintaining reliability of concrete structures. II: Optimum inspection/repair. *Journal of Structural Engineering, 120*, 846–862.

NCHRP. (2006). *Manual on service life of corrosion-damaged reinforced concrete bridge superstructure elements* (NCHRP-Report 558). Washington, DC: Transportation Research Board, National Cooperative Highway Research Program.

Okasha, N. M., & Frangopol, D. M. (2010). Redundancy of structural systems with and without maintenance: An approach based on lifetime functions. *Reliability Engineering & System Safety, 95*, 520–533.

Onoufriou, T. & Frangopol, D. M. (2002). Reliability-based inspection optimization of complex structures: A brief retrospective. *Computers & Structures, 80*, 1133–1144.

Parmee, I. C., Cvetkovic, D., Watson, A. H., & Bonham, C. R. (2000). Multiobjective satisfaction within an interactive evolutionary design environment. *Evolutionary Computation, 8*, 197–222.

Purshouse, R. C., & Fleming, P. J. (2007). On the evolutionary optimization of many conflicting objectives. *Evolutionary Computation, 11*, 770–784.

Saxena, D. K., Duro, J. A., Tiwari, A., Deb, K., & Zhang, Q. (2013). Objective reduction in many-objective optimization: linear and nonlinear algorithms. *Evolutionary Computation, 17*, 77–99.

Singh, H. K., Isaacs, A., & Ray, T. (2011). A Pareto corner search evolutionary algorithm and dimensionality reduction in many-objective optimization problems. *Evolutionary Computation, 15*, 539–556.

Soliman, M., & Frangopol, D. M. (2014). Life-cycle management of fatigue-sensitive structures integrating inspection information. *Journal of Infrastructure Systems, 20*, 04014001.

Stewart, M. G., & Rosowsky, D. V. (1998). Time-dependent reliability of deteriorating reinforced concrete bridge decks. *Structural Safety, 20*, 91–109.

Teytaud, O. (2007). On the hardness of offline multi-objective optimization. *Evolutionary Computation, 15*, 475–491.

Thoft-Christensen, P., & Sørensen, J. D. (1987). Optimal strategy for inspection and repair of structural systems. *Civil Engineering and Environmental Systems, 4*, 94–100.

Torres-Acosta, A. A., & Martinez-Madrid, M. (2003). Residual life of corroding reinforced concrete structures in marine environment. *Journal of Materials in Civil Engineering, 15*, 344–353.

Val, D. V., & Melchers, R. E. (1997). Reliability of deteriorating RC slab bridges. *Journal of Structural Engineering, 123*, 1638–1644.

Verel, S., Liefooghe, A., Jourdan, L., & Dhaenens, C. (2011, January 17–21). Analyzing the effect of objective correlation on the efficient set of MNK-landscapes. *Proceedings of the 5th conference on learning and intelligent optimization (LION 5)*, Rome, Italy.

Wallbank, E. J., Tailor, P., & Vassie, P. R. (1999). Strategic planning of future maintenance needs. In P. C. Das (Ed.), *Management of highway structures* (pp. 163–172). London: Thomas Telford.

Probabilistic assessment of an interdependent healthcare–bridge network system under seismic hazard

You Dong and Dan M. Frangopol

ABSTRACT

Strong earthquakes can destroy infrastructure systems and cause injuries and/or fatalities. Therefore, it is important to investigate seismic performance of interdependent infrastructure systems and guarantee their abilities to cope with earthquakes. This paper presents an integrated probabilistic framework for the healthcare–bridge network system performance analysis considering spatial seismic hazard, vulnerability of bridges and links in the network, and damage condition of a hospital at component and system levels. The system level performance is evaluated considering travel and waiting time based on the damage conditions of the components. The effects of correlation among the seismic intensities at different locations are investigated. Additionally, the correlations associated with damage of the investigated structures are also incorporated within the probabilistic assessment process. The conditional seismic performance of the hospital given the damage conditions of the bridge network and the effect of bridge retrofit actions are also investigated. The approach is illustrated on a healthcare system located near a bridge network in Alameda, California.

1. Introduction

Strong earthquakes can destroy infrastructure systems and cause injuries and/or fatalities. Therefore, it is important to investigate seismic performance of healthcare systems to guarantee immediate medical treatment after earthquakes. The 1971 San Fernando, California, earthquake caused a severe damage to hospitals associated with about 53 million US$ loss in building damage and more than 1 million US$ loss in equipment damage (Murphy, 1973). The World Health Organization (WHO, 2007) stated that healthcare systems "must be physically resilient and able to remain operational and continue providing vital health services" after disasters. Thus, healthcare systems need to be resilient enough to cope with earthquakes and to provide timely medical treatment. In this paper, the seismic performance assessment of a healthcare system located near a bridge network is investigated considering both component and system performance levels.

The assessment of healthcare–bridge network system performance depends on the seismic vulnerability of bridges located in a bridge network and hospital, as well as on the ground motion intensity. After a destructive earthquake, the functionality of a highway network can be affected significantly; this, in turn, may lead to hinder the emergency management. Additional travel time would result due to the damaged bridges and links; consequently, injured persons may not receive treatment in time. Thus, it is important to account for the effects of damage condition associated with highway bridge network on the healthcare system performance. In this paper, the extra travel time of injured persons through the damaged bridge network to a hospital under the seismic hazard is investigated.

Myrtle, Masri, Nigbor, and Caffrey (2005) carried out a series of surveys on performance of hospitals during several earthquakes to identify the important components; Yavari, Chang, and Elwood (2010) investigated performance levels for interacting components (i.e. structural, nonstructural, lifeline and personnel) using data from past earthquakes; Achour, Miyajima, Kitaura, and Price (2011) investigated the physical damage of structural and nonstructural components of an hospital under seismic hazard; and Cimellaro, Reinhorn, and Bruneau (2011) introduced a model to describe the hospital performance under earthquake considering waiting time. However, the damage conditions associated with bridge networks have not been incorporated within the healthcare system performance assessment process. Additionally, the correlation effects have also not been addressed in these studies.

After an earthquake, it is common to experience a sudden increase in the number of patients for a period of time, which in turn can bring delay in treating them. The estimation of hospital capacity after an extreme event is of vital importance to determine the waiting time of the injured persons. Hospital functionality may be disrupted by damage associated with structural and nonstructural components or medical equipment. A proposed approach considering both structural and nonstructural components (e.g. medical equipment) is presented in this paper to investigate the hospital performance under a given seismic scenario. The relationship between structural and nonstructural seismic demands (e.g. peak inter story drift ratio and peak floor acceleration) is considered. Additionally, the correlations among the damages of structural and nonstructural components are also considered in the hospital functionality assessment process. Finally, the effects of correlation on the healthcare–bridge network system performance at a system level are investigated.

The performance of interdependent infrastructure systems under seismic hazard could be incorporated within the framework of life-cycle analysis of structural systems under extreme events. By considering the probability of occurrence of seismic hazard and structural deteriorations, the seismic performance of interdependent systems in a life-cycle context could be investigated. This will help to (a) promote the concept of interdependency in life-cycle design and assessment of structural systems under extreme events and (b) improve the risk-informed decision-making process for maintenance actions on interdependent infrastructures. It is clear that a life-cycle framework considering interdependent systems under seismic hazard is more realistic than the approach based on independent systems, since it involves the consideration of complex interaction processes among deteriorating structures during their lifetime. Overall, this study provides a novel approach to compute the seismic performance of interdependent healthcare–bridge network systems.

This paper aims to assess probabilistically an interdependent healthcare–bridge network system under seismic hazard and to aid the emergency preparedness to cope with the sudden increase of patients. The damage conditions of the bridges, links and hospital are considered in the overall system performance assessment. Fragility curves are employed to identify the components vulnerability under seismic hazard. The effects of disruption associated with transportation system on the emergency management are investigated. Additionally, the correlations among structural damages and the effect of bridge retrofit actions are considered in the assessment process. The system level performance indicators are expressed in terms of the extra travel and waiting time of the injured persons from the damaged region to the hospital given the occurrence of the earthquake. The approach is illustrated on a healthcare system located near a bridge network in Alameda, California.

2. Earthquake scenarios

The first step in seismic performance assessment of a healthcare system located near a bridge network is to identify the seismic scenarios at the location of the system. The seismic scenarios associated with an earthquake fault are introduced herein. The earthquake rupture is given as a characteristic magnitude distribution, modelled as a Gaussian distribution using the mean, a coefficient of variation of .12, and a truncation at ±2 standard deviations of magnitude above and below the mean (USGS, 2003). Generally, the mean magnitude M associated with an earthquake rupture can be computed as (Hanks & Bakun, 2002):

$$M = \begin{cases} 3.98 + \log_{10}(A_F) & A_F \leq 468 \text{ km}^2 \\ 3.09 + \frac{4}{3}\log_{10}(A_F) & A_F > 468 \text{ km}^2 \end{cases} \quad (1)$$

where A_F is the total area of the fault segment (km^2). Given the seismic scenario, the following step is to predict the ground motion intensity at the location of the structure. The attenuation relation, expressed in a logarithmic form, is used to predict the ground motion intensity at a certain site (Campbell & Bozorgnia, 2008).

The seismic intensities at different sites caused by the same earthquake are correlated. It is necessary to consider the spatial correlation of ground motion intensities within the seismic performance assessment of interdependent hospital – bridge network system. Several studies (e.g. Jayaram & Baker, 2009; Wang & Takada, 2005) revealed that the peak ground acceleration (PGA) associated with a given seismic scenario at different sites is spatially correlated and this correlation is higher for closer sites. Accordingly, the correlation among the seismic intensities at different locations is modelled as an exponential decay function (Wang & Takada, 2005):

$$\rho\left(IM_i, IM_j\right) = \exp\left(-h/l_{co}\right) \quad (2)$$

where h is the distance between two sites i and j (km); IM is the ground motion intensity measure (e.g. PGA); and l_{co} is the correlation length (km). The value of l_{co} can be estimated based on the statistical analysis of the past earthquake data. Given the distribution types associated with the ground motion intensities and correlation coefficients in Equation (2), the correlated ground motion intensities could be generated using straightforward numerical procedures, such as Monte Carlo simulation. In this paper, the effects of ground motion correlation are accounted for within the seismic assessment of spatially distributed interdependent healthcare–bridge network system.

3. Bridge, link and hospital seismic damage assessment

3.1. Bridge seismic vulnerability

A transportation network is composed of nodes, links and bridges. Nodes describe the locations of highway intersections, while links represent the highway segments connecting two nodes. Generally, bridges are considered to be the most vulnerable components in a transportation network (Dong, Frangopol, & Saydam, 2014; Lee & Kiremidjian, 2007; Liu & Frangopol, 2006). Fragility curves are commonly used to quantify structural performance under seismic hazard and are defined by the exceedance probability of a damage state under a given ground motion intensity (Dong, Frangopol, & Saydam, 2013; Mander, 1999). The fragility curve of a structural system under a specific ground motion intensity can be expressed as (Mander, 1999):

$$P_{S \geq DS_i \mid IM} = \Phi\left(\frac{\ln(IM) - \ln(m_i)}{\beta_i}\right) \quad (3)$$

where $\Phi(.)$ is the standard normal cumulative distribution function; β_i is the standard deviation of the logarithm of ground motion intensity associated with damage state i; and m_i is the median value of the intensity measure of damage state i. For the seismic performance assessment of a transportation network, it is necessary to develop the fragilities of all bridges. The correlations

161

among the seismic performance of bridges in terms of fragility have also to be considered. Specifically, bridges in a transportation network can be classified into different subgroups to characterise their fragilities considering structural characteristics, such as number of spans and material types (e.g. steel, concrete). The median values of ground motion intensities with different damage states of highway bridges have been investigated by Mander (1999) and Shinozuka, Feng, Kim, Uzawa, and Ueda (2001). For a given ground motion intensity, the probability $P_{DSi|IM}$ of a bridge being in a damage state i is given by the difference between the probabilities of exceedance of damage states i and $i+1$.

The expected bridge damage index BDI is obtained by multiplying the probability of a bridge being in each damage state with the corresponding bridge damage state index $BDDI$. Accordingly, the BDI of bridge k under a certain ground motion intensity is (Shiraki et al., 2007):

$$BDI_k = \sum_{i=1}^{n_{BDS}} BDDI_i \cdot P_{DSki|IM} \qquad (4)$$

where $BDDI_i$ is the damage index associated with the damage state i; n_{BDS} is the number of damage states; and $P_{DSki|IM}$ is the probability of a bridge k being in a damage state i. In this paper, the considered damage states for a highway bridge are none, slight, moderate, major and complete.

3.2. Link seismic damage assessment

A link is considered as an element connecting the nodes of a network. The performance of network links is related to individual bridge located on the link. The performance of the link after an earthquake can be expressed in terms of link damage index LDI, which depends on the $BDIs$ of the bridges on the link, as (Chang, Shinozuka, & Moore, 2000):

$$LDI = \sqrt{\sum_{j=1}^{n_b} \left(BDI_j \right)^2} \qquad (5)$$

where n_b is the number of the bridges located on the link. Traffic flow in the damaged link can be different and speed limits might be reduced. The level of link traffic flow capacity and speed for a damaged link depends on damage states of the link. The intact state LDS_1, slight LDS_2, moderate LDS_3 and major damage LDS_4 states represent $LDI \le .5$, $.5 < LDI \le 1.0$, $1.0 < LDI \le 1.5$ and $LDI > 1.5$, respectively (Chang et al., 2000). The correlation coefficient $\rho(BDDI_{Bk}, BDDI_{Bj})$ between the damage indices of bridge k and j is considered in the assessment process.

3.3. Hospital functionality assessment

The effects of damage states associated with structural and nonstructural components on the damage performance of a hospital are introduced in this section. The damage assessment of the hospital should determine the capacity of how many patients it can handle to provide timely treatment to the injured people (Cimellaro et al., 2011). The functionality of a hospital could be assessed based on its components (e.g. structural, nonstructural) (Yavari et al., 2010). Building components can be categorised into structural, nonstructural and content (Mitrani-Reiser, 2007). The

performance associated with different components of a hospital system should be investigated using different seismic demand indicators (e.g. story drift, floor acceleration). As beams and columns can provide support for the structure's weight and the strength needed to resist lateral forces, the beams and columns are usually classified as structural components. Nonstructural components usually include architectural, mechanical and electrical components within a building (FEMA, 2009; HAZUS, 2003). The damage to structural components is usually related to structural drift and ground motion acceleration, while the damage of nonstructural components is generally associated with the floor acceleration and structural drift (Dong & Frangopol, 2016; Mitrani-Reiser, 2007).

Given the ground motion intensity, the seismic performance of a hospital is investigated using fragility curves. Based on HAZUS (2003), the PGA is adopted to predict the performance of structural components under earthquakes. The probability of the structural components being in different damage states could be computed accordingly. The peak floor acceleration (PFA) acts as seismic demand for the damage assessment of nonstructural components. The peak floor acceleration amplification Ω (i.e. PFA/PGA) is adopted herein to compute the PFA, which in turn could be utilised for the vulnerability analysis of nonstructural components. The peak floor acceleration amplification factor Ω is (Chaudhuri & Hutchinson, 2004):

$$\Omega = \left(1.0 + \alpha_1 \sqrt{h_{nor}} \right)\left(1.0 - h_{nor} \right) + \left(\alpha_2 h_{nor}^2 \right) h_{nor} \qquad (6)$$

where α_1 and α_2 are empirical constants and h_{nor} is normalised height computed as the floor height divided by the total building height. Given the detailed information of the investigated hospital and seismic inputs, the PFA could also be obtained using nonlinear time history or incremental dynamic analysis (Dong & Frangopol, 2016).

The expected damage indices associated with structural and nonstructural components can be expressed, respectively, as follows:

$$D_{SC} = \sum_{i=1}^{n_{SD}} HCDI_{SC,i} \cdot P_{SCi|IM} \qquad (7)$$

$$D_{NSC} = \sum_{i=1}^{n_{NSD}} HCDI_{NSC,i} \cdot P_{NSCi|IM} \qquad (8)$$

where $HCDI_{SC,i}$ and $HCDI_{NSC,i}$ are the damage indices associated with state i of structural and nonstructural components of the hospital, respectively; n_{SD} and n_{NSD} are the numbers of damage states associated with structural and nonstructural components, respectively; and $P_{SCi|IM}$ and $P_{NSCi|IM}$ are the probabilities of the structural and nonstructural components being in damage state i, respectively. These probabilities are obtained based on the fragility curves considering different seismic demands. The correlation coefficient between the damage indices of structural and nonstructural components $\rho(HCDI_{SC}, HCDI_{NSC})$ is considered in the assessment process.

A comprehensive performance assessment of a hospital should be conducted based on a system level approach. Herein, a composite damage index of a hospital based on the structural and nonstructural damages is developed. Given the weighting factors associated with structural and nonstructural components, the composite expected hospital damage index HDI is

$$HDI = r_{SC} \cdot D_{SC} + r_{NSC} \cdot D_{NSC} \qquad (9)$$

where r_{SC} and r_{NSC} are the weighting factors associated with structural D_{SC} and nonstructural D_{NSC} damage indices, respectively. Given the probability density function (PDF) associated with hospital damage index HDI and threshold values (i.e. lower and upper bounds), the probability of a hospital being in different functionality levels $HFLs$ can be identified. Holmes and Burkett (2006) suggested classifying structural and nonstructural damages into four levels: none, minor, affecting hospital operations and temporary closure. Yavari et al. (2010) presented an approach considering the overall facility as fully functional, functional, affected functionality and not functional. In this paper, the four functionality levels for a hospital are none, slight, moderate and major (Yavari et al., 2010).

The capacity of a hospital depends on the classified hospital functionality levels $HFLs$. The waiting time is an important parameter to evaluate the capacity of a hospital during normal and extreme event conditions (Yi, 2005). When the number of injured persons is larger than the number of patients treated, additional waiting time is necessary. The waiting time associated with hospital functional level HFL_i under daily patient rate λ is (Paul, George, Yi, & Lin, 2006):

$$WT_i(\lambda) = \exp\left(A_i + B_i \lambda\right) \qquad (10)$$

$$A_i = \frac{\lambda_U \ln\left(WT_L\right) - \lambda_L \ln\left(WT_{Ui}\right)}{\lambda_U - \lambda_L} \qquad (11)$$

$$B_i = \frac{\ln\left(WT_{Ui}\right) - \ln\left(WT_L\right)}{\lambda_U - \lambda_L} \qquad (12)$$

where A_i and B_i are constants associated with the hospital functionality level i; λ_L is the pre-disaster average daily patient arrival number; WT_L is the waiting time associated with the normal hospital operation; λ_U is the maximum daily arrival number; and WT_{Ui} is the waiting time associated with functionality level i under the maximum arrival rate λ_U. Equation (10) is used herein to compute the waiting time; given additional data, other functions could also be adopted.

4. System level performance assessment

In this section, the effects of the damage states associated with a bridge network and a hospital on system level performance of an interdependent healthcare–bridge network are investigated. The extra travel and waiting time at the system level are computed. Based on the configuration of the investigated bridge network, travel time is defined as the time it takes to transfer the injured people to the hospital immediately after the earthquake. With respect to waiting time analysis, the healthcare system performance is measured by the waiting time needed to get the injured persons treated.

The extra travel time experienced by an injured person is due to the damages of bridges and links in a bridge network. The travel time is representative of the functionality of a bridge network; large travel time reveals a high reduction of functionality associated with a bridge network (Bocchini & Frangopol, 2011; Dong, Frangopol, & Sabatino, 2015; Frangopol & Bocchini, 2011). The extra daily travel time $EDTT$ for the injured persons in a bridge network can be expressed as (Dong et al., 2015):

$$EDTT = \sum_{j=1}^{n_l} \sum_{i=1}^{n_{LD}} P_{LDSj,i|IM} \left[ADR_{ij} \cdot \left(\frac{l_j}{S_{Dj,i}} - \frac{l_j}{S_{0j}} \right) + ADT_{ij} \cdot \frac{D_j}{S} \right] \qquad (13)$$

where n_l is the number of links in a bridge network; n_{LD} is the number of damage states associated with link damage; $P_{LDSj,i|IM}$ is the conditional probability of the jth link being in damage state i; ADT_{ij} is average daily number of injured persons that follow detour due to damage state i of the jth link; D_j is the length of the extra detour of jth link (km); S is the detour speed (km/h); ADR_{ij} is the average daily number of injured persons that remain on the jth link under damage state i; l_j is the length of link j (km); S_{0j} is the traffic speed on intact link j (km/h); and $S_{Dj,i}$ is the traffic speed on link j associated with damage state i (km/h).

The waiting time is related to the hospital functionality levels under a given seismic scenario. Given the limited functionality associated with the hospital, the extra waiting time of the injured people could be computed. Based on the theorem of total probability, the extra daily waiting time $EDWT$ can be computed as:

$$EDWT = \sum_{i=1}^{n_H} \left[P_{HFi|IM} \cdot WT_i(ATV) \right] \cdot ATV - ATV \cdot WT_0(ATV) \qquad (14)$$

where ATV is the total number of injured persons transferred though a bridge network to a hospital; WT_i is the waiting time associated with functionality level i given ATV; n_H is the number of functionality levels of a hospital under investigation; $WT_0(ATV)$ is the waiting time associated with the intact hospital under ATV; and $P_{HFi|IM}$ is the conditional probability of hospital being in functionality level i under IM. The flowchart of the computation associated with the extra travel and waiting time of a healthcare system under seismic hazard is shown in Figure 1. As indicated, the seismic scenarios should be identified first. Then, the vulnerability analyses of a bridge network and a hospital are conducted. Finally, based on the damage conditions and functionality levels, the extra travel and waiting time are computed as system level performance indicators.

During the system level performance assessment process, the correlations associated with the ground motion intensities and seismic damage indices are considered. For example, the correlations among the IMs at different locations are computed using Equation (2). Then, using Monte Carlo simulation, these correlated IMs could be generated. Overall, given the correlation coefficients, the correlated random variables used in the functionality assessment procedure could be generated. The flowchart of generating these random variables using Monte Carlo simulation is shown in Figure 2. Finally, the system level performance

163

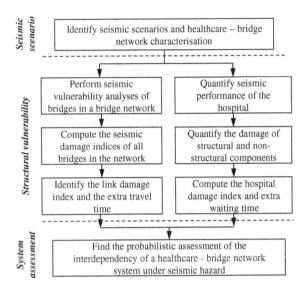

Figure 1. Flowchart of component and system levels functionality assessment of an interdependent healthcare–bridge network system under seismic hazard.

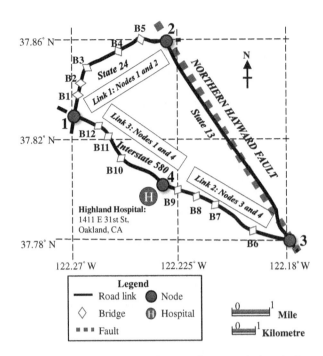

Figure 3. Layout of the healthcare–bridge network system under investigation.

indicators (e.g. extra travel and waiting time) could be computed using Equations (13) and (14).

Due to the correlations among the random variables of ground motion intensities and seismic damage indices associated with bridges and hospital, the seismic performance of a bridge network and a hospital is correlated. The correlation effects on the conditional seismic performance of a hospital given the damage state of a link are investigated herein. The conditional probability of a hospital being in functionality level j given the link in damage state i is:

$$P\left(HFL_{j|IM}\middle|LDS_{i|IM}\right) = \frac{P\left(HFL_{j|IM} \cap LDS_{i|IM}\right)}{P\left(LDS_{i|IM}\right)} \quad (15)$$

where $HFL_{j|IM}$ is the event that a hospital is in functionality level j given IM and $LDS_{i|IM}$ is the event that a link is in damage state i given IM. The probability $P(HFL_{j|IM} \cap LDS_{i|IM})$ could be computed by considering the events $HFL_{j|IM}$ and $LDS_{i|IM}$ as a parallel system. Then, given the correlation coefficients among the random variables (e.g. ground motion intensities, seismic damage indices), the probability $P(HFL_{j|IM} \cap LDS_{i|IM})$ is computed. When the events $HFL_{j|IM}$ and $LDS_{i|IM}$ are

independent, $P(HFL_{j|IM} \cap LDS_{i|IM}) = P(HFL_{j|IM}) \times P(LDS_{i|IM})$. Finally, given $P(HFL_{j|IM} \cap LDS_{i|IM})$ and $P(LDS_{i|IM})$, the conditional probability of the hospital being in functionality level j is computed according to Equation (15).

5. Illustrative example

The proposed approach is illustrated on an interdependent healthcare–bridge network system located in Alameda, California. The schematic layout of the hospital and bridge network is shown in Figure 3. The Highland Hospital, constructed in 1954, is investigated herein. This is a public hospital and is operated by the Alameda Health system. A set of locations is connected to the hospital via a bridge network composed of 12 bridges. Link 1 connects the nodes 1 and 2, link 2 connects nodes 3 and 4 and link 3 connects nodes 1 and 4. The hospital investigated herein is treated as a mid-rise steel moment frame building and designed based on moderate-code seismic provision.

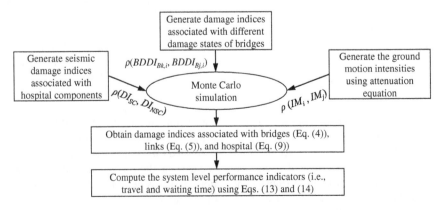

Figure 2. Flowchart of generating correlated random variables using Monte Carlo simulation to compute the system level performance indicators.

5.1. Seismic performance of bridges and links

In order to quantify the seismic performance of the bridges and links, the probabilistic earthquake scenarios should be identified. The seismic scenarios are selected based on the seismic rupture sources in the San Francisco bay area (USGS, 2003). The investigated earthquake magnitudes are associated with the Northern Hayward Fault as the healthcare–bridge network system is located in this area. The segment length and width of the fault are 50 and 14 km, respectively. Using Equation (1), the expected magnitude associated with the investigated rupture is 6.88. Using Monte Carlo Simulation, 100,000 samples of seismic scenarios are generated. Then, based on the attenuation equation (Campbell & Bozorgnia, 2008), the ground motion intensity at the location of the healthcare–bridge network system is predicted.

The PGA is utilised as ground motion intensity measure and other ground motion intensities (e.g. spectral acceleration) could also be used (Campbell & Bozorgnia, 2008). The PGA is assumed lognormal. Its expected value at the location of the hospital is .865 g, and the standard deviation is .51 g using the attenuation equation (Campbell & Bozorgnia, 2008). The probabilistic PGAs at the locations of the bridges and hospital are generated using Monte Carlo simulation. The correlation among the PGAs at the locations of the hospital and bridges is computed based on Equation (2). The exponential decay function has been widely adopted in the assessment of the spatial correlation of PGA (Esposito & Iervolino, 2011; Goda & Hong, 2008; Jayaram & Baker, 2009). Then, the correlated random variables associated with ground motion intensities are generated using Monte Carlo simulation as indicated in Figure 2.

The parameters of the fragility curves associated with the bridges located on the network are based on Shinozuka et al. (2001). Bridges 2, 7 and 9 are single-span concrete bridges; the remaining bridges are multiple-span continuous concrete bridges. The fragility curves of the basic single and multiple-span

continuous concrete bridges are shown in Figures 4(a) and (b). Given the skew angle and soil condition of the specific bridges, the fragility curves could be updated accordingly (Shinozuka et al., 2001).

Given the ground motion intensity and fragility curves, the probabilities of the bridges being in different damage states are computed. Then, using Equation (4), the bridge damage index *BDI* under the investigated ground motion intensity is obtained. The damage state index *BDDI* is considered lognormal with a coefficient of variation of .5 (HAZUS, 2003; Shinozuka et al., 2008). The expected values of the damage index associated with slight, moderate, major and complete damage states are .1, .3, .75 and 1, respectively (Shiraki et al., 2007). Monte Carlo simulation is adopted to generate these random variables. The bridge damage states associated with bridges 1 (multiple-span continuous concrete) and 2 (single-span concrete bridge) are shown in Figure 4(c) and (d), respectively. The bridge damage indices associated with bridges 1 and 2 are best fitted by a gamma distribution with mean values .338 and .196, and standard deviations .272 and .187, respectively. Subsequently, the damage indices of links are computed using Equation (5).

The correlation among the random damage indices *BDDI* of different seismic damage states is now considered. Herein, the correlation coefficients are assumed to be 0, .5 and 1, representing uncorrelated, partially and fully correlated random variables, respectively. These values are adopted to investigate the correlation effects on the network performance under seismic hazard. Additionally, the damage indices of different bridges are also correlated. The probabilistic damage index associated with link 1 under different correlation coefficients among the damage indices *BDDI* is shown in Figure 5(a). As indicated, the standard deviation of the link damage index increases as the correlation coefficient among the random variables increases. Given the threshold associated with link performance, the probabilities of the link 1 being in different performance levels (i.e. from no

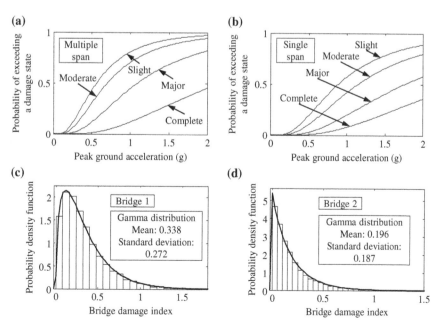

Figure 4. Fragility curves of (a) multiple-span concrete bridge, (b) single-span bridges; probability density functions of damage indices of (c) bridge 1 (B1) and (d) bridge 2 (B2).

(a)

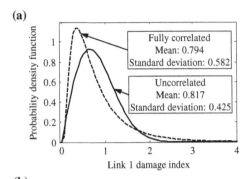

(b)

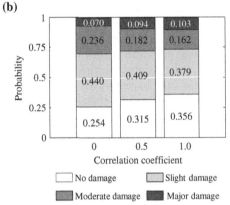

Figure 5. (a) Effects of correlation among the damage indices *BDDI* on the link 1 damage index and (b) probabilities of link 1 being in different damage states using 0, .5 and 1.0 correlation coefficients.

damage to major damage) are shown in Figure 5(b). As indicated, without considering the correlation effect, the probabilities of link being in none and major damage states would be underestimated. Consequently, the correlation coefficients among the damage indices *BDDI* have a significant effect and should be carefully evaluated. The probabilistic damage indices associated with links 2 and 3 could also be obtained. The probabilities of

the links 2 and 3 being in different damage states are shown in Table 1.

5.2. Hospital damage assessment

In order to quantify the hospital damage index *HDI*, the seismic intensity and vulnerability of structural and nonstructural components should be identified. The PGA is adopted as seismic demand indicator for the structural components, while the PFA is used to investigate the seismic performance of nonstructural components (HAZUS, 2003). The relationship between PFA and PGA is indicated in Equation (6). For the mid-rise hospital building, α_1 and α_2 are assumed 1.63 and 1.53, respectively (Chaudhuri & Hutchinson, 2004). The maximum PFA occurs at $h_{nor} = 1$ for the investigated hospital. The parameters of the fragility curves associated with structural and nonstructural components are based on HAZUS (2003) and shown in Table 2. Given moderate-code seismic provision, the median values and standard deviations of fragility curves associated with structural and nonstructural components are obtained. Subsequently, given the PGA and PFA, the seismic vulnerabilities of the structural and nonstructural components are computed.

Herein, the damage indices of damage states associated with structural and nonstructural components are based on Aslani and Miranda (2005). The mean values of slight, moderate, major and complete damage states of structural components are .025, .12, .6 and 1.2, respectively. The mean values of slight, moderate, major and complete damage states of nonstructural components are .02, .12, .36 and 1.2, respectively. The coefficients of variation of these random variables are .7. Then, based on Equations (7) and (8), the damage indices of structural and nonstructural components are computed. The PDFs of damage indices of structural and nonstructural components are indicated in Figure 6. As indicated, the expected damage index of the structural component is almost 2.05 times of that associated with the nonstructural component.

The correlation coefficient among the damage indices of structural and nonstructural components $\rho(HCDI_{SC}, HCDI_{NSC})$ is considered in the assessment process. The three correlation coefficients 0, .5 and 1 are considered. Monte Caro simulation is used to generate the random variables considering correlation as indicated in Figure 2. Given the damage index of structural and nonstructural components, the composite building damage index is computed using Equation (9). This equation is used to compute the hospital damage index. Herein, r_{SC} and r_{NSC} are assumed .5. The expected value and standard deviation of *HDI*

Table 1. Probabilities of links 2 and 3 in Figure 3 being in different damage states considering correlation among the random damage indices *BDDI*.

Link	Correlation coefficient	No damage	Slight damage	Moderate damage	Major damage
Link 2	.0	.3514	.4509	.1643	.0334
	.5	.4136	.4001	.1356	.0507
	1.0	.4498	.3573	.1277	.0652
Link 3	.0	.4472	.4223	.1111	.0194
	.5	.4963	.3666	.1051	.0320
	1.0	.5327	.3239	.0977	.0457

Table 2. Median and standard deviation associated with fragility curves of structural and nonstructural components (adapted from HAZUS, 2003).

Design level	Slight damage		Moderate damage		Major damage		Complete damage	
	Median	Standard deviation	Median	Standard deviation	Median	Standard deviation	Median	Standard deviation
Structural component								
Moderate code	.16	.64	.28	.64	.6	.64	1.27	.64
Low code	.15	.64	.23	.64	.42	.64	.73	.64
High code	.17	.64	.34	.64	.85	.64	2.1	.64
Nonstructural component								
Moderate code	.38	.67	.75	.67	1.5	.67	3	.67
Low code	.3	.65	.6	.67	1.2	.67	2.4	.67
High code	.45	.66	.9	.67	1.8	.68	3.6	.66

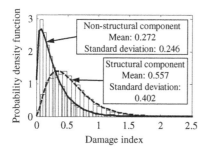

Figure 6. Probability density functions of structural and nonstructural components damage indices.

Table 3. Probabilities of the hospital having different functionality levels under different correlation coefficients among the damage indices of structural and nonstructural components.

Correlation coefficient	No damage	Slight damage	Moderate damage	Major damage
.0	.3323	.4946	.1526	.0205
.5	.4116	.3903	.1569	.0412
1.0	.4625	.3234	.1527	.0614

associated with $\rho(HCDI_{SC}, HCDI_{NSC}) = 0$ are .416 and .221, respectively. Given $\rho(HCDI_{SC}, HCDI_{NSC}) = 1$, the expected value and standard deviation of HDI are .416 and .319, respectively. As indicated, the correlation has a significant effect on the standard deviation of the hospital damage index.

For the hospital functionality level analysis, none, slight, moderate and major damage are represented by the values of the hospital damage index $HDI \leq .3$, $.3 < HDI \leq .6$, $.6 < HDI \leq 1$ and $HDI > 1$, respectively. Given the hospital functionality criterion, the probabilities of the hospital being in different functionality levels are identified and shown in Table 3. As revealed, the probabilities of being in none and major damaged functionality levels increase when considering correlation effects. Given the threshold values associated with hospital damage indices determined by a decision-maker, the probabilities of the hospital being in different functionality levels could be updated.

5.3. System level performance

The seismic performance of a healthcare–bridge network system is investigated considering two indicators: (a) travel time and (b) waiting time. After the earthquake, the injured persons from nodes 1, 2 and 3 in Figure 3 are transferred to node 4, as the hospital is near this node. The extra travel and waiting time associated with the daily patient volume are investigated. If the damage state of the link is slight, the remaining patient volume and the flow speed are 100% and 75% of those for the intact link, respectively. In moderate damage state, the remaining volume and the flow speed are 75% and 50% of those for the intact link. In major damage state, the remaining volume and the flow speed are 50% and 50% of those for the intact link (Chang et al., 2000; Dong et al., 2014). The extra detour length of links is 3.5 km. The daily patient volumes after the earthquake from node 1 to 4, 2 to 4 and 3 to 4 are 60, 120 and 60, respectively. The total number of daily injured persons transferred to the hospital is 240 (i.e. 60 + 120 + 60).

Then, given the probabilities of links being in different damage states as computed previously, the extra daily travel time

considering the number of injured persons through the damaged transportation network is computed using Equation (13). The correlation coefficients $\rho(BDDI_{Bk,i}, BDDI_{Bj,i})$ and $\rho(PGA_i, PGA_j)$ are denoted as ρ_1 and ρ_2, respectively. Given $\rho_1 = \rho_2 = 0$, the extra daily travel time is 15.13 h; this value reduces to 12.02 h given $\rho_1 = \rho_2 = 1$. As indicated, the correlation among the random variables could affect the extra travel time significantly. Compared with the uncorrelated case (i.e. $\rho_1 = \rho_2 = 0$), the extra travel time associated with fully correlated random variables ($\rho_1 = \rho_2 = 1$) decreases by almost 26%. Furthermore, the correlation among the ground motion intensities has a larger effect on the extra travel time. Given $\rho_1 = \rho_2 = 0$, the extra daily travel time is 15.13 h; this value reduces to 13.22 h when $\rho_1 = 0$ and $\rho_2 = 1$ and to 14.09 h when $\rho_1 = 1$ and $\rho_2 = 0$.

Given the hospital being in the performance levels as shown in Table 3, the extra daily waiting time is evaluated using Equation (14). The pre-earthquake average patient arrival per day λ_L is 80. The waiting time WT_L associated with the normal operation condition is 20 min. Given more information (e.g. number of beds, number of operating rooms) of the investigated hospital, the arrival rate and waiting time could be updated. The maximum arrival per day λ_U is assumed 450. The waiting times associated with none, slight, moderate and major damage levels under maximum arrival rate are 40, 60, 80 and 120 min, respectively. Then, using Equations (10) – (12), the waiting times associated with different hospital functional levels are shown in Figure 7(a). The waiting time is expressed using the exponential function (Paul et al., 2006). With additional data, other models could also be incorporated within the assessment process.

The correlation effects are considered in the extra waiting time assessment process. Herein, $\rho(HCDI_{SC}, HCDI_{NSC})$ is denoted as ρ_3. Given $\rho_3 = 0$, the extra daily waiting time of the injured person

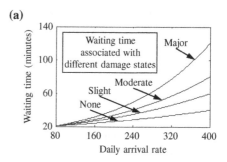

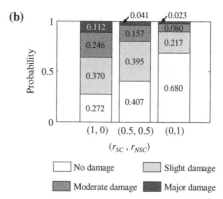

Figure 7. (a) Waiting time associated with hospital having different functionality levels (b) effects of r_{SC} and r_{NSC} on the probabilities of hospital being in different functionality levels.

(a)

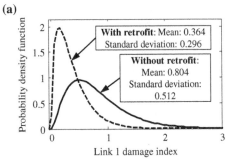

(b)

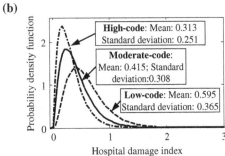

Figure 8. Effects of (a) retrofit actions on link 1 damage index and (b) high- and low-design codes on the hospital–bridge network system under seismic hazard.

and nonstructural components under low- and high-code design provisions are shown in Table 2. Given the fragility curves associated with these design provisions, the extra daily waiting time with low-, moderate- and high-design codes are 25.41, 16.06 and 10.59 h, respectively. The PDF of the hospital damage index is shown in Figure 8(b). As revealed, the seismic design code has a significant effect on the hospital functionality levels.

Moreover, the effects of correlation on the conditional probabilities of hospital being in different functionality levels given the seismic damage of the link are investigated using Equation (15). Given $\rho(BDDI, HCDI) = \rho(PGA_i, PGA_j) = .5$, the conditional probabilities of the hospital being in different functionality levels are shown in Table 4. The case without considering the correlation effects is also shown in this table. As indicated, the correlation effects could affect the conditional performance of the hospital significantly. For example, given the link 1 being in damage state 3 (i.e. moderate damage state), the conditional probability of the hospital being in moderate damaged functionality level is .2555, while this value is reduced to .1526 without considering the correlation effects.

is 16.84 h; this value reduces to 15.46 h when $\rho_3 = 1$. The extra daily waiting time of the injured people is slightly larger than the extra travel time. The effects of r_{SC} and r_{NSC} on the hospital functionality levels are indicated in Figure 7(b). The two parameters (i.e. r_{SC} and r_{NSC}) have a significant effect on the hospital functionality assessment and should be carefully assessed.

The effects of retrofit actions associated with the bridges on extra travel time are also studied. The fragility enhancement of bridges retrofitted by steel jacketing is investigated based on the approach presented in Shinozuka et al. (2008). The enhancement ratios associated with fragility median values are 55%, 75%, 104% and 145% considering slight, moderate, major and complete damage states, respectively (Shinozuka et al., 2008). Given $\rho(BDDI_{Bk,i}, BDDI_{Bj,i}) = .5$ and Equation (2), the extra daily travel time with and without retrofit actions are 3.12 and 13.01 h, respectively. As revealed, the seismic retrofit actions have a profound effect on the extra travel time and can improve the healthcare–bridge network performance significantly. The probabilistic damage index associated with link 1 under these two scenarios (i.e. with and without retrofit actions) is shown in Figure 8(a).

Additionally, the effects of seismic vulnerability of hospital on the system level performance are investigated. Medians and standard deviations associated with fragility curves of structural

6. Conclusions

This paper presents an approach for assessment of an interdependent healthcare–bridge network system under seismic hazard considering uncertainties and correlation effects. The functionalities associated with the individual bridges, bridge networks and hospital are investigated and combined for the system performance assessment. Rather than focusing only on structural damage, the extra travel and waiting time are investigated. The correlation among the damage indices is considered in the performance assessment process in addition to the correlation among the spatial seismic intensities. The approach is illustrated on a healthcare system near a bridge network in Alameda, California. The approach presented can be extended to systems with multiple hospitals, in which the interdependencies among hospitals and the interdependencies healthcare facilities–bridge network systems could be considered.

The following conclusions are drawn:

(1)   The proposed performance assessment of an interdependent healthcare–bridge network system under seismic hazard provides system level probabilistic measures that can aid the emergency management process. In addition to the consideration of bridge damage assessment, the damages associated with structural and nonstructural components of a hospital are considered in the functionality assessment.

Table 4. Conditional probabilities of the hospital being in different functionality levels given the seismic damage of the link 1 considering the correlations among the ground motion intensities and damage indices of bridges and hospital.

Correlation coefficient	Hospital functional level	$P(HFL_{i|M}|LDS_{1|M})$	$P(HFL_{i|M}|LDS_{2|M})$	$P(HFL_{i|M}|LDS_{3|M})$	$P(HFL_{i|M}|LDS_{4|M})$
$\rho = .5$	No	.6460	.4032	.2115	.0823
	Slight	.2978	.4397	.4681	.3273
	Moderate	.0520	.1367	.2555	.3906
	Major	.0042	.0204	.0649	.1998
$\rho = 0$	No	.3323	.3323	.3323	.3323
	Slight	.4946	.4946	.4946	.4946
	Moderate	.1526	.1526	.1526	.1526
	Major	.0205	.0205	.0205	.0205

(2) The correlation among the random damage indices has a significant effect on the probabilities of links being in different damage states and should be carefully evaluated. Without considering the correlation effect, the probabilities of link being in none and major damage states would be underestimated.

(3) It is necessary to consider the correlation coefficients among the spatial ground motion intensities and component-to-component damage indices for the healthcare–bridge network system performance assessment. The correlation coefficients have a significant effect on the standard deviations of the damage indices of bridges, links and hospital.

(4) Regarding the system level performance assessment, the extra travel and waiting time decrease when the correlation coefficients (e.g. correlations among the ground motion intensities and seismic damage indices) are accounted for. Additionally, the correlation among the ground motion intensities has a larger effect on the extra travel time than that among the damage indices.

(5) The seismic performance of interdependent healthcare–bridge network system depends highly on the seismic performance of both bridge network and hospital. The effects of retrofit and seismic strengthening associated with bridges and hospital are significant. Bridge retrofit actions could result in a profound improvement of the performance of healthcare–bridge network system. Given the specific bridge retrofit actions and correlation coefficients considered in this paper, the extra travel time of the injured persons associated with the case when bridges are retrofitted is about 24% (i.e. 3.12/13.01) of the case when bridges are not retrofitted.

(6) Based on the damage conditions of the link, the functionality of the hospital under the investigated seismic scenarios could be predicted considering correlation effects. Overall, it is of vital importance to incorporate correlation effects within the seismic performance assessment of interdependent infrastructure systems.

Acknowledgements

The support from the National Science Foundation, the Commonwealth of Pennsylvania, Department of Community and Economic Development, through the Pennsylvania Infrastructure Technology Alliance (PITA) and the U.S. Federal Highway Administration Cooperative Agreement is gratefully acknowledged. The opinions and conclusions presented in this paper are those of the authors and do not necessarily reflect the views of the sponsoring organisations.

Disclosure statement

No potential conflict of interest was reported by the authors.

Funding

This work was supported by the National Science Foundation [grant number CMS-0639428], [grant number CMMI-1537926]; the Commonwealth of Pennsylvania, Department of Community and Economic Development, through the Pennsylvania Infrastructure Technology Alliance (PITA); the U.S. Federal Highway Administration Cooperative Agreement [award number DTFH61-07-H-00040].

References

Achour, N., Miyajima, M., Kitaura, M., & Price, A. (2011). Earthquake-induced structural and nonstructural damage in hospital. *Earthquake Spectra, 27*, 617–634.

Aslani, H., & Miranda, E. (2005). *Probabilistic earthquake loss estimation and loss disaggregation in building* (Report No. 157). Stanford, CA: Department of Civil and Environmental Engineering, Stanford University.

Bocchini, P., & Frangopol, D. M. (2011). A stochastic computational framework for the joint transportation network fragility analysis and traffic flow distribution under extreme events. *Probabilistic Engineering Mechanics, 26*, 182–193.

Campbell, K. W., & Bozorgnia, Y. (2008). NGA ground motion model for the geometric mean horizontal component of PGA, PGV, PGD and 5% damped linear elastic response spectra for periods ranging from 0.01 to 10 s. *Earthquake Spectra, 24*, 139–171.

Chang, S. E., Shinozuka, M., & Moore, J. E. (2000). Probabilistic earthquake scenarios: Extending risk analysis methodologies to spatially distributed systems. *Earthquake Spectra, 16*, 557–572.

Chaudhuri, S., & Hutchinson, T. (2004). *Distribution of peak horizontal floor acceleration for estimating nonstructural element vulnerability*. Proceedings of the 13th World Conference on Earthquake Engineering, Vancouver, BC, Canada, paper No. 1721.

Cimellaro, G. P., Reinhorn, A. M., & Bruneau, M. (2011). Performance-based metamodel for healthcare facilities. *Earthquake Engineering and Structural Dynamics, 40*, 1197–1217.

Dong, Y., Frangopol, D. M., & Saydam, D. (2013). Time-variant sustainability assessment of seismically vulnerable bridges subjected to multiple hazards. *Earthquake Engineering and Structural Dynamics, 42*, 1451–1467.

Dong, Y., Frangopol, D. M., & Saydam, D. (2014). Sustainability of highway bridge networks under seismic hazard. *Journal of Earthquake Engineering, 18*, 41–66.

Dong, Y., Frangopol, D. M., & Sabatino, S. (2015). Optimizing bridge network retrofit planning based on cost-benefit evaluation and multi-attribute utility associated with sustainability. *Earthquake Spectra, 31*, 2255–2280.

Dong, Y., & Frangopol, D. M. (2016). Performance-based seismic assessment of conventional and base-isolated steel buildings including environmental impact and resilience. *Earthquake Engineering and Structural Dynamics, 45*, 739–756.

Esposito, S., & Iervolino, I. (2011). PGA and PGV spatial correlation models based on European multievent datasets. *Bulletin of the Seismological Society of America, 101*, 2532–2541.

FEMA. (2009). *NEHRP recommended provisions for seismic regulations for new buildings and other structures, FEMA P-750*. Washington, DC: Federal Emergency Management Agency.

Frangopol, D. M., & Bocchini, P. (2011, April, 14–16). *Resilience as optimization criterion for the rehabilitation of bridges belonging to a transportation network subjected to earthquake*. Proceedings of the SEI-ASCE 2011 Structures Congress, Las Vegas, NV, USA.

Goda, K., & Hong, H. P. (2008). Spatial correlation of peak ground motions and response spectra. *Bulletin of the Seismological Society of America, 98*, 354–365.

Hanks, T. C., & Bakun, W. H. (2002). A bilinear source-scaling model for M-log observations of continental earthquakes. *Bulletin of the Seismological Society of America, 92*, 1841–1846.

HAZUS. (2003). *Multi-hazard loss estimation methodology, earthquake model*. Technical Manual. Washington, DC: Department of Homeland Security Emergency Preparedness and Response Directorate, FEMA, Mitigation Division.

Holmes, W. T., & Burkett, L. (2006). *Seismic vulnerability of hospitals based on historical performance in California*. Proceedings of the 8th U.S. National Conference on Earthquake Engineering, San Francisco, CA, USA.

Jayaram, N., & Baker, J. W. (2009). Correlation model for spatially distributed ground-motion intensities. *Earthquake Engineering and Structural Dynamics, 38*, 1687–1708.

Lee, R. G., & Kiremidjian, A. S. (2007). *Uncertainty and correlation in seismic risk assessment of transportation systems* (PEER Report 2007/05). Berkeley, CA: Pacific Earthquake Engineering Research Center College of Engineering, University of California.

Liu, M., & Frangopol, D. M. (2006). Probability-based bridge network performance evaluation. *Journal of Bridge Engineering, 11*, 633–641.

Mander, J. B. (1999). *Fragility curve development for assessing the seismic vulnerability of highway bridges* (Technical Report). Buffalo, NY: University at Buffalo, State University of New York.

Mitrani-Reiser, J. (2007). *Probabilistic loss estimation for performance-based earthquake engineering* (PhD Dissertation) (pp. 1–155). Pasadena, CA: California Institute Technology.

Murphy, L. (1973). *San Fernando, California, Earthquake of February 9 1971*. Washington, DC: U.S Department of Commerce.

Myrtle, R. C., Masri, S. E., Nigbor, R. L., & Caffrey, J. P. (2005). Classification and prioritization of essential systems in hospitals under extreme events. *Earthquake Spectra, 21*, 779–802.

Paul, J. A., George, S. K., Yi, P., & Lin, L. (2006). Transient modelling in simulation of hospital operations for emergency response. *Prehospital and Disaster Medicine, 21*, 223–236.

Shinozuka, M., Feng, M. Q., Kim, H., Uzawa, T., & Ueda, T. (2001). *Statistical analysis of fragility curves* (Technical Report MCEER). Los Angeles, CA: Department of Civil and Environmental Engineering, University of South California.

Shinozuka, M., Zhou, Y., Kim, S. H., Murachi, Y., Banerjee, S., Cho, S., & Chung, H. (2008). *Social-economic effect of seismic retrofit implemented on bridges in Los Angeles highway network* (Report CA06-0145). Irvine, CA: Department of Civil and Environmental Engineering, University of California.

Shiraki, N., Shinozuka, M., Moore, J. E., Chang, S. E., Kameda, H., & Tanaka, S. (2007). System risk curves: probabilistic performance scenarios for highway networks subject to earthquake damage. *Journal of Infrastructure Systems, 13*, 43–54.

USGS. (2003). *Earthquake probabilities in the San Francisco Bay Region: 2002–2031* (Open File Report 03-214). Menlo Park, CA: United States Geological Survey.

Wang, M., & Takada, T. (2005). Macrospatial correlation model of seismic ground motions. *Earthquake Spectra, 21*, 1137–1156.

World Health Organization (WHO). (2007). *Risk Reduction in the Health Sector and Status of Progress*. Proceedings of Disaster Risk Reduction in the Healthcare Sector- Thematic Workshop, World Health Organization (WHO), Geneva, Switzerland.

Yavari, S., Chang, S., & Elwood, K. J. (2010). Modeling post-earthquake functionality of regional health care facilities. *Earthquake Spectra, 26*, 869–892.

Yi, P. (2005). *Real-time generic hospital capacity estimation under emergency situations*. Buffalo, NY: State University of New York at Buffalo.

Reliability-based structural design framework against accidental loads – ship collision

Hyun-Moo Koh, Jeong-Hyun Lim, Hyunjoong Kim, Jiwoon Yi, Wonsuk Park and Junho Song ⓘ

ABSTRACT

As civil infrastructures become highly complex and interconnected, accidental events such as fire, explosion, dropped objects and collision tend to occur more frequently, and are causing critical structural damage and socio-economic costs. Therefore, accidental loads should be considered appropriately during the structural design. To achieve this goal, the paper presents a reliability-based structural design framework against accidental loads. The proposed framework employs a scenario-based simulation in which the occurrence of an accidental load event and the resulting structural loads are modelled probabilistically. The probabilistic distribution of accidental loads is then determined based on the occurrence probabilities of identified scenarios and the conditional probability of structural loads given each accident scenario. In order to estimate the structural responses under accidental loads, an appropriate type of structural analysis is carried out for the given limit states of the structural resistance at the critical sections. The proposed framework is illustrated by an example of a cable-stayed bridge whose piers are under ship collision risk. To consider the dynamic amplification of bridge response caused by ship–bridge collision, time history analysis is performed for the equivalent static ship collision loads. The paper also identifies research needs and challenges in applying the proposed framework to other accidental loads.

1. Introduction

Up to 80–90% of catastrophic structural failures arise from accidental events caused by human errors rather than uncertainties in loads and strengths (Ellingwood, 2001). Among those accidental events, fire, explosion, dropped objects and collision of vehicles or vessels are considered critical ones that may result in serious structural damage and huge socio-economic costs. The risk of such accidental events tends to increase worldwide due to the urbanisation which requires highly complex and interdependent civil infrastructures for robust service of gas, oil, water and electricity and for effective large-size transportation. Moreover, the unexpected failure of a structure in a complex and interdependent infrastructure environment may induce a chain reaction that can make the impacts of the structural failure even more disastrous.

As the society and urban system continuously develop and change, a structure might experience unexpected accidents that were not originally considered during the original design process. For example, in 1980, Sunshine Skyway Bridge collapsed due to a ship collision. The main cause of the collapse was the failure of a pier column located 13 m above the waterline where the overhang of the ship's bow contacted the structure at a weak point. This type of accident was not properly considered when the bridge was originally designed and constructed (Knott, 1998).

A main challenge in dealing with such accidental events is that it is difficult to estimate the severity and the likelihood of accident scenarios. This is because many man-made factors contributing to the risk are highly uncertain and thus difficult to quantify or predict, and the knowledge base of structural behaviour under such extreme loading conditions is limited. Despite this challenge, the accidental events need to be considered properly during the design and rehabilitation of critical infrastructures such as bridges and industrial facilities for the sake of public safety and sustainability.

Current design approaches for fire, explosion and ship collision can be considered as *performance-based* design (American Bureau of Shipping [ABS], 2013; Det Norske Veritas [DNV], 2010; Ellingwood, 2000). That is, the design criteria are usually determined by the owner or the agency in charge of the infrastructure. Therefore, a proper definition of the design criteria related to accidental events is a critical issue in efforts to optimise the design loads from the trade-off between safe and cost-effective design options. However, it is difficult to determine the capacity and redundancy of a structure that would ensure a proper level of structural reliability against extreme loads. This is the reason why accidental events are characterised by various uncertainties, e.g. possible locations of critical accidents, the severity and likelihood of the accidental loads and the mechanism of chain reaction. Therefore, the society has a pressing need for implementing reliability-based decision framework for accidental loads in the process of planning and designing an infrastructure.

As an effort to satisfy this important research need, this paper presents a reliability-based design framework for accidental loads based on findings and outcomes of recent research efforts on bridge designs under ship collision risks, which was performed through the Core Technology Development Program of Super Long Span Bridge R&D Center of Korea (Lim, Park, Kim, & Koh, 2012a, 2012b; Lim, Park, & Koh, 2013; Park, Lim, & Koh, 2013a, 2013b). After a brief review of existing approaches, the paper discusses the proposed framework and main procedures. The proposed framework is then illustrated step by step using an example of reliability analysis of a cable-stayed bridge whose piers are subject to ship collision risk. The paper is concluded by identifying future research needs and challenges in applying the proposed design framework to other accidental loads.

2. Existing probabilistic approaches for accidental loads

2.1. Current design practice for accidental loads

In current design practices regarding accidental loads, the structural design criteria are often determined through hazard analysis consisting of accidental loading scenario evaluation and risk assessment (ABS, 2013; British Standards Institution [BSI], 2006). In this approach, the acceptance criteria for all affected members and systems should be defined by the owner so that the health and safety, environment and facility risk levels are properly characterised (ABS, 2013). Since this approach determines the structural design based on the target performance of the owner, it can be considered as a performance-based design approach.

In such a performance-based design, the hazard and class of the structure subject to the risk are classified into several categories. For example, ASCE Standard 7-2 provides the categories of buildings according to performance requirement. According to the standard, it is also the responsibility of owners to classify the grade of the building (Structural Engineering Institute [SEI], 2006). However, it may be challenging for the owner to decide the categories of structures based on quantitative assessment of the hazard.

For this reason, probabilistic approaches have been proposed for proper quantification of such risk. For example, Ellingwood (2001) compared the risk of accidental loads with the probability of structural failure caused by natural hazards. An appropriate level of target reliability could be also determined based on the *de minimis* risk (Ellingwood, 2006; Paté-Cornell, 1994), which is the risk level below which the society does not impose regulatory guidance. Recently, a scenario-based design approach was proposed for ship collision using an approach similar to those

developed for natural hazard assessment (Park et al., 2013a). The details of this approach will be presented later in this paper.

2.2. Probabilistic method for bridge design against ship collision

The ship collision problem has been handled by probabilistic approaches, and design specifications have adopted probabilistic approaches for ship collision risks. Current design practices such as AASHTO guidelines (American Association of State Highway and Transportation Officials [AASHTO], 2009) or Eurocode (BSI, 2006) adopt probabilistic design approaches, which aim to manage the quantified risk of bridge collapse below a prescribed level. AASHTO LRFD employs the 'annual frequency of bridge collapse (AF)' concept to estimate the risk of bridge collapse. Similarly, Eurocode suggested to estimate the 'probability of overall collapse of bridge', whose detailed procedure is mainly dependent on engineer's judgment.

In detail, the annual frequency (AF) of structural collapse by vessel collision suggested by AASHTO is calculated as follows:

$$AF = N \cdot PA \cdot PG \cdot PC \tag{1}$$

where N denotes the annual number of ships passing by the bridge, PA denotes the aberration probability of each ship, PG is the geometric collision probability of the aberrated ship and PC denotes the probability of bridge collapse due to a collision with an aberrant vessel. Based on the available data concerning risk comparisons and consultation with risk analysis experts, AASHTO introduced safety criteria using annual frequency concept as follows.

- Critical/essential bridges: The acceptable annual frequency of collapse of critical/essential bridge should be equal to, or less than, .01 in 100 years, i.e. $AF \le .0001$.
- Typical bridges: The acceptable annual frequency of collapse of typical bridges should be equal to, or less than, .1 in 100 years, i.e. $AF \le .001$.

Table 1 summarises uncertain quantities affecting the probabilistic terms in Equation (1) and methods suggested for estimating the quantities.

Although current design processes for ship collision are employing probabilistic approach, they are not considered legitimate reliability-based frameworks yet, thus require further developments. For example, the occurrence of ship aberrancy and collision needs to be determined systematically based on site-dependent characteristics, human factor, bad weather and so on. In addition, collision loads need to be calculated with full consideration of uncertainties of collision event and collision mechanism. Finally, design criteria need to be determined

Table 1. Uncertain quantities affecting the probabilistic terms used in ship collision risk analysis and estimation methods.

Terms	Uncertain quantities	Estimation method
Vessel frequency (N)	Characteristics of vessel traffic under the bridge (size, type, etc.)	Prediction of future vessel traffic based on statistical analysis of traffic demands and environments
Probability of aberration (PA)	Human error, mechanical failures, adverse environmental conditions (poor visibility, wind squalls, etc.)	Statistical analysis of data or traffic simulation
Geometric probability (PG)	Vessel traffic distribution, human intervention	Statistical analysis of data or mathematical model
Probability of collapse (PC)	Vessel impact force, ultimate bridge resistance	Evaluate ratio of vessel impact force to ultimate bridge resistance

based on a probabilistic approach which can deal with various uncertainties of the ship collision event. Such desirable further developments for reliability-based bridge design framework against ship collision are identified in terms of three main tasks: hazard modelling, performance evaluation and decision-making (see Figure 1).

In order to develop such reliability-based design framework, researchers at the Core Technology Development Program of Super Long Span Bridge R&D Center of Korea have recently performed state-of-the-art research on probabilistic assessment of bridge–ship collision (Lim et al., 2012a, 2012b, 2013; Park et al., 2013a, 2013b). Based on the experience, a reliability-based structural design framework is proposed as follows, which could be extended to manage the risk of structural failures caused by other accidental loads.

3. Reliability-based structural design framework against accidental loads

3.1. Design framework and main tasks

As described in the previous section, a probabilistic approach may facilitate the structural design process based on the quantified risk caused by accidental loads. Therefore, a probabilistic bridge design framework is herein proposed to deal with ship collision risk effectively during the design process.

3.1.1. Probabilistic design framework

The proposed approach employs scenario-based simulations in which critical scenarios of accidental loads are identified and analysed to predict the resulting structural loads in a probabilistic manner. A variety of uncertainties are fully considered throughout the process from the occurrence of the accident to the resulting load on the structure. Using the total probability theorem, the probability of events that the structural load Q exceeds a threshold level q is described as follows:

$$P(q < Q) = \sum_{i=1}^{n} P\left(q < Q|E_i\right) \cdot P\left(E_i\right) \qquad (2)$$

where n is the total number of critical accidental event scenarios during a certain period, E_i is the ith accident scenario, $P(E_i)$ is the probability of the ith scenario and $P(q < Q|E_i)$ is the conditional

probability that the load exceeds the threshold given the ith scenario. Equation (2) implies that the identified scenario events are mutually exclusive to each other and collectively exhaustive, and the probabilistic distribution of the load is statistically dependent on the scenario. From Equation (2), the complementary cumulative distribution function (CCDF) of the load Q is described as follows:

$$G_Q(q) = \sum_{i=1}^{n} G_{Q|E_i}(q|E_i)P(E_i) \qquad (3)$$

where $G_{Q|E_i}(q|E_i) = P(q < Q|E = E_i)$ denotes the conditional CCDF of the load Q given the ith scenario event E_i.

Equation (3) shows that the probabilistic analysis requires the following: (1) *the event-scenario analysis* to determine $P(E_i)$ and (2) *the probabilistic structural analysis* to find the conditional CCDF $G_{Q|E_i}(q|E_i)$. It should be noted that these two analyses are closely related to each other because both the structural load Q and an accidental event E_i are affected by common environmental and structural design variables.

Suppose a target exceedance probability P_E is given for a certain period. From Equation (3), it is seen that the design load Q_D corresponding to the target exceedance probability can be determined by

$$Q_D = G_Q^{-1}(P_E) \qquad (4)$$

where $G_Q^{-1}(\cdot)$ denotes the inverse CCDF of the load Q. This means Q_D is the minimum load capacity of the structure to control the exceedance probability below the target level P_E.

In summary, the proposed framework facilitates finding a design load corresponding to the target exceedance probability by combining the results of three main tasks that were identified during the ship collision analysis (Figure 1): (1) **hazard modelling**: identifying critical ship collision accident scenarios and computing the probabilities, (2) **performance evaluation**: computing conditional CCDF of accidental loads given each collision scenario and (3) **decision-making**: determining a proper level of risk acceptance criteria in terms of annual exceedance probability. These three main tasks of the probabilistic design framework are explained in more detail as follows.

3.1.2. Task 1: hazard modelling

This task aims to identify critical scenarios of the accidental events, and to develop their probabilistic models. In order to calculate the probabilities of accidental events, potential accident scenarios need to be first identified in terms of their cause and mechanism. Potential risk factors should be identified through a brainstorming process of experienced experts. In addition, the statistical analysis of historical accidents can help estimate the probability of such accidents. If the proposed framework is applied to gas explosion hazards, for example, the gas explosion rate per gas supply in dwellings should be considered in computing the probability of gas explosion (Ellingwood, 2006). In addition, the effect of the devices for mitigating the risk, e.g. leak alarms of inflammables, automatic lock and collision protection facilities, should be considered to avoid overestimating the risk.

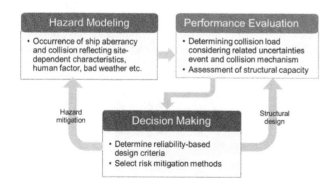

Figure 1. Reliability-based bridge design framework against ship collision.

3.1.3. Task 2: performance evaluation

This task determines accidental load considering related uncertainties and collision mechanism. Thus, solid understanding about the behaviour of the structure under the identified accident scenarios is required. The structural response caused by impact forces from explosion or collision is affected by the duration and magnitude of the impact force (Chopra, 1995). This means that a proper dynamic analysis is needed for accurate estimation of the structural response caused by the accidental loads. In this case, it is challenging to determine the representative value of accidental loads because the loads depend on the particular time history. Nevertheless, in order to calculate the probability distribution of loads, a representative value of the dynamic load is still needed. For example, ship collision loads are represented by equivalent static loads in the approaches by AASHTO LRFD and Eurocode 1–7. In particular, Eurocode 1–7 employs dynamic amplification factor to determine the equivalent static loads representing the dynamic response of the bridge.

3.1.4. Task 3: decision-making

In current design practices, the design loads against accidental events are determined based on the owner's risk acceptance. In such cases, the owner should make a reasonable decision on design criteria by considering both safety and cost-effectiveness of the structure. However, if the owner has lack of experience in the accidental loads and cannot achieve reasonable design criteria, the society may take an intolerable level of risk or pay unreasonable cost for an excessively conservative design.

In current design practice for earthquake and a strong wind loads, the design load is determined by using exceedance probability or return period as a design criteria (AASHTO, 2012). Thus, we have experience in the design load by observing data from natural hazard. Therefore, the annual exceedance probability of the artificial hazard can be compared with that of the natural hazard which has similar intensity or consequence.

The final task for the reliability-based design framework aims to determine a proper level of risk acceptance criteria in terms of annual exceedance probability. In contrast to previous approaches, structural reliability analysis can be performed for more precise interpretation of the exceedance probability. The reliability analysis method requires the probabilistic distribution of accidental loads as well as dead weight and live load. Many research efforts have been made to determine the load factor of LRFD against some accidental loads such as drift ice, vehicle collision and ship collision, which are configured in Extreme Event II in AASHTO LRFD (AASHTO, 2012; Ghosn, Moses, & Wang, 2003). Section 4 of this paper shows an example of such

reliability analysis to determine the design criteria of exceedance probability for ship collision load cases.

3.2. Important features of the proposed framework

An accidental load is a *mixed* random variable, i.e. the probabilistic distribution is described by the combination of the probability mass function (PMF) and the probability density function (PDF). This is because the load has a finite probability of being zero ($Q = 0$), i.e. no load impact despite an accident, and shows a continuous probabilistic distribution for non-zero load values ($Q = q \in (0, \infty)$). Thus, the average and the standard deviation of the total distribution may not represent the relevant characteristics of accidental load distribution function. Instead, the shape of the PDF where the design loads may set on is important. Therefore, if the probability of accidents is already less than the design criteria of annual exceedance probability, the examination of accidental loads is unnecessary.

However, it is difficult to consider all possible accidental events for the purpose of accurate estimation of the probability distribution. In the proposed framework, one needs to identify *critical* accident scenarios that make significant contributions to the probability distribution. Therefore, it is desirable to develop advanced reliability algorithms or methods to identify failure mechanisms that contribute to larger accidental loads for the given type of the structure and hazard.

4. Application example of the proposed bridge design framework – ship collision

In this section, the proposed reliability-based structural design framework under accidental loads is demonstrated by a ship collision example investigated by the Core Technology Development Program of Super Long Span Bridge R&D Center of Korea. The design accidental load of a bridge is determined using the framework and tasks described in Section 3.

4.1. Problem description

The structure of interest in this example is the 'E1' pylon of the Incheon Bridge, a cable-stayed bridge in Korea (Figure 2). The 800 m main span of the Incheon Bridge provides two-way navigable channel between the pylons. The water depth is about 20 m, which allows for passage of 100,000 dead weight tonnage (*DWT*) cargo ship. The pylon is 225.5 m high and is supported by a cast-in-place pile foundation. The foundation has a relatively flexible behaviour for lateral loads such as ship impact. Thus, the

Figure 2. Layout and pier notation of Incheon Bridge.

174

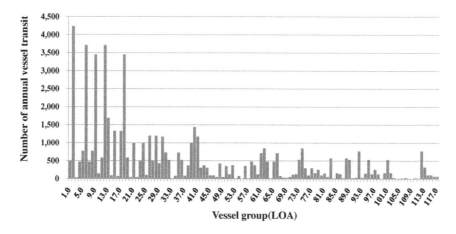

Figure 3. Annual transit data of vessels passing the Incheon Bridge.

dynamic behaviour of the bridge should be analysed for ship collision loads.

Annually, about 50,000 vessels transit under the bridge for the arrivals and departures at the nearby ports. The vessels are categorised into 117 groups, whose annual traffic statistics are reported in Figure 3. Important parameters related to geometrical information of the route are the following: the width of collision part (25 m), the distance from centreline to pier (243.5 m), the waterway width (313 m), the maximum impact speed (5.14 m/s) and the minimum impact speed (.72 m/s). According to the AASHTO LRFD bridge design specifications (2012), the width of collision part and the maximum and minimum impact speed are determined as the width of the pier foundation of the pylon near the transit path and the speed limits on vessel sailing and the average tide speed of nearby seaway respectively. These parameters are used to calculate the collision probability and the ship collision load according to the procedure in Section 3.14 of AASHTO LRFD bridge design specifications (AASHTO, 2012).

4.2. Hazard modelling: probability of bridge–ship collision

Based on AASHTO LRFD bridge design specification, the vessel transit distribution is determined as a normal distribution whose mean value and standard deviation have been assumed equal to the centreline of the vessel transit path and the LOA (length of overall) of the vessels, respectively (see Figure 4). The annual probability of collision for a vessel is determined based on the vessel transit distribution whose area meets with the bridge pier or the foundation. Therefore, the probability of collision can be computed by integrating the area in the normal distribution bounded by the pier width and the width of the vessel on each side of the pier. In this example, the annual probability of collision between transit vessels and the pylon E1 is computed (see Figure 5). Each vessel group has a separate set of collision scenarios, and the conditional ship collision load distribution is calculated for each scenario.

4.3. Performance evaluation: conditional CCDF of ship collision load given collision scenario

The representative value of the collision load is defined as the equivalent static load that the bridge can experience during the collision. The conditional probability distribution of ship collision loads given a scenario is assumed as the symmetric triangular distribution as suggested by Frandsen and Langso (1980). This distribution model was proposed based on Woisin's experiments in 1971, which was conducted using a scaled ship's bow model. According to the triangular distribution model, the conditional PDF of the collision load Q for the ships in the ith collision scenario event E_i is defined as follows:

$$f_{Q|E_i}(q|E_i) = \begin{cases} 4(q - 0.5Qs)/Qs^2 & 0.5Q_s \leq q \leq Q_s \\ -4(q - 1.5Qs)/Qs^2 & Q_s \leq q \leq 1.5Q_s \\ 0 & \text{otherwise} \end{cases} \quad (5)$$

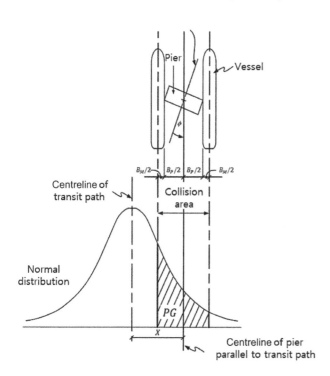

Figure 4. Geometric probability of pier collision.

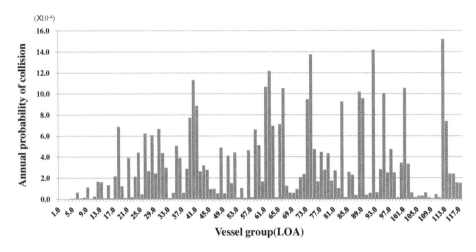

Figure 5. Annual probability of collision by transit vessels.

where Q_s is the mean collision force given in the following equation (unit: MN):

$$Q_S = 0.88 \sqrt{DWT} \frac{V}{8.2} \qquad (6)$$

where V denotes the velocity of the colliding ship (m/s). This value is derived from the speed and weight of a vessel without considering the behaviour of collided structure. The impact speed and weight of vessel are calculated by the procedure in AASHTO LRFD bridge design specifications (AASHTO, 2012). The triangular distribution is derived based on the shape, structure and stiffness of ship's bow. For more precise calculations of the collision load distribution, Lim et al. (2013) proposed a method that considers the impact speed, weight of vessel and the colliding angle as random variables.

Using Equation (3) and the probabilities computed from the previous steps, the total CCDF of the collision load distribution is derived as shown in Figure 6. In this process, it is assumed that all collision scenarios are mutually exclusive and (almost) collectively exhaustive. This graph helps determine the collision

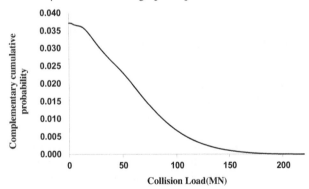

Figure 6. Complementary CDF of the total vessel collision load distribution.

design load for the target annual exceedance probability of catastrophic events by ship collisions.

4.4. Decision-making: selection of annual exceedance probability for bridge design

In this example, in order to prove the appropriateness of the design load, the section properties of the bridge pylon are designed by the load combination of extreme event II in AASHTO LRFD bridge design specifications (2012) which contain the ship collision design load. Then, the structural reliability analysis of the bridge pylon is conducted. The critical section against ship impact was defined as the most vulnerable part of the top of the piles. The limit state for reliability analysis is defined by use of the P-M interaction diagram of the top of the piles because the axial force of self-weight and the moment of ship collision simultaneously affect the piles of the bridge pylon (Kim, Lee, Paik, & Lee, 2013). The statistical parameters of the dead load, live load and the resistance of nominal strength are defined as in Table 2. In this way, the reliability index and annual failure probability of bridge pier can be compared with the design load and the annual exceedance probability. Then, the appropriateness of the design load can be validated by the corresponding reliability index.

In order to examine the dynamic effects of the ship collision load, dynamic analysis is performed by finite element analysis (see Figure 7). The time history of ship collision load computed by a method proposed by Lim et al. (2013), as well as static collision load, was applied to the foundation of the pylon to derive the moment responses of design section, M_{CV}. Plotting the results as in Figure 7, the amplification factor can be computed by dividing the maximum moment from dynamic analysis by that by static analysis. A total of 22 collision scenarios established in terms of load magnitudes were used in the analysis. As a result, the upper limit of dynamic amplification factor about 1.3 is set from the result (see Figure 8). The first-order reliability method (FORM)

Table 2. Statistical parameters of the dead load, live load and the resistance (Nowak, 1999).

Random variable	Bias factor	COV	Distribution type
Dead load	1.050	.100	Normal
Live load	1.000	.180	Lognormal
Resistance	1.165	.206	Lognormal

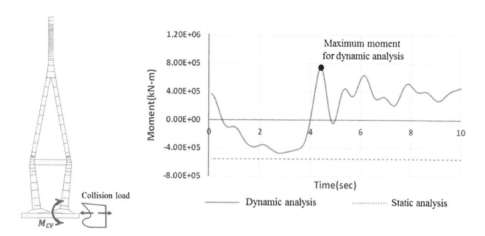

Figure 7. Finite element model for collision analysis (left) and the result (right).

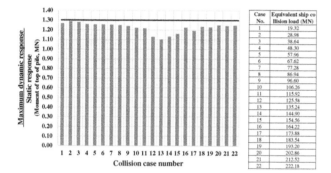

Case No.	Equivalent ship collision load (MN)
1	19.32
2	28.98
3	38.64
4	48.30
5	57.96
6	67.62
7	77.28
8	86.94
9	96.60
10	106.26
11	115.92
12	125.58
13	135.24
14	144.90
15	154.56
16	164.22
17	173.88
18	183.54
19	193.20
20	202.86
21	212.52
22	222.18

Figure 8. Dynamic amplification factors of 22 vessel collision scenarios.

is used to conduct reliability analysis using the probability distributions of loads and the resistance.

The results of reliability analysis for vessel collision load acted on the critical section of pylon are shown in Table 3. This table can provide a basis for the owner or designer's decision-making process during the bridge design against ship collision risk. For example, the design load for the reliability index of 3.0 is the load 193.2MN corresponding to annual exceedance probability .0001.

When the risk analysis results confirm that the bridge of concern does not have sufficient capacity against the estimated ship collision loads, decision-makers need to identify a proper risk mitigation method. One of the most fundamental methods is to place piers on land to avoid ship collision completely. Alternatively, piers can be designed such that they are strong enough to withstand the estimated impact of direct collisions. However, these approaches may not be feasible because of the construction cost or site characteristics. Therefore, the most prevailing approach in practice is installing independent protective structures, such as fender, artificial island and dolphin protections around bridge piers to reflect or stop the colliding ships. It is noted that the reliability-based design framework proposed in this paper can be used to predict the effectiveness of such risk mitigation methods as well.

5. Research needs for applying the proposed framework to other accidental loads

In the same manner with ship collision events, other accidental events such as fire, explosion and dropping object can be dealt with reliability-based design framework shown in Figure 1. In order to apply the design framework, scenarios for each event and corresponding probabilistic models shall be developed first reflecting site dependency (hazard modelling). Then, the load has to be determined using conditional probability distribution reflecting the uncertainties and mechanisms (performance evaluation). After establishing an acceptance criterion, a designer makes decision on risk mitigation methods which meet probabilistic level of safety (decision-making). The reliability-based design framework is illustrated in Figure 9.

As an effort to facilitate the actual applications of the above procedure, uncertainties related to Task 1 and Task 2 and related research issues are summarised in Tables 4 and 5, respectively. Hazard modelling, performance evaluation and decision-making can be accomplished for various accidental events through research efforts related to the issues. A schematic outline of scenarios-based design for accidental events summarising contents on tables is also given in Figure 10.

Table 3. Results of reliability analysis for E1 pylon of Incheon Bridge against ship collision.

Annual exceedance probability of design load, P_E	Vessel collision design load (MN)	Reliability index β	Annual failure probability P_f
.001	159.7	2.7483	.002995
.0002	185.7	2.9555	.001561
.0001	193.2	3.0124	.001296
.00002	210.7	3.1437	.000834

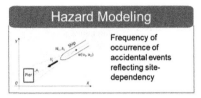
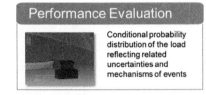

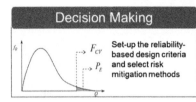

Figure 9. Reliability-based design framework against accidental events.

Table 4. Uncertainties and issues related to Task 1: identifying critical accident scenarios and computing the probabilities (AASHTO, 2009; ABS, 2013).

Accidental event	Uncertainties and research issues
Fire	• Release condition of the inflammables (leakage of pipe/valve/tank, stack of the inflammables bulk such as wood, coal, chemicals)
	• Probability of the ignition (by human, machinery)
	• Environmental conditions (wind, temperature, pressure, etc.)
Explosion (blast)	• Release condition of the explosive materials (terrorism not considered)
	• Probability of the trigger action (by human, machinery)
	• Environmental conditions (wind, temperature, pressure, etc.)
Dropped objects	• Frequency of the working (lifting, drilling, etc.)
	• Probability of errors (by human, mechanical, environment)
Ship collision	• Frequency of vessel transit
	• Probability of errors (by human, mechanical)
	• Geometry of the channel (or fairway)
	• Location of structure (fixed)
	• Traffic regulations of the site
	• Environmental conditions (the speed of fluid, the height of a wave, etc.)
Common factors	• Scenario definition

Table 5. Uncertainties and issues related to Task 2: computing conditional CCDF of accidental loads given each scenario (AASHTO, 2009; ABS, 2013).

Accidental event	Uncertainties and research issues
Fire	• Thermal loads/structural load
	• Fire type: pool or jet
	• Flame geometry with respect to time (width, length, shape, etc.)
	• Intensity (radiant/convective heat flux, momentum pressure load)
Explosion (blast)	• Blast definition (overpressure exceedance curve, relationship between event duration and overpressure, inclusion of rebound blast, presence of projectiles)
	• Rise time
	• Maximum pressure
	• Pulse duration
Dropped objects	• Ongoing operation, e.g. lifting, drilling or severe environment
	• Dropped object (including structure description and mass)
	• Drop characteristics (including location of anticipated impact and height of drop)
	• Velocity at impact
	• Impact angle
	• Location of impact
	• Environmental conditions, e.g. weather and sea state
Ship collision	• Mass
	• Characteristic properties, e.g. hull shape, form and draft
	• Velocity of impact
	• Approach angle
	• Location of contact
	• Environmental conditions, e.g. current, wind and sea state, during event
Common factors	• Structural configuration
	• Material properties

$$P_E = P[q < Q] = \sum_{i=1}^{n} P[q < Q|E_i]P[E_i]$$

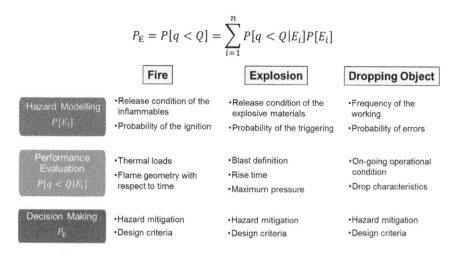

Figure 10. Scenario-based design for accidental events.

6. Conclusions

In this paper, a new reliability-based framework is presented to determine the structural design load appropriately based on the quantified risk of accidental loads. In the proposed framework, the probabilistic distributions of the accidental loads are derived based on the probabilities of accidental scenarios and the conditional probability distribution of the load given each scenario. From the target annual exceedance probability determined based on the socio-economic impact of the catastrophic events and comparison with other hazards, the corresponding design load can be calculated from the accidental load distribution. The proposed framework and related tasks were demonstrated by a design example of a design example of a bridge under the risk of ship collision. Through dedicated research efforts in the future, the proposed framework can be extended to manage the risk of other catastrophic events such as fire, explosion, dropped objects and ship collision.

In order to extend the proposed framework to structural design under other types of accidental loads, both probabilistic data of accidental loads and the mechanical reaction of structures under extreme conditions, such as high temperature and pressure, and impact force should be investigated in detail. For more reliable design loads, various uncertainties related to accident scenarios and structural analysis under extreme conditions should be identified and understood by research efforts. The probabilistic analysis of historical accidents and mechanical experiments in extreme condition would be particularly helpful. As an effort to gain better understanding of the behaviour of structures under various extreme conditions, structural performance testing facilities for extreme conditions are currently under construction at Seoul National University in Korea, and are expected to provide useful information that would promote a wide application of the proposed framework to various problems.

One important question we need to answer regarding structural design against accidental loads is 'how safe is safe enough?' Efforts to answer this question would help select the reasonable design criteria of accidental loads. The decisions related to design and management of infrastructures entail significant impacts on public safety, economy and culture. Therefore, a multidisciplinary research is needed to make the best decision on developing the design criteria. For example, concepts such as 'quality of life index' (Ditlevsen, 2003; Rackwitz, 2002) could be used as helpful indicators for structural designers to develop such design criteria. It is also important to compare the quantified risk of accidental loads with other risks such as traffic accidents and cancer. Future research in this area is expected to promote engineers' active participation in the society's effort for public happiness through the development of sustainable and resilient infrastructure against various man-made hazards.

Disclosure statement

No potential conflict of interest was reported by the authors.

Funding

The work was supported by the grant [15CCTI-A052531-08-000000] from the Ministry of Land, Infrastructure and Transport of Korean government as the Core Engineering Technology Development Project at Seoul National University through Super Long Span Bridge R&D Center; the Institute of Construction and Environmental Engineering (ICEE) at Seoul National University; and National Research Foundation of Korea (NRF) Grant [number 2015R1A5A7037372], funded by the Korean Government (MSIP).

ORCID

Junho Song http://orcid.org/0000-0003-4205-1829

References

American Association of State Highway and Transportation Officials. (2009). *Guide specification and commentary for vessel collision design of highway bridges*. Washington, DC: Author.

American Association of State Highway and Transportation Officials. (2012). *AASHTO LRFD bridge design specifications* (6th ed.). Washington, DC: Author.

American Bureau of Shipping. (2013). *Guidance notes on accidental load analysis and design for offshore structures*. Houston: Author.

British Standards Institution. (2006). *Eurocode 1 - Actions on structures Part 1–7: General actions – Accidental actions*. London: Author.

Chopra, A. K. (1995). *Dynamics of structures*. Englewood Cliffs, NJ: Prentice Hall.

Det Norske Veritas. (2010). *Design against accidental loads – Recommended practice*. DNV-RP-C204. Hovik: Author.

Ditlevsen, O. (2003). Decision modeling and acceptance criteria. *Structural Safety, 25*, 165–191.

Ellingwood, B. R. (2000). *Performance-based design: Structural reliability considerations*. Advanced Technology in Civil Engineering, Proceedings of Structures Congress, Philadelphia, PA.

Ellingwood, B. R. (2001). Acceptable risk bases for design of structures. *Progress in Structural Engineering and Materials, 3*, 170–179.

Ellingwood, B. R. (2006). Mitigating risk from abnormal loads and progressive collapse. *Journal of Performance of Constructed Facilities, 20*, 315–323.

Frandsen, A., & Langso, H. (1980). Ship collision problems: I. Great Belt Bridge and II. International enquiry. *Proceedings of IABSE Periodica, 2*, 81–108.

Ghosn, M., Moses, F., & Wang, J. (2003). *Design of highway bridges for extreme events*. Washington, DC: Transportation Research Board.

Kim, J.-H., Lee, S.-H., Paik, I., & Lee, H. S. (2013). *Reliability assessment of reinforced concrete column under axial-flexural loading*. Proceeding of Annual Conference of the Korean Society of Civil Engineers. Jeongseon, Korea: Korean Society of Civil Engineers.

Knott, M. A. (1998). Vessel collision design codes and experience in the United States. In Gluver, H. & Olsen, D. (Eds.), *Ship collision analysis* (pp. 75–84). Rotterdam: A.A. Balkema.

Lim, J.-H., Park, W., Kim, H.-J., & Koh, H.-M. (2012a). *Probabilistic load model for ship-bridge collision using Monte Carlo simulation*. Proceeding of 18th IABSE Congress, Seoul, Korea.

Lim, J.-H., Park, W., Kim, H.-J., & Koh, H.-M. (2012b). *Simulation based probability model of impact load due to ship-bridge collision*. Proceeding of 2012 IABSE Conference, Sharm El Sheikh.

Lim, J.-H., Park, W., & Koh, H.-M. (2013). *Dynamic vessel collision load for long-span cable-supported bridges*. Proceeding of The 7th International Symposium on Steel Structure, Jeju, Korea.

Nowak, A. S. (1999). *Calibration of LRFD bridge design code* (Final Rep. NCHRP 12-33). Ann Arbor: Department Of Civil and Environmental Engineering, University of Michigan.

Park, W., Lim, J.-H., & Koh, H.-M. (2013a). Estimation of probabilistic scenario-based design load for extreme events. *KSCE Journal of Civil Engineering, 17*, 594–601.

Park, W. L., Lim, J.-H. & Koh, H.-M. (2013b). *Vessel collision design method for long-span bridges*. Proceeding of 36th IABSE Symposium, Kolkata, India.

Paté-Cornell, M. E. (1994). Quantitative safety goals for risk management of industrial facilities. *Structural Safety, 13*, 145–157.

Rackwitz, R. (2002). Optimization and risk acceptability based on the life quality index. *Structural Safety, 24*, 297–331.

Structural Engineering Institute. (2006). *Minimum design loads for buildings and other structures*. Reston, VA: ASCE Publications.

Target reliability for existing structures considering economic and societal aspects

Miroslav Sykora [iD], Dimitris Diamantidis, Milan Holicky and Karel Jung

ABSTRACT

Specification of target reliability levels is one of the key issues of the assessment of existing structures. ISO 13822:2010 and ISO 2394:2015 indicate procedures for specification of target reliability levels by optimisation of the total cost related to an assumed remaining working life of a structure. These approaches are critically compared with human safety criteria, with target levels based on a marginal life-saving costs principle, and with recommendations of present standards. Optimal target reliability levels are then derived in the representative case study for an existing structural member. It appears that the requirement to reach the same target reliability levels for existing and new structures is uneconomical. Decisions made in the assessment can result in the acceptance of the actual state or in the upgrade of an existing structure. Two reliability levels are thus needed – the minimum level below which the structure is unreliable and should be upgraded, and the target level indicating an optimum upgrade strategy. It is recommended that these levels be established using economic optimisation and the marginal life-saving costs principle, as both the approaches take into account the costs of safety measures and the failure consequences.

Introduction

At present, about 50% of the investments in construction are related to existing structures. Regarding reliability verifications, the assessment of existing structures differs from the design of a new structure in a number of ways, including:

- Increased costs of upgrading in order to achieve the same safety level.
- A remaining working life shorter than the standard design working life of 50–100 years, depending on the type of structure (residential or industrial building, bridge, etc.).
- Additional information on actual structural conditions that may be available for the assessment of an existing structure due to inspections, tests and measurements.

In addition, existing structures are mostly verified using conservative procedures based on the partial factor method for structural design. Such assessments may then lead to expensive upgrades. A more realistic verification of the actual performance of existing structures can be achieved by probabilistic methods when relevant actual uncertainties are described by appropriate probabilistic models. Specification of the target reliability levels is needed for probabilistic assessments, as well as modifications of partial factors (Caspeele, Sykora, Allaix, & Steenbergen, 2013; Sykora, Holicky, & Markova, 2013). Target reliability levels used in the verification process are mainly associated with the failure of components and not with the collapse of a whole structure or its main part.

The target reliability levels recommended in (EN 1990, 2002) are primarily intended for new structures, while (ISO 2394, 2015) gives more detailed and, to some extent, different recommendations. The target reliability levels provided in these documents are partly based on calibrations relating to previous practice and should be considered as indicative only. Furthermore, (ISO 13822, 2010) indicates the possibility of specifying the target reliability levels for existing structures by optimisation of the total cost related to an assumed remaining working life. However, the standard does not provide further guidance for reducing the target reliabilities, e.g. for a shorter residual lifetime.

Amongst others, (Vrouwenvelder & Scholten, 2010) recognised that the requirement to reach the same reliability levels for existing and new structures would be uneconomical. More recently, the background document for the European standard on assessment of existing structures (Luechinger et al., 2015) recommended the same reliability for an existing structure as that required in the design of this structure. However, in many cases such reliability levels are unknown. (Diamantidis & Bazzurro, 2007) argued that lower target reliabilities for existing structures were unavoidable. (Bigaj-van Vliet & Vrouwenvelder, 2013) concluded that further calibrations and improvements of the methodology for reliability differentiation would still be needed.

The present contribution attempts to derive appropriate target reliability levels for the structural member dominating the system failure mode of an existing structure – 'key structural member' – when ultimate limit states are decisive and cost optimisation and

Table 1. List of symbols.

Symbol	Description
A	Constant in the societal risk criterion – frequency of the occurrence of an event with $N \geq 1$ fatalities
A_{col}	Collapsed area
A_{floor}	Floor area
C_0	Cost independent of the decision parameter
C_1	Cost dependent on decision parameter
C_f	Consequences of failure expressed in monetary units, i.e. failure cost
C_{str}	Present value of structural cost
C_{tot}	Expected total costs
G	Permanent load
G_x	Societal Willingness to Pay per life saved
K	Constant in the LQI acceptance criterion according to (ISO 2394, 2015)
K_E	Load effect uncertainty
K_R	Resistance uncertainty
N	Expected number of fatalities given collapse of the structure
Q	Time factor
R	Resistance
R_d	Design value of resistance
S	Snow load
V_X	Coefficient of variation of X
X	Basic variable
c	Cost C related to C_{str}, $c = C/C_{str}$
d	Decision parameter
d_0	Value of decision parameter before upgrade
d_{0lim}	Limiting value of d_0 – for $d_0 < d_{0lim}$, the reliability level of an existing structure is too low and the structure should be upgraded, while from an economic point of view, the present state is accepted for $d_0 > d_{0lim}$
d_{opt}	Optimum value of decision parameter
$p_{d/f}$	Conditional probability of an occupant fatality given failure of the structure
p_f	Failure probability related to t_0, typically to one year
t_0	Basic periods for which failure events are independent
t_{ref}	Reference period for reliability analysis
q	Annual discount rate
x_k	Characteristic value
Φ, Φ^{-1}	Cumulative distribution function of the standardised normal distribution and its inverse
a	Constant in the societal risk criterion – risk aversion factor
β	Target reliability index without the distinction between β_0 and β_{up} levels
β_0	Minimum reliability index below which the considered structural member is unreliable and should be upgraded
β_{up}	Target reliability index corresponding to an optimum upgrade strategy
γ_s	Interest rate of relevance for decision-making on behalf of society
μ_X	Mean of X
ω	Annual rate of obsolescence

(Sykora, Diamantidis, Holicky, & Jung, 2015). The symbols used in the paper are listed in Table 1.

Target reliability levels in codes

In (EN 1990, 2002) and (ISO 2394, 2015), reliability index β is generally used as a measure of the reliability. This index is related to failure probability through the inverse of the standardised normal cumulative distribution. It is noted that the target reliability levels in codes of practice provide criteria for limit states that do not account for human errors, i.e. the target levels should be compared with the so-called notional reliability indicators (ISO 2394, 2015). (Melchers, 2001) argued that the optimum solutions should be hardly affected by considerations of human errors. The target levels are differentiated with respect to various parameters. It is shown here that the target reliability can be specified by taking into account.

Costs of safety measures

These costs should reflect the efforts needed to improve structural reliability considering the properties of the construction materials and the characteristics of the failure modes investigated. The relative cost of safety measures depends significantly on the variability of load effects and resistances (Vrouwenvelder, 2002).

Failure consequences

Herein failure consequences are understood to cover all direct and indirect (follow-up) consequences related to failure. When specifying these costs, the distinction between ductile or brittle failure (warning factor), redundancy and possibility of progressive collapse should be taken into account. In this way it would be possible to consider system failure in component design. However, such implementation is not always feasible in practice, and therefore consequence classes are specified with respect to the use of a structure (EN 1990, 2002), and to the number of persons at risk (ASCE 7-10, 2013).

Time parameters

Target levels are commonly related to a reference period or a design working life. The reference period is understood as a chosen period of time, used as a basis for statistically assessing time-variant basic variables, and the corresponding probability of failure. The design working life is considered here as an assumed period of time for which a structure is to be used for its intended purpose without any major repair work being necessary. Therefore, the concept of reference period is fundamentally different from the concept of design working life. Obviously, the target reliability should always be specified together with a reference period considered in reliability verification. (ISO 2394, 2015) indicates that the remaining working life can be considered as a reference period for the serviceability and fatigue limit states, while a shorter reference period might be reasonable for the ultimate limit states. When related to failure costs, it is proposed here to refer to lifetime probabilities if economic consequences

criteria for human safety are applied. In (EN 1991-1-7, 2006), a key element of the structure is defined as a structural member upon which the stability of the remainder of the structure depends. The European standard also provides guidance on how to identify such important members on the basis of expected damaged floor area. Exterior columns and load-bearing walls represent examples of these members. In practice, key elements are often identified on the basis of the experience of the designer.

A representative existing building with a remaining working life of 15 years, moderate costs of safety measures and moderate failure consequences – hereafter 'the representative structure' – is considered throughout the paper to clarify and illustrate the general methodology. Such a structure may be represented by an existing residential, office, public or industrial building, or a common bridge with relatively inexpensive strengthening, for example, a reinforced concrete or masonry structure strengthened by fibre-reinforced polymers, with no restrictions regarding its use. The paper is an extended revision of the recent contribution

Table 2. Target reliability indices according to selected standards – different reference periods and comparable failure consequences.

Standard	Failure conseq.	Reference period	β in standard	β represent. str.
(EN 1990, 2002)	Medium	50 y. (1 y.)	3.8(4.7)	4.1[*]
ISO 2394:1998	Moderate	Life-time	2.3/3.1/3.8[**]	3.1
(ISO 2394, 2015) (economic optimisation)	Class 3	1 y.	3.3/4.2/4.4[**]	3.5[*]
(ISO 2394, 2015) (LQI)	–	1 y.	3.1/3.7/4.2[**]	2.9[*]
(ISO 13822, 2010)	Moderate	Minimum standard period for safety	3.8	3.8
(NEN 8700, 2009)	Moderate	Rem. work. life	2.5	2.5

[*]Recalculated for a reference period of 15 years; [**]High/moderate/low relative costs of safety measures, respectively.

dominate. When human safety is endangered, a reference period of one year is normally accepted.

In (EN 1990, 2002), it is recommended that the target reliability index β for two reference periods (1 and 50 years) should be used; see the example for medium consequences of failure in Table 2. These target reliabilities are intended to be used primarily for the design of members of new structures. The two β-values given in Table 2 are provided for two reference periods used for reliability verification and should correspond approximately to the same reliability level:

$\beta = 3.8$ should be used when probabilistic models of basic variables are related to a reference period of 50 years.

The same reliability level should be approximately reached when $\beta = 4.7$ is applied using models related to one year and failure events in individual yearly intervals – basic reference periods for variable loads (see the following section) – are independent.

Considering an arbitrary reference period t_{ref}, the reliability level is derived from an annual target according to (EN 1990, 2002) as follows:

$$\beta(t_{ref}) = \Phi^{-1}\left\{ \left[\Phi(\beta(1 \text{ y.})) \right]^{t_{ref}} \right\} \tag{1}$$

where $\Phi(\cdot)$ = cumulative distribution function of the standardised normal distribution, Φ^{-1} being its inverse; and $\beta(1 \text{ y.})$ = target reliability index related to $t_{ref} = 1$ year. Thus, $\beta = 4.1$ should be accepted for $t_{ref} = 15$ years in the case of the representative structure.

When compared to (EN 1990, 2002), a more detailed and substantially different recommendation is provided by a former edition of ISO 2394 dating from 1998. The target reliability index is given for the working life and related not only to the consequences but also to the relative costs of safety measures, as is exemplified in Table 2. The consideration of costs of safety measures is particularly important for existing structures. The target reliability $\beta = 3.1$ may be selected for the representative structure. According to ISO 2394:1998, the target level for existing structures apparently decreases as it takes relatively more effort to increase the reliability level compared to the design of a new structure. Consequently, for an existing structure one may use the values of one category higher, i.e. instead of 'moderate' consider 'high' relative costs of safety measures (Vrouwenvelder, 2002). This is in agreement with recommendations of (fib, 2013).

Similar recommendations are provided in (JCSS, 2001a) and (ISO 2394, 2015). The target reliability indices are also related to both the consequences and to the relative costs of safety measures, for the reference period of one year, however. (ISO 2394, 2015) gives the target levels based on economic optimisation and

acceptance criteria using the Life Quality Index (LQI) (Nathwani, Pandey, & Lind, 2009), as is briefly explained in Section Life Quality Index of this paper. The β-values, given in Table 2 for the representative structure, are obtained as follows:

- Economic optimisation: annual $\beta = 4.2$ for medium relative cost of safety measures and economic losses expected for most residential buildings or common pedestrian and road bridges is recalculated to $\beta = 3.5$ for a 15-year reference period.
- LQI: annual $\beta = 3.7$ for the medium relative lifesaving cost is recalculated to $\beta = 2.9$ for $t_{ref} = 15$ years.

In (ASCE 7-10, 2013) buildings and other structures are classified into four risk categories according to the number of persons at risk. Category I is associated with few persons at risk and category IV with tens of thousands. For all loads except for earthquake, (ASCE 7-10, 2013) aims to reach the annual reliability levels reported by (Hamburger, 2013), i.e. from 3.7 for category I up to 4.4 for category IV. By including additional factors such as inspectability, the Canadian Standards Association uses a different and slightly more detailed approach for bridges than the aforementioned documents (Allen, 1993).

On the other hand, (ISO 13822, 2010) indicates four target reliability levels for the ultimate limit states and different consequences of failure: small – 2.3, some – 3.1, moderate – 3.8 and high – 4.3. The related reference period is 'a minimum standard period for safety (e.g. 50 years)'. $\beta = 3.8$ may thus be adopted for the representative structure associated with moderate consequences of failure, assuming that a 15-year reference period can be considered as a minimum standard period for safety (Table 2).

Recommendations on the target reliability levels are also provided in several national standards (Diamantidis & Bazzurro, 2007). The results obtained mostly indicate that the requirement to reach the same reliability levels for existing and new structures is uneconomical. For instance, the Dutch standard for the rules of the assessment of existing structures (NEN 8700, 2009) provides β-values for the remaining working life in the case of strengthening and disapproval. Reliability indices 1.8, 2.5 and 3.3 are considered for low, moderate and high consequences of failure, respectively, and a minimum reference period of 15 years. The target reliability $\beta = 2.5$ may thus be accepted for the representative structure associated with moderate consequences of failure (Table 2).

For bridges, the reduction is 0.5 in Canada and 1.7 in the U.S.A (Casas & Wisniewski, 2013). These reductions agree with the observations obtained in the case study presented in this paper. However, it must be emphasised that the same target

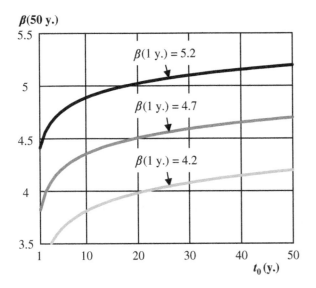

$\beta(50 \text{ y.})$

$\beta(1 \text{ y.}) = 5.2$

$\beta(1 \text{ y.}) = 4.7$

$\beta(1 \text{ y.}) = 4.2$

$t_0 \text{ (y.)}$

Figure 1. Variation of β(50 y.) with t_0 and selected β(1 y.) values (failures during t_0 basic periods mutually independent).

reliabilities are required for existing and new structures in many other countries.

It follows from Table 2 that the target reliability levels obtained for the representative structure according to the selected standards are considerably different. The difference between (EN 1990, 2002), deemed to provide recommendations particularly for structural design, and (NEN 8700, 2009) focused on existing structures is significant, $\Delta\beta = 1.6$. Obviously, further insights into the specification of target levels are needed to clarify appropriate concepts and reach the desired harmonisation in code provisions.

In general, (ISO 2394, 2015) and (JCSS, 2001a, 2001b) seem to provide a more appropriate reliability differentiation for existing structures than (EN 1990, 2002) and (ISO 13822, 2010), since costs of safety measures are taken into account. It is worth noting that other standards consider additional factors such as type of failure and expected warning before failure that do, however, affect failure consequences. In addition, several empirical models for the assessment of target reliabilities have been proposed in previous studies. A brief overview of these models was provided by (Sykora, Holicky, & Markova, 2011).

Influence of reference period

Equation (1) is based on the assumption of independent failure events in subsequent years. (Vrouwenvelder, 2002) suggested that the target level of 3.8 for 50 years in Table 2 could better be interpreted as corresponding to $\beta_1 = 4.5$ (and not to 4.7) as the full independence of failure events in subsequent years is hardly realistic.

To illustrate this, it is assumed that the basic periods t_0 for which failure events are independent can be identified. Equation (1) can then be reformulated as follows:

$$\beta\left(t_{\text{ref}}, t_0\right) = \Phi^{-1}\left\{ \left[\Phi\left(\beta\left(1 \text{ y.}\right)\right)\right]^{t_{\text{ref}}/t_0} \right\} \qquad (2)$$

where $t_{\text{ref}}/t_0 \geq 1$ is the expected number of events that may lead to failure. The following indications regarding t_0 can be useful:

- One year can be accepted in many cases, e.g. when climatic or traffic actions govern structural reliability.
- 5–10 years can be considered for structures dominated by a sustained component of the imposed load in office buildings (JCSS, 2001a).
- $t_0 = t_{\text{ref}}$ can be used for the cases in which the reliability is insignificantly affected by time-variant phenomena, e.g. for the structures subjected to dominating permanent actions, masonry or geotechnical structures.

Variation of reliability index $\beta(t_{\text{ref}})$ with t_0 is shown in Figure 1. It can be observed that for $t_{\text{ref}} = 50$ y. and $\beta(1 \text{ y.}) = 4.7$, the following target values should be considered:

- $\beta(50 \text{ y.}) = 3.8$ for $t_0 = 1$ y.
- $\beta(50 \text{ y.}) = 4.4$ for $t_0 = 10$ y.
- $\beta(50 \text{ y.}) = 4.7$ for $t_0 = 50$ y.

Apparently, the assumption of independent failure events needs to be carefully verified before selecting appropriate target levels.

Cost optimisation

According to (ISO 13822, 2010), lower target reliability levels can be used if justified on the basis of societal, cultural, economical and sustainable considerations. (ISO 2394, 2015) indicates that the target level of reliability should depend on the balance between consequences of failure and costs of safety measures. From an economic point of view, the objective is to minimise the total working-life cost.

The expected total costs C_{tot} may be generally considered as the sum of the expected costs of inspections, maintenance, upgrades and costs related to failure (malfunction) of a structure. It is noted that life-cycle cost as a key parameter in decision-making at the design phase include in addition end-of-life expenses that normally cover decommissioning, deconstruction, demolition or recycling costs (ISO 15686-5, 2008). However, these costs do not affect decision-making regarding the reliability of existing facilities, and thus are not taken into account in the following analysis.

The decision parameter(s) d to be optimised in the assessment may influence resistance, serviceability, durability, maintenance, inspection, upgrade strategies, etc. Examples of d include shear and flexural resistances or stiffness of a girder to control deflections. In the present study, the decision parameter is assumed to concern structural resistance affecting ultimate limit states with a need for an immediate upgrade, while inspection, maintenance and future upgrade strategies are influenced marginally. Moreover, the benefits related to the use of a structure that in general should be considered in the optimisation are assumed to be independent of a decision parameter. These are reasonable assumptions in many practical cases.

In general, the immediate upgrade costs consist of:

- Cost C_0 independent of the decision parameter, such as costs related to surveys, design, administration and

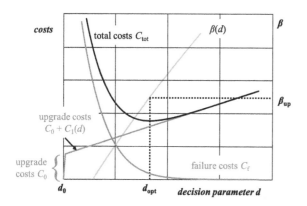

Figure 2. Variation of upgrade and failure costs with decision parameter and optimal upgrade strategy based on minimisation of total costs.

management, economic losses due to business interruption or replacement of users.

• Cost $C_1(d)$ dependent on the decision parameter; normally a linear relationship with respect to d is appropriate.

In general, the former cost significantly exceeds the latter, $C_0 \gg C_1(d)$, (Rackwitz, 2000). For more details see (ISO 15686-5, 2008).

Depending on the subject concerned, the failure cost C_f may include (ISO 2394, 2015):

• Cost of upgrade or replacement.
• Economic losses due to non-availability or malfunction of the structure.
• Societal consequences covering costs of injuries and fatalities, expressed commonly in terms of compensations or insurance costs.
• Unfavourable environmental effects including CO_2 emissions, energy use or release of dangerous substances.
• Other costs related to the loss of reputation or to undesirable suboptimal changes in design practice.

It is noted that the consequences of failure of a specific structure may differ when estimated by public authorities or private owners (Faber, 2015). For instance, the public owner might consider business losses and increased insurance costs only, while the public authority should duly consider all the economic and societal costs related to the failure.

Figure 2 illustrates the variation of upgrade and failure costs with the decision parameter. For consistency, the upgrade and failure costs need to be expressed on a common basis. The upgrade cost is normally given according to its present value. All the expected failure costs that may occur within a reference period should thus be estimated likewise according to the present worth. This is achieved by converting the expected failure costs to a present value (Holicky, 2013):

$$E\left[C_f\left(t_{\text{ref}}, d\right)\right] = C_f p_f(d) Q_f\left(t_{\text{ref}}, d\right) \quad (3)$$

where C_f = present value of the failure cost; $p_f(d)$ = failure probability related to a basic reference period t_0; and Q = time factor. The time factor Q obtained for the sum of a geometric sequence is (Holicky, 2013, 2014):

$$Q\left(t_{\text{ref}}, d\right) = \frac{1 - \left[\left(1 - p_f(d)\right)/(1+q)\right]^{\frac{t_{\text{ref}}}{t_0}}}{1 - \left[\left(1 - p_f(d)\right)/(1+q)\right]} \quad (4)$$

where q = annual discount rate for which (ISO 15686-5, 2008) assumes values between 0 and 0.04. (ISO 2394, 2015) then suggests 0.01–0.03 for most Western economies and 0.05–0.08 for most Asian economies. (Lentz, 2007) discussed in detail discounting for private and public sectors, indicating values around 0.05 for a long-term average of q, while (Lee & Ellingwood, 2015) suggested lower values for investments over multiple generations.

The expected total costs are now expressed as:
In case of upgrade:

$$E\left[C_{\text{tot}}\left(t_{\text{ref}}; d\right)\right] = C_0 + C_1(d) + C_f p_f(d) Q\left(t_{\text{ref}}, d\right) \quad (5a)$$

In case of no upgrade – accepting a present state:

$$\text{E}\left[C_{\text{tot}}\left(t_{\text{ref}}; d_0\right)\right] = C_f p_f\left(d_0\right) Q\left(t_{\text{ref}}, d_0\right) \quad (5b)$$

where d_0 = value of the decision parameter before the upgrade.

Previous experience shows that it is more convenient to analyse costs related to the present value of the structural cost C_{str} – the present cost of a new structural member identical to an existing member to be assessed, hence corresponding to d_0. The cost C_{str} is considered here as including all costs of labour, materials, equipment, construction site management and quality control. Equations (5a) and (5b) can then be rewritten as follows:
Upgrade:

$$\begin{aligned}
C_{\text{tot}}\left(t_{\text{ref}}; d\right)/C_{\text{str}} \\
= c_{\text{tot}}\left(t_{\text{ref}}; d\right) = C_0/C_{\text{str}} + C_1(d)/C_{\text{str}} \\
+ C_f/C_{\text{str}} p_f(d) Q\left(t_{\text{ref}}, d\right) \\
= c_0 + c_1(d) + c_f p_f(d) Q\left(t_{\text{ref}}, d\right)
\end{aligned} \quad (6a)$$

No upgrade:

$$c_{\text{tot}}\left(t_{\text{ref}}\right) = c_f p_f\left(d_0\right) Q\left(t_{\text{ref}}, d_0\right) \quad (6b)$$

The costs related to C_{str} are hereafter denoted by the small letters c; the symbol of expectation is omitted. Apparently, the total costs C_{tot} and c_{tot} attain a minimum for the same value of the decision parameter. From Equation (6a), the optimum value of the decision parameter d_{opt} – an optimum upgrade strategy – can be assessed as:

$$\text{minimum}_d c_{\text{tot}}\left(t_{\text{ref}}; d\right) = c_{\text{tot}}\left(t_{\text{ref}}; d_{\text{opt}}\right) \quad (7)$$

It follows from Equations (6a) and (7) that d_{opt} is independent of c_0. The optimum upgrade strategy should aim at the target reliability corresponding to d_{opt}:

$$\beta_{\text{up}} = \Phi^{-1}\left\{\left[1 - p_f\left(d_{\text{opt}}\right)\right]^{t_{\text{ref}}/t_0}\right\} \quad (8)$$

The identification of the optimal upgrade strategy is indicated in Figure 2. From an economic point of view, no upgrade is

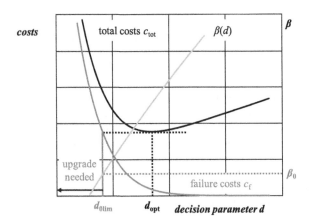

Figure 3. Decision on upgrade using economic optimisation.

undertaken when the total cost according to Equation (6b) is less than the total cost of the optimum upgrade as indicated in Figure 3. The limiting value $d_{0\text{lim}}$ of the decision parameter before the upgrade is then found as follows:

$$c_f p_f\left(d_{0\text{lim}}\right)Q\left(t_{\text{ref}},d_{0\text{lim}}\right) = c_0 + c_1\left(d_{\text{opt}}\right) \\ + c_f p_f\left(d_{\text{opt}}\right)Q\left(t_{\text{ref}},d_{\text{opt}}\right) \tag{9}$$

For $d_0 < d_{0\text{lim}}$, the reliability level of an existing structure is too low, failure costs become high, and the decision is to upgrade the structure or its member, as the optimum upgrade strategy yields a lower total cost. For $d_0 > d_{0\text{lim}}$, the present state is acceptable from an economic point of view, since a no upgrade strategy leads to a lower total cost than the costs expected when no upgrade is taken. The minimum reliability index β_0 below which the considered structural member is unreliable, and should be upgraded, is thus obtained as follows:

$$\beta_0 = \Phi^{-1}\left\{\left[1 - p_f\left(d_{0\text{lim}}\right)\right]^{t_{\text{ref}}/t_0}\right\} \tag{10}$$

An application of these principles in decision-making concerning road bridges was shown by (Lenner & Sýkora, 2016).

Note that the decision may also be made to demolish the structure and build a new one. The optimisation then needs to account for:

- Costs of the upgrade and construction of a new structure
- Additional benefits associated with a new structure, possibly better suited for the intended purpose

However, demolition might be the optimum decision, particularly when the structure is inappropriate for the intended new use, or has a very low reliability level. A detailed discussion of these related aspects is rather beyond the scope of this paper.

Requirements on human safety

As indicated in (ISO 2394, 2015), the target failure probabilities may be selected solely on the basis of economic optimisation, only if no risk to human life is associated with structural failure.

Otherwise, individual or societal risk criteria, or the marginal life-saving costs principle should apply. Such criteria essentially represent social requirements, whereas economic optimisation reflects the preferences of the owner of a particular structure. Acceptable risks to occupants or users of the structure should be assured, as compared to other daily activities, or to other industries. Basically, these criteria are associated with the global failure of a structure, or of the main parts of it, but can also be related to structural components dominating the system failure, as is considered here.

Individual risk criterion

General guidelines for the assessment of target reliabilities with respect to human safety are provided in (ISO 2394, 2015). Based on the concept of individual risk, the annual target failure probability depends on the conditional probability of an occupant fatality $p_{d|f}$, given the failure of the structure:

$$p_f\left(\text{per year}\right) \leq 10^{-5}\left(\text{per year}\right)/p_{d|f} \tag{11}$$

where the value 10^{-5} per year is assumed to be optimal for existing structures (JCSS, 2001b; Tanner & Hingorani, 2015), reflecting somewhat indirectly the differences between existing and new structures, and is accepted here. ISO 2394:1998 indicated a value of 10^{-6} for the design of new structures, a value that was recently re-examined and justified (Tanner & Hingorani, 2015).

With respect to the loss of human life, (EN 1990, 2002) distinguishes between low, medium and high consequences by the Consequence Classes CC1, CC2 and CC3, respectively. Based on the study (Steenbergen & Vrouwenvelder, 2010) and limited statistical data (Sykora, Holicky, Markova, & Senberger, 2016), probabilities $p_{d|f}$ might be approximated by the values 0.0025, 0.025 and 0.25 for CC1 to CC3, respectively (Table 3). The evaluation of the excessive database of structural failures revealed probability of fatality given the structural failure provided in Table 3 (Eldukair & Ayyub, 1991). It must be emphasised that these values should be considered as an upper bound on $p_{d|f}$. Given the structural failure, the probability of fatality of an individual occupant whose risk is being assessed is obviously lower than the probability of at least one fatality of any occupant.

The annual target failure probabilities of a dominating structural component, or of the structural system, become, from the individual risk criterion (11):

$$\begin{aligned} \text{CC1:} \quad & p_f \leq 4 \times 10^{-3}(\beta \geq 2.7) \\ \text{CC2:} \quad & p_f \leq 4 \times 10^{-4}(\beta \geq 3.4) \\ \text{CC3:} \quad & p_f \leq 5 \times 10^{-5}(\beta \geq 3.9) \end{aligned} \tag{12}$$

Apparently, it is highly desired to improve the estimates of $p_{d|f}$ on the basis of additional empirical data. For small probabilities, the target failure probabilities related to a reference period are $t_{\text{ref}} > t_0$ obtained as follows for:

$$\begin{aligned} \text{CC1:} \quad & p_f\left(t_{\text{ref}}\right) \leq 4 \times 10^{-3} \times t_{\text{ref}}/t_0 \\ \text{CC2:} \quad & p_f\left(t_{\text{ref}}\right) \leq 4 \times 10^{-4} \times t_{\text{ref}}/t_0 \\ \text{CC3:} \quad & p_f\left(t_{\text{ref}}\right) \leq 5 \times 10^{-5} \times t_{\text{ref}}/t_0 \end{aligned} \tag{13}$$

Table 3. Probabilities of fatality given structural failure (in %).

Consequence class	CC1	CC2	CC3	
Probability	Low	Medium	High	References
Condit. probab. occupant fatal. $p_{d|f}$	0.1	3	30	[b]
$p_{d|f}$	NA[a]	NA	0.03, 0.8, 6.9, 19–45, 20	[c]
Prob. fatality given failure (>$p_{d|f}$)	0.5	1–3	3–5.5	[d]
$p_{d|f}$ accepted here	0.25	2.5	20	–

[a]Not available.
[b](Steenbergen & Vrouwenvelder, 2010) and background to (NEN 8700, 2009).
[c](Stewart, 2010a): for the major U.S. terrorist attacks – World Trade Center (1993), Alfred P. Murrah Federal Building (1995), World Trade Center (2001) and Pentagon (2001); (Stewart, 2010b): I35W bridge collapse in Minneapolis (2007), respectively.
[d](Eldukair & Ayyub, 1991).

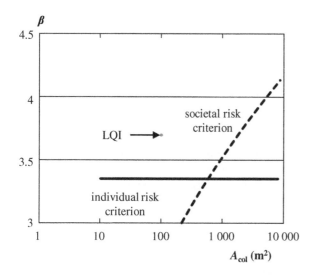

Figure 4. Annual target reliability index based on the individual and societal risk criteria, and the LQI approach (A_{col} = 100 m², A_{floor} = 1000 m²) as a function of the collapsed area for CC2.

Societal risk criterion

As indicated in (ISO 2394, 1998), in many cases authorities explicitly require the avoidance of accidents wherein a large number of people may be killed. The additional societal risk criterion is then applied:

$$p_f \leq A N^{-\alpha} \qquad (14)$$

where N = the expected number of fatalities given structural failure; and A and α = constants for which (ISO 2394, 1998) provides examples A = 0.01 or 0.1 and α = 2.

The number of fatalities should be estimated for a failure mode for which the target level is to be provided, considering the possibility of progressive collapse and its consequences. In general, the number of fatalities depends on several factors, including the number of occupants, the probability of exposure, warning factors, the probability of self- or assisted rescue, etc.

Based on the analysis of more than one hundred building collapses, empirical relationships were provided for the number of fatalities N and a collapsed area A_{col} in m² (Tanner & Hingorani, 2010, 2015):

CC1: no recommendation
CC2: $N = 0.27 A_{col}^{0.5} - 1 \geq 0$ $\qquad (15)$
CC3: $N = 0.59 A_{col}^{0.56} - 1 \geq 0$

Apparently, this must be understood as a first indication of the expected number of fatalities that should always be assessed on a case-specific basis; for further guidance see (FEMA, 2009; Janssens, O'Dwyer, & Chryssanthopoulos, 2012; Smit & Barnardo-Vijloen, 2013).

Considering A = 0.01 and α = 2, annual target reliability indices based on the individual – Equation (11) – and societal – Equations (14) and (15) – risk criteria are shown as a function of the collapsed area in Figure 4 for CC2 and Figure 5 for CC3. It follows from the figures that the individual risk criterion dominates over the societal risk criterion for, approximately, A_{col} < 1 000 m². Note that the conditional probability $p_{d|f}$ is assumed to be independent of A_{col} (Smit & Barnardo-Vijloen, 2013); their mutual relationship should, however, be investigated in further studies.

Moreover, A = 0.01 and α = 2 should be considered as mere indications and should be carefully specified taking into consideration a reference system – a group of buildings or an individual structural member – and other structure-specific and industry-specific parameters (Hingorani & Tanner, 2014; Tanner & Hingorani, 2015). Based on the results obtained for individual risks (Tanner & Hingorani, 2015), it is foreseen that an optimal value of the coefficient A should be somewhat greater for existing structures when compared to the design of new structures. α > 1 represents a risk-averse decision-making which is understood as the essential source of conservatism in structural reliability-related decisions (Cha & Ellingwood, 2012). In principle, α = 1 leads to a fair decision-making process, irrespective of whether it concerns new or existing structures.

Life Quality Index

The LQI (Nathwani et al., 2009) and several other metrics were derived to support decisions related to allocations of available public resources between and within various societal sectors and industries. The LQI is an indicator of the societal preference and capacity for investments into life safety expressed as a function of GDP, life expectancy at birth and ratio between leisure and working time (ISO 2394, 2015).

The ISO standard provides detailed guidance on how preferences of the society in regard to investments into health and life safety improvements can be described by the LQI concept. The target level is derived by considering the costs of safety measures, the monetary equivalent of societal willingness to save one life and the expected number of fatalities in the event of structural

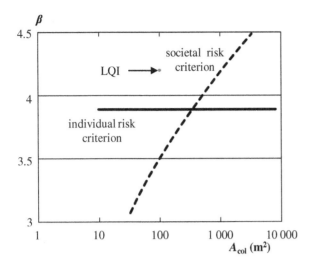

Figure 5. Annual target reliability index based on the individual and societal risk criteria, and the LQI approach (A_{col} = 100 m², A_{floor} = 1000 m²) as a function of the collapsed area for CC3.

failure. Essentially, this approach combines economic and human safety aspects. Compared with economic optimisation, it should lead to lower target reliability indices, as only the human consequences of structural failure are taken into account, while other losses such as economic and environmental costs are disregarded.

In principle, the LQI approach weighs the value of the expected casualties of a certain activity against the economic value of the activity. In such an analysis, the danger to which the people are subjected might vary on an individual basis within the group of people affected, which may be deemed unethical (Steenbergen, Sykora, Diamantidis, Holicky, & Vrouwenvelder, 2015). The example given in Annex G to (ISO 2394, 2015) is used here to compare the LQI target levels with the other approaches under consideration. The approach outlined in the ISO standard for structural design is adjusted for the assessment of existing structures by assuming a greater relative cost of risk reduction measures than for the design of a new structure. The annual values indicated in Figures 4 and 5 are obtained by considering:

- Moderate relative cost of risk reduction measure, $c_1(d)/c_0 = 0.01$,
- N/m^2 – input parameter in Table G.6, per m² of floor area of the building A_{floor} - derived from N given in Equation (15):

$$\text{CC1:} \quad N/A_{floor} = \left(0.27A_{col}^{0.5} - 1\right)/A_{floor} \geq 0$$
$$\text{CC2:} \quad N/A_{floor} = \left(0.59A_{col}^{0.56} - 1\right)/A_{floor} \geq 0 \qquad (16)$$

assuming a known damage extent, expressed by A_{col} and A_{floor}, that may depend on:

- A type of exposure – for earthquake it is likely greater than for a local impact.
- Type of structural member subjected to damage.
- Robustness and size of the structure as, e.g. total collapse may be less likely for larger systems with more redundancies and alternative load paths.

Detailed discussion on the damage extent for different exposures and structural systems is beyond the scope of this study. As an example, values of β for the LQI approach are obtained from Annex G to (ISO 2394, 2015) considering A_{col} = 100 m² and A_{floor} = 1000 m². It follows from Figures 4 and 5 that β(LQI) leads to similar target levels to the individual risk criterion, the difference in terms of reliability index being about 0.25.

Considering that the societal criterion is conservative due to risk aversion, recommendations concerning the target levels for human safety may be derived from the individual risk criterion and the LQI approach. Based on the results in Equation (12) and Figures 4 and 5, the following annual target reliabilities could thus be accepted as a first approximation:

- β(1 y.) = 2.7 for CC1
- β(1 y.) = 3.5 for CC2
- β(1 y.) = 4.0 for CC3

It is observed that these values are lower than those given in (EN 1990, 2002); for instance, β(1 y.) = 4.7 for CC2. This indicates that a design strategy, as represented by the EN 1990, 2002 values, is mostly driven by economic arguments, since it is affordable and economically optimal to design for higher reliability than that required by the human safety criteria. Furthermore, based on the comparison of the β-values, it is recommended to increase the reliability index by 0.5 when the consequence class shifts one level up, for instance from CC2 to CC3. This suggestion is consistent with the current recommendations of (EN 1990, 2002).

Case study

The implementation of the cost optimisation procedure and the human safety criteria including the LQI approach is illustrated in the reliability assessment of a dominant member of the representative structure with a remaining working life of 15 years. The example is an extension of a previous study (Steenbergen et al., 2015). The reference period equal to the remaining working life is taken into account for economic optimisation. For comparative purposes, the target reliabilities based on human safety and LQI criteria are recalculated from annual to 15-year values using Equation (1).

The structural member under consideration is exposed to permanent and snow loads. As degradation effects are significantly case-dependent, deterioration is not taken into account in order to obtain and provide results with wide validity. The decision parameter is the ratio of resistance of the member over resistance required by Eurocodes for design. Initially, the reliability of the member is verified by the partial factor method. Deterministic verification reveals insufficient reliability of the member – d_0 = 0.65, i.e. 65% of the design resistance, according to Eurocodes.

Probabilistic reliability analysis

The failure probability is obtained as follows:

$$p_f(d) = P\left[K_R R(d) - K_E(G + S) < 0\right] \qquad (17)$$

Table 4. Models for basic variables.

Variable	X	Dist.	x_k	μ_X/x_k	V_X
Resistance before upgrade	R	LN	0.77	1.29	0.15
Permanent load	G	N	0.37	1	0.05
Snow load (1 y.)	S	GU	0.37	0.35	0.7
Resistance uncertainty	K_R	LN	1	1	0.05
Load effect uncertainty	K_E	LN	1	1	0.1

x_k = characteristic value; μ_X = mean; V_X = coefficient of variation; LN = lognormal; N = normal; and GU = Gumbel distribution (maximum values).

The characteristic values and probabilistic models of the basic variables given in Table 4 are based on the recommendations of (Holicky & Sykora, 2011; JCSS, 2001a). The basic reference period for the snow load is $t_0 = 1$ year. Using the FORM method, reliability index 1.9 is low and the reliability of the member seems to be insufficient.

Input data for the cost optimisation

The total cost optimisation is based on the following assumptions:

(1) The failure costs can be estimated using the available data (Kanda & Shah, 1997) as follows: $c_f = 1$ to 3 for CC1, $c_f = 5$ to 20 for CC2 and $c_f = 20$ to 50 for CC3, (Vrouwenvelder, 2002) suggested lower values: $c_f = 2$ for the class of minor consequences, $c_f = 4$ for moderate consequences, and finally $c_f = 8$ for the class of large consequences. For $c_f > 10$ a full cost–benefit analysis was recommended. However, previous numerical experience indicated that cost optimisation based on such low c_f-values leads to unrealistically low target levels (Sykora & Holicky, 2012).

(2) The upgrade cost independent of the decision parameter c_0 is assumed to be dominated by (A) the losses due to business interruptions and (B) costs of structural rehabilitation independent of the decision parameter d, related to surveys, design, administration and management: (A) A two-week period is needed to upgrade a structure considering the data provided by (Eldukair & Ayyub, 1991) and general experience with structural upgrading. The return period for the investments into the structure – C_{str} – is about 30 years. The economic loss due to the upgrade is obtained as the ratio of the period of business interruption over the return period of investments, $1/26 \times C_{str}/30 = C_{str}/780$. This loss is estimated to be 10% of the total loss due to business interruption, and thus $c_{0(A)} = C_{0(A)}/C_{str} = 10/780 = 0.013$, (B) (Ang & De Leon, 1997) proposed a model for estimating the cost of rehabilitation after failure on the basis of the damage extent. $c_{0(B)} = 0.083$ corresponding to a low value of the damage extent – $A_{col}/A_{floor} = 0.05$ – is considered here to represent the costs of structural rehabilitation independent of d. Therefore, a value of $c_0 = 0.1$ is taken into account in the case study.

(3) Based on the authors' experience with upgrades of steel members exposed to compression, bending moments and shear forces (Sykora et al., 2016), the upgrade cost $c_1(d)$ can be estimated as follows:

$$c_1(d) = 0.25d/d_0 - 0.23; \quad \text{for:}1 < d/d_0 < 1.5 \qquad (18)$$

(4) The discount rate is $q = 3\%$.

Concerning the second assumption, c_0 is considered herein as independent of the consequence class. Obviously, this consideration can hardly be generalised and needs to be specified for the actual conditions of an assessed structure. General economic experience indicates that the return period for the investments into buildings is about 10 years for industrial structures, 15 years for office buildings and 30 years for residential houses. Consideration of the longest period is conservative, as this reduces the value of c_0, and consequently increases β_0. Similarly, a conservatively short upgrade time is selected. Moreover, Equation (18) likely leads to a lower estimate of c_1 for most structures, which again yields conservative estimates of β_0 and β_{up}.

Optimum upgrade strategy and decision on upgrade based on cost optimisation

The total cost given in Equation (6a) is optimised with respect to the decision parameter d, considering the failure probability obtained from Equation (17). Figures 6–8 show the variation of the total costs c_{tot} and of the reliability index β with the parameter d for the consequence classes CC1 to CC3, respectively. Considering purely economic criteria, it can be concluded that:

- CC1: No upgrade is advisable since the total cost for the no upgrade strategy is lower than the total cost for the optimum upgrade strategy. The optimum upgrade strategy would be to strengthen by up to 10% as $d_{opt}/d_0 = 1.02$–1.11 (for $c_{f,min}$ and $c_{f,max}$, respectively) that corresponds to about $0.7R_d$.
- CC2: The present state is accepted for $c_{f,min}$ while the upgrade to achieve $\beta_{up} = 3.1$ is carried out for $c_{f,max}$. The optimum upgrade strategy is $d_{opt}/d_0 = 1.15$–1.26 (0.75–$0.8\,R_d$).
- CC3: The upgrade to achieve $\beta_{up} = 3.1$ or 3.3 is carried out for $c_{f,min}$ or $c_{f,max}$. The optimum upgrade strategy is $d/d_0 = 1.26$–1.32 (0.8–$0.85\,R_d$). Particularly in this case, a small difference in c_{tot} is observed for the optimum strategy (d_{opt}) and Eurocode design ($d = 1$). That is why the strategy for upgrade based on design approaches, for instance using the partial factors for structural design, may be acceptable.

Using Equations (9) and (10), the following minimum target reliability indices β_0 are obtained for $c_{f,min}$ and $c_{f,max}$:

- $\beta_0 = 1.0$–1.5 for CC1 (0.55–$0.6\,R_d$),
- $\beta_0 = 1.7$–2.3 for CC2 (0.6–$0.7\,R_d$),

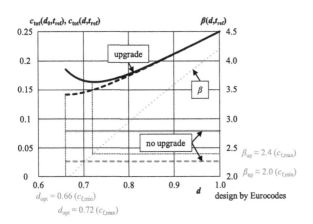

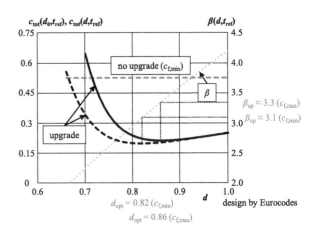

Figure 6. Variation of the total costs and reliability index with the parameter d for CC1 – $1 < c_f < 3$; dashed curves – $c_{f,min}$, solid – $c_{f,max}$).

Figure 8. Variation of the total costs and reliability index with the parameter d for CC3 – $20 < c_f < 50$; dashed curves – $c_{f,min}$, solid – $c_{f,max}$.

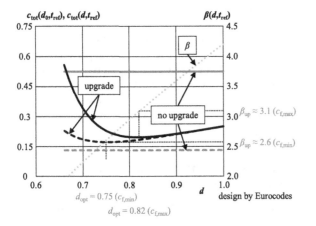

Figure 7. Variation of the total costs and reliability index with the parameter d for CC2 – $5 < c_f < 20$; dashed curves – $c_{f,min}$, solid – $c_{f,max}$.

- $\beta_0 = 2.3$–2.6 for CC3 (0.7–0.75 R_d).

Human safety criteria and LQI

The target reliabilities β_0 and β_{up} based on economic optimisation are rather low when compared to the reliability levels obtained from relationships (12) and (13). Figure 9 shows the variation of the target reliability indices with relative failure costs c_f for $t_{ref} = 15$ years. It appears that the human safety criteria lead to lower levels than economic optimisation that essentially takes into account economic and societal consequences and should thus lead to higher target reliabilities. It must be emphasised that the individual risk criterion in Equation (11) is here adjusted for existing structures by modifying the overall individual fatal accident rate to 10^{-5}. For the value of 10^{-6} indicated in ISO 2394:1998, the individual risk criterion yields higher target reliabilities than the economic optimisation (Sykora et al., 2016).

In addition, a minimum reliability level based on the LQI approach is derived following the guidance and indicative values given in Annex G to (ISO 2394, 2015). Using Equation (G.7) in (ISO 2394, 2015), the LQI reliability index β_{up} is obtained using the following relationship:

$$-\frac{d\,p_f(d)}{d\,d} \leq K_{up} = \frac{c_1(d)C_{str}\left(\gamma_s + \omega\right)}{G_x\left(N/A_{floor}\right)} \quad (19)$$

where the economic parameters are adopted from the example given in Annex G to (ISO 2394, 2015): $C_{str} = 2000$ CHF/m² (related to floor area); $\gamma_s = 0.03$ (interest rate of relevance for decision-making on behalf of society, moderately selected rate of economic growth); $\omega = 0.02$ (annual rate of obsolescence); and $G_x = 5,100,000$ CHF (Societal Willingness to Pay per life saved). The expected number of fatalities per m² of floor area, N/A_{floor}, is obtained from Equation (16) for a specified damage extent A_{col}/A_{floor}, collapsed area A_{col} and consequence class. The derivative of failure probability with respect to decision parameter is obtained numerically considering the limit state function in Equation (17). Figure 10 shows the identification of a minimum value of the decision parameter in the case of upgrade for CC2 and CC3 and corresponding reliability indices β_{up}(LQI) obtained for $A_{col} = 1000$ m², $A_{floor} = 10,000$ m² and $t_{ref} = 15$ years.

The values of β_{up}(LQI) are also shown in Figure 9. It appears that they correspond well to β_{up}-levels based on economic optimisation and, for CC3 structures, to the individual risk criterion. For CC2, β_{up}(LQI) is higher than the level based on the individual risk criterion. This is consistent with the results plotted in Figures 4 and 5, where the LQI approach provides somewhat greater values than the individual and societal risk criteria.

The structural member should be upgraded whenever total costs related to an optimally upgraded member – d_{opt}(LQI) – are lower than the expected failure costs for a no upgrade strategy. This can be expressed on the basis of Equation (19) as follows:

$$\left[c_0 + c_1\left(d_{opt}\right)\right]C_{str}\left(\gamma_s + \omega\right) + G_x N/A_{floor}p_f\left(d_{opt}\right) < G_x N/A_{floor}p_f\left(d_0\right)$$
$$K_0 = \left[c_0 + c_1\left(d_{opt}\right)\right]C_{str}\left(\gamma_s + \omega\right)/\left(G_x N/A_{floor}\right) < \Delta p_f = p_f\left(d_0\right) - p_f\left(d_{opt}\right) \quad (20)$$
$$\text{CC2:}K_0 = 0.0044 > \Delta p_f = 0.0031; \quad \text{CC3:}K_0 = 0.0013 < \Delta p_f = 0.0031$$

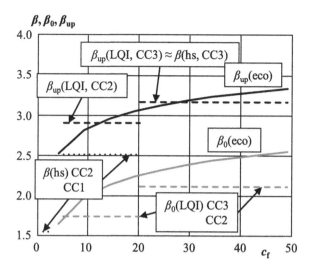

Figure 9. Variation of the target reliability indices with the failure costs c_f (t_{ref} = 15 years, hs – human safety).

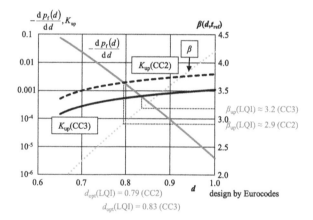

Figure 10. Identification of d_{opt} in the case of upgrade for CC2 and CC3 and corresponding reliability indices β_{up} based on the LQI approach (A_{col} = 1000 m², A_{floor} = 10,000 m² and t_{ref} = 15 years).

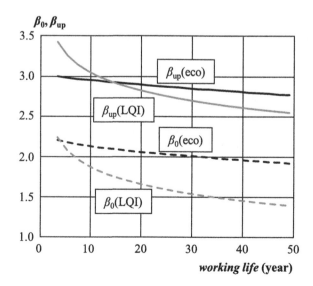

Figure 11. Variation of target reliabilities with t_{ref} for CC2 (c_f = 12.5).

Therefore, the LQI criterion suggests upgrading for CC3, while the present state should be accepted for CC2. The limiting value d_{0lim} for which $K_0 = \Delta p_f$ can be obtained from relationship (20). The LQI target reliability for the decision concerning upgrade, $\beta_0(d_{0lim})$, is shown in Figure 9. It clearly yields lower reliability levels compared to β_0 based on economic optimisation.

The comparison of the reliability indices in Figure 9 indicates that the economic optimisation and LQI approach lead to similar target reliability levels for an optimum upgrade strategy. As these approaches take into account costs of safety measures and failure consequences, they should in principle be preferred to the human safety criteria for establishing target reliabilities for existing structures. For a minimum level below which upgrade is needed (β_0), economic optimisation yields slightly greater target reliabilities than the LQI approach, with the difference less than 0.5. Both the β_0-levels based on economic optimisation and LQI are considerably lower than the individual risk criterion, which can be explained as follows:

- For existing structures, upgrades are expensive and economic criteria yield lower optimum reliabilities than those required to assure acceptable human safety.
- For new structures, costs of safety measures are relatively low and it is affordable – economically optimal – to design for greater reliabilities than those required for human safety.

Assuming that a reference period is equal to the working life of the structure, Figure 11 illustrates the variation of the target reliabilities based on economic optimisation and LQI with a working life for CC2. It appears that:

- Levels based on economic optimisation are nearly independent of a working life for the relatively low value of the discount rate considered here. For the range of a working life from 3 to 50 years, the β_0 and β_{up} levels decrease by less than 0.5.
- Economic optimisation is decisive for the β_0-level and working lifetimes over 5 years, and for the β_{up}-level and working lifetimes over 15 years. For a shorter working life, the LQI approach should be applied to verify acceptable human safety.

Similar trends are observed for CC3. Complementary studies indicate that the target reliabilities are also insignificantly affected by the discount rate for $0 < q < 0.04$ (Holicky, 2013).

It is noted that the LQI values depend on the economy of a country. In particular, the G_x value adopted here from Annex G to (ISO 2394, 2015) is rather high compared to average economies and, consequently, the LQI target levels are slightly lower in common cases.

Comparison of target reliabilities

Table 5 provides a comparison of target reliabilities estimated for the representative structure; the target levels are based on different standards, total cost optimisation and the LQI approach (t_{ref} = 15 y.). A considerable scatter is observed, for instance the reliability index varies within the range of 2.9 to 4.1 for the CC2

Table 5. Overview of target reliabilities for the representative structure (t_{ref} = 15 years).

	Consequence class		
	CC1	CC2	CC3
Code, method	$c_f = 3$	$c_f = 20$	$c_f = 50$
(EN 1990, 2002)	3.6	4.1	4.6
(ISO 13822, 2010)	2.3/3.1*	3.8	4.3
(ISO 2394, 2015)**	3.1	3.3	3.7
β_{up} economic optimisation	2.4	3.1	3.3
β_{up} LQI	–	2.9	3.2
β_0 economic optimisation	1.5	2.3	2.6
β_0 LQI	–	1.7	2.1

*Small or some failure consequences; **economic optimisation – large relative costs of safety measures, for consequence classes comparable to those in (EN 1990, 2002).

member when the β_0 level is disregarded. The following observations can be made from the results in Table 5:

(1) (EN 1990, 2002) recommends considerably larger values that seem to be intended primarily for new structures. Economic optimisation and the LQI approach suggest $\beta_{up} = \beta_{EN\,1990} - 1$.

(2) Similarly, $\beta_0 = \beta_{EN\,1990} - 2$ may be considered.

(3) (ISO 13822, 2010) provides rather high reliability levels when compared to (ISO 2394, 2015) – *economic optimisation*; the latter might be more appropriate for existing structures when large relative costs of safety measures are taken into account.

Discussion of the results

Difficulties in establishing target levels in practical applications

The present study intends to provide unambiguous recommendations regarding specification of target reliabilities for existing structures. In all cases (ISO 2394, 2015) suggest that the target reliabilities for verifications of structural members should be calibrated against well-established cases, proven by past experience to be of adequate reliability. This recommendation is supported by the following list of deficiencies that may need to be overcome in practical applications:

(1) *Difficult generalisation of the input data.* Input data for economic optimisation are of considerable uncertainty. In particular, the estimates of failure costs may be difficult in real situations. Applications of human safety criteria may become complicated due to the need to estimate probability $p_{d|f}$ and a collapsed area A_{col}.

(2) *Difficulties in the specification of a reference period.* The target reliabilities should always be defined for a specified reference period. In the case of economic optimisation, a reference period is commonly equal to a required remaining working life that can be lower for private owners – as related to specific business plans – than for public authorities, since infrastructures are intended to serve their purposes over long periods of time. Uncertainties in specifying an adequate working life may then also affect the target reliabilities. There is an on-going debate

concerning what changes in target reliabilities should be accepted in the case of a shorter working life or in the case of transient design situations. (EN 1990, 2002) is not sufficiently instructive in this regard, partly due to insufficient attention paid to this issue, partly due to contradictory opinions amongst experts (Vrouwenvelder, 2012).

Another difficulty could arise from the fact that the relative cost of safety measures depends on the variability of the influencing loads and resistances (Vrouwenvelder, 2002). This variability is affected by the level of information about an existing structure. When detailed information on the actual condition of the existing structure is available, variability reduces and upgrading increases the reliability level more significantly than is the case with vague information, thereby decreasing the relative costs of safety measures. This effect is difficult to generalise and should be assessed on a case-specific basis. In addition, the effect of the level of information on target reliability is expected to be small when compared to uncertainties in specifying the consequences of failure. Hence, this effect is not treated in this study.

Practical implications for member and system design

Since designers are mainly applying safety formats at a component – structural member – level, the definition of targets for members and associated limit states is preferred (EN 1990, 2002) and (ISO 2394, 2015). Structural members designed using such a procedure may have sufficient reliability beyond local failure conditions due to sectional and internal redundancies, possible alternative load paths, and/or the formation of plastic hinges. Therefore, failures of individual members often have significantly lower consequences compared to the failure of a main part of the structure – system failure. It thus appears important to reflect the differences between member failure and structural collapse in target levels (Bhattacharya, Basu, & Ma, 2001).

When deriving target reliability levels by considering failures resulting in large collapsed areas and many fatalities, it is obvious that these levels should be related more to system failure, i.e. global failure, or the failure of a main part of the structure. This distinction has been implemented in standards reflecting performance objectives of structures; see (Diamantidis, 2008) for a review.

For instance, (ASCE 7-10, 2013) requires the target reliability index to be about 0.5 greater for progressive collapse than for a member or connection failure (Hamburger, 2013). This difference was also proposed for bridges to distinguish between single-path and multiple-path failures (Kaszynska & Nowak, 2005). A difference of 0.4 reflects, in practice, the fact that in the case of system failure the number of expected fatalities is greater than is the case with member failure (Allen, 1981; Rackwitz, 2002).

It seems appropriate to increase the target reliability index by 0.5 when the reliability of a system or its key component is verified (as compared to the design of a member). This corresponds to shifting one consequence class higher, as discussed above. This recommendation relates to ductile systems with alternative load paths – parallel systems, as is relevant to many civil engineering structures. For a series system, the target level for components is higher than with a system target. As the system behaviour of structures cannot easily be classified into idealised types of systems such as series, parallel ductile or parallel brittle, detailed discussion on system targets requires a separate study.

Conclusions

The target reliability levels recommended in various documents for new and existing structures are inconsistent in terms of the values and criteria according to which optimum levels are to be selected. In general, optimum reliability levels can be obtained by considering both the relative costs of safety measures and failure consequences over a specified working life of the structure; the minimum reliability for human safety should also be verified, if relevant. The following conclusions are drawn from the present study:

- It is uneconomical to require the same reliability levels for existing and new structures.
- Decisions made in assessment can result in the acceptance of an actual state or in the upgrade of a structure; two reliability levels are thus needed – the minimum level below which the structure is considered unreliable and should be upgraded – reliability index β_0; and the target level indicating an optimum upgrade strategy – β_{up}.
- Economic optimisation can be used to derive target reliabilities; thereby minimum human safety levels should not be exceeded.
- The difference of 0.5 in β is appropriate to distinguish between two subsequent consequence classes such as CC2 to CC3; β should likewise be increased by about 0.5 – as compared to verification of a common member – when reliability of a system or its key component is verified.
- It is recommended that the target levels for existing structures be established using economic optimisation and the LQI approaches, as these take into account both costs of safety measures and failure consequences; the former aspect is important for existing structures and is difficult to be covered by the individual and societal risk criteria.
- (EN 1990, 2002) recommends considerably greater target reliabilities as compared to those obtained in this study for existing structures; as a first approximation it can be considered that: $\beta_{up} = \beta_{EN\ 1990} - 1$, and $\beta_0 = \beta_{EN\ 1990} - 2$.
- Human safety levels and LQI criteria are commonly related to a period of one year and need to be translated to a reference period accepted for the analysis using Equation (1); the recalculation of the target levels amongst different reference periods may be complicated by mutual dependencies of failure events.
- When using economic optimisation, target reliability levels primarily depend on failure and upgrade costs; the influence of the working life is less significant.
- (ISO 13822, 2010) for existing structures provides quite high reliability levels when compared to (ISO 2394, 2015) – *economic optimisation* that might be more appropriate for existing structures when large relative costs of safety measures are taken into account.

Disclosure statement

No potential conflict of interest was reported by the authors.

Funding

This work was supported by the Czech Science Foundation [grant number P105/12/G059]; the Ministry of Education, Youth and Sports of the Czech Republic [grant number LG14012], [grant number LD15037]; the Regensburg Center of Energy and Resources (RCER) of the Ostbayerische Technische Hochschule Regensburg (OTH Regensburg); the Technology and Science Network Oberpfalz (TWO); the European Commission [grant number CZ/13/LLP-LdV/TOI/134014]. The contribution reflects the views of the authors, and the Commission cannot be held responsible for any use of the information contained therein.

ORCID

Miroslav Sykora (iD) http://orcid.org/0000-0001-9346-3204

References

Allen, D. E. (1993). Safety criteria for the evaluation of existing structures. *Proceedings IABSE Colloquium on Remaining Structural Capacity,* Copenhagen, Denmark.

Allen, D. E. (1981). Criteria for design safety factors and quality assurance expenditure. *Proceedings of 3rd International Conference on Structural Safety and Reliability* (pp. 666–678). Trondheim.

Ang, A. H. S., & De Leon, D. (1997). Determination of optimal target reliabilities for design and upgrading of structures. *Structural Safety, 19,* 91–103. doi: 10.1016/S0167-4730(96)00029-X

ASCE 7-10. (2013). *Minimum design loads for buildings and other structures.* Reston, VA: American Society of Civil Engineers. doi:10.1061/9780784412916

Bhattacharya, B., Basu, R., & Ma, K. (2001). Developing target reliability for novel structures: The case of the mobile offshore base. *Marine Structures, 14,* 37–58. doi: 10.1016/S0951-8339(00)00024-1

Bigaj-van Vliet, A., & Vrouwenvelder, T. (2013). Reliability in the performance-based concept of fib Model Code 2010. *Structural Concrete, 14,* 309–319. doi: 10.1002/suco.201300053

Casas, J. R., & Wisniewski, D. (2013). Safety requirements and probabilistic models of resistance in the assessment of existing railway bridges. *Structure and Infrastructure Engineering, 9,* 529–545. doi: 10.1080/15732479.2011.581673

Caspeele, R., Sykora, M., Allaix, D. L., & Steenbergen, R. (2013). The design value method and adjusted partial factor approach for existing structures. *Structural Engineering International, 23,* 386–393. doi: 10.2749/10168661 3X13627347100194

Cha, E. J., & Ellingwood, B. R. (2012). Risk-averse decision-making for civil infrastructure exposed to low-probability, high-consequence events. *Reliability Engineering and System Safety, 104,* 27–35. doi: 10.1016/j. ress.2012.04.002

Diamantidis, D., & Bazzurro, P. (2007). Safety acceptance criteria for existing structures. *Special Workshop on Risk Acceptance and Risk Communication*, Stanford University, CA, USA.

Diamantidis, D. (2008). Current safety acceptance criteria in codes and standards – A critical review. *Proceedings of ASCE Structures Congress*, Vancouver, Canada. doi: 10.1061/41016(314)76

Eldukair, Z. A., & Ayyub, B. M. (1991). Analysis of recent U.S. structural and construction failures. *Journal of Performance of Constructed Facilities, 5*, 57–73.

EN 1990. 2002. *Eurocode – Basis of structural design*. Brussels: CEN.

EN 1991-1-7. (2006). *Eurocode 1: Actions on structures – Part 1-7: General actions – Accidental actions*. Brussels: CEN.

Faber, M. H. (2015). Codified risk informed decision making for structures. *Proceedings of Symposium on Reliability of Engineering Systems SRES'2015*, Hangzhou, China.

FEMA. (2009). *Handbook for rapid visual screening of buildings to evaluate terrorism risks*. No. FEMA 455. Federal Emergency Management Agency, USA.

fib. (2013). *fib model code for concrete structures 2010*. Lausanne: Author.

Hamburger, R. O. (2013). *Provisions in present U.S. building codes and standards for resilience* (NIST technical note 1795). Gaithersburg, MD: NIST.

Hingorani, R., & Tanner, P. (2014). Structural safety requirements based on risks to persons. *Proceedings of Jornadas Internacionales Conmemorativas Del 80 Aniversario Del IETcc*, Madrid. 12.

Holicky, M. (2013). Optimisation of the target reliability for temporary structures. *Civil Engineering and Environmental Systems, 30*, 87–96. doi: 10.1080/10286608.2012.733373

Holicky, M. (2014). Optimum reliability levels for structures. In M. Beer, S.-K. Au & J. W. Hall (Eds.), *Vulnerability, uncertainty, and risk: Quantification, mitigation, and management* (pp. 184–193). Reston, VA: American Society of Civil Engineers. doi: 10.1061/9780784413609.019

Holicky, M., & Sykora, M. (2011). Conventional probabilistic models for calibration of codes. *Proceedings of ICASP11* (pp. 969–976). ETH Zurich, Switzerland.

ISO 2394. (1998). *General principles on reliability for structures* (2nd ed.). Geneve: ISO.

ISO 13822. (2010). *Bases for design of structures – Assessment of existing structures*. Geneve: ISO TC98/SC2.

ISO 15686-5. (2008). *Buildings and constructed assets – Service-life planning – Part 5: Life-cycle costing* (1st ed.). Geneve: ISO.

ISO 2394. (2015). *General principles on reliability for structures* (4th ed.). Geneve: Autor.

Janssens, V., O'Dwyer, D. W., & Chryssanthopoulos, M. K. (2012). Assessing the consequences of building failures. *Structural Engineering International, 22*, 99–104. doi: 10.2749/101686612X13216060213473

JCSS. (2001a). *JCSS probabilistic model code (periodically updated, online publication)*. Lyngby. Retrieved from www.jcss.byg.dtu.dk

JCSS. (2001b). *Probabilistic assessment of existing structures*. D. Diamantidis (Ed.), Joint Committee on Structural Safety. Bagneux: RILEM Publications S.A.R.L.

Kanda, J., & Shah, H. (1997). Engineering role in failure cost evaluation for buildings. *Structural Safety, 19*, 79–90. doi: 10.1016/S0167-4730(96)00039-2

Kaszynska, M., & Nowak, A. S. (2005). Target reliability for design and evaluation of bridges. *Management 5: Inspection, maintenance, assessment and repair* (pp. 401–408). London. doi: 10.1680/bmf.33542.0051

Lee, J. Y., & Ellingwood, B. R. (2015). Ethical discounting for civil infrastructure decisions extending over multiple generations. *Structural Safety, 57*, 43–52. doi: 10.1016/j.strusafe.2015.06.001

Lenner, R., & Sýkora, M. (2016). Partial factors for loads due to special vehicles on road bridges. *Engineering Structures, 106*, 137–146. doi: 10.1016/j.engstruct.2015.10.024

Lentz, A. (2007). *Acceptability of civil engineering decisions involving human consequences* (Ph.D. thesis). TU Munich, Munich, Germany.

Luechinger, P., Fischer, J., Chrysostomou, C., Dieteren, G., Landon, F., Leivestad, S., & Tanner, P. (2015). *New european technical rules for the assessment and retrofitting of existing structures (JRC science and policy report)*. (No. JRC94918). Luxembourg: JRC. doi: 10.2788/095215

Melchers, R. E. (2001). *Structural reliability analysis and prediction* (2nd ed.). Chichester: Wiley.

Nathwani, J. S., Pandey, M. D., & Lind, N. C. (2009). *Engineering decisions for life quality: How safe is safe enough?*. London: Springer-Verlag.

NEN 8700. (2009). *Grondslagen van de beoor deling van de constructieve veiligheid van een bestaand bouwwerk* [Assessment of existing structures in case of reconstruction and disapproval – basic rules; in Dutch]. Delft: Nederlands Normalisatie Instituut.

Rackwitz, R. (2000). Optimization – The basis of code-making and reliability verification. *Structural Safety, 22*, 27–60. doi: 10.1016/S0167-4730(99)00037-5

Rackwitz, R. (2002). Optimization and risk acceptability based on the Life Quality Index. *Structural Safety, 24*, 297–331. doi: 10.1016/S0167-4730(02)00029-2

Smit, C. F., & Barnardo-Vijloen, C. (2013). Reliability based optimization of concrete structural components. *Proceedings of 11th International Probabilistic Workshop* (pp. 391–404). Brno.

Steenbergen, R. D. J. M., & Vrouwenvelder, A. C. W. M. (2010). Safety philosophy for existing structures and partial factors for traffic loads on bridges. *Heron, 55*, 123–139.

Steenbergen, R. D. J. M., Sykora, M., Diamantidis, D., Holicky, M., & Vrouwenvelder, A. C. W. M. (2015). Economic and human safety reliability levels for existing structures. *Structural Concrete, 16*, 323–332. doi: 10.1002/suco.201500022

Stewart, M. (2010a). Risk-informed decision support for assessing the costs and benefits of counter-terrorism protective measures for infrastructure. *International Journal of Critical Infrastructure Protection, 3*, 29–40. doi: 10.1016/j.ijcip.2009.09.001

Stewart, M. G. (2010b). Acceptable Risk Criteria for Infrastructure Protection. *International Journal of Protective Structures, 1*, 23–40. doi: 10.1080/15732470902726023

Sykora, M., Diamantidis, D., Holicky, M., & Jung, K. (2015). *Target reliability levels for assessment of existing structures considering economic and societal aspects. Proceedings of IALCCE 2014* (pp. 838–845). Tokyo, Japan.

Sykora, M., & Holicky, M. (2012). Target reliability levels for the assessment of existing structures – case study. *Proceedings of IALCCE 2012* (pp. 813–820). Vienna, Austria.

Sykora, M., Holicky, M., & Markova, J. (2011). Target reliability levels for assessment of existing structures. *Proceedings of ICASP11* (pp. 1048–1056). ETH Zurich, Switzerland.

Sykora, M., Holicky, M., & Markova, J. (2013). Verification of existing reinforced concrete bridges using the semi-probabilistic approach. *Engineering Structures, 56*, 1419–1426. doi: 10.1016/j.engstruct.2013.07.015

Sykora, M., Holicky, M., Markova, J., & Senberger, T. (2016). *Probabilistic reliability assessment of existing structures focused on industrial heritage buildings*. Prague: Ceska technika (the Publishing House of CTU in Prague).

Tanner, P., & Hingorani, R. (2010). Development of risk-based requirements for structural safety. *Joint IABSE – fib Conference Codes in Structural Engineering* (pp. 379–386). Dubrovnik, Croatia.

Tanner, P., & Hingorani, R. (2015). Acceptable risks to persons associated with building structures. *Structural Concrete, 16*, 314–322. doi: 10.1002/suco.201500012

Vrouwenvelder, A. C. W. M. (2002). Developments towards full probabilistic design codes. *Structural Safety, 24*, 417–432. doi: 10.1016/S0167-4730(02)00035-8

Vrouwenvelder, A. C. W. M. (2012). Target reliability as a function of the design working life. *Proceedings of 6th IFED Forum*, Lake Louise, Canada.

Vrouwenvelder, A. C. W. M., & Scholten, N. (2010). Assessment criteria for existing structures. *Structural Engineering International, 20*, 62–65.

Vibration monitoring via spectro-temporal compressive sensing for wireless sensor networks

Roman Klis and Eleni N. Chatzi

ABSTRACT

The reliable extraction of structural characteristics, such as modal information, from operating structural systems allows for the formation of indicators tied to structural performance and condition. Within this context, reliable monitoring systems and associated processing algorithms need be developed for a robust, yet cost-effective, extraction. Wireless Sensor Networks (WSNs) have in recent years surfaced as a promising technology to this end. Currently operating WSNs are however bounded by a number of restrictions relating to energy self-sustainability and energy data transmission costs, especially when applied within the context for vibration monitoring. The work presented herein proposes a remedy to heavy transmission costs by optimally combining the spectro-temporal information, which is already present in the signal, with a recently surfaced compressive sensing paradigm resulting in a robust signal reconstruction technique, which allows for reliable identification of modal shapes. To this end, this work outlines a step-by-step process for response time-series recovery from partially transmitted spectro-temporal information. The framework is validated on synthetic data generated for a benchmark structure of the American Society of Civil Engineers. On the basis of this example, this work further provides a cost analysis in comparison to fully transmitting wireless and tethered sensing solutions.

1. Introduction

Wireless Sensor Networks (WSNs) were introduced in civil engineering applications by Straser et al. (1998), via fusion of a radio with a sensing unit in order to remotely connect to a data logging device. Since then important work has been delivered in this field, among others by Hsu et al. (2011), Lynch et al. (2001), Nagayama et al. (2009), Wang (2007), Zhu et al. (2010). Lynch and Spencer in particular present a broad review of currently available wireless sensor solutions and corresponding algorithms in their works (Lynch, 2006; Spencer et al., 2004).

However, one major issue affecting applicability of WSNs is the battery life of the wireless nodes. Once the wireless sensor drains the available power source, the measurement node remains inoperational until battery replacement is carried out by the system operator. Such an action is associated with additional costs that need be accounted for by the infrastructure operator. The frequency at which such a maintenance action needs to be performed is directly related to the average power consumption of the wireless node. It has already been identified that the most energy consuming operation the wireless nodes perform lies in data communication (Feltrin et al., 2009; Sichitiu & Dutta, 2007). This problem is even more pronounced when large data blocks need be communicated across the network as in the case of vibration-based monitoring. In the latter case, the data volume handled by the sensors is large; thus, energy transmission costs are bound to be high. Efforts on tackling this issue have been organised by the research community on two levels, firstly on a hardware level, and secondly on a software level. On the hardware level a significant body of research is focused on the usage of various energy harvesting techniques, such as solar panels for charging up the wireless sensor battery (Klis & Chatzi, 2013; Le Cam et al., 2012; Miller et al., 2010). On the software level, effort is allocated in lowering the transmitted samples, and thereby the associated power consumption. In this direction, data compression schemes, which are also the topic of the current work, have come to play an important role.

Several works in the area of WSN implementations for the monitoring of civil structures provide a discussion within a more general context (Sim et al.,2011; Rice et al., 2011; Wang et al., 2007). For some purposes a decentralised approach is followed where the processing ability of the wireless nodes is exploited. In such a scheme, the acquired time history data can be processed in the node level to compute the envelope of the vibration amplitude and the dominant peaks of the Fourier spectrum, which serve as indicators of structural condition (Feltrin et al., 2011). These are then transmitted to a central processing unit as lower cost information. Nonetheless, it is oftentimes of the essence to transmit a richer information content from the reference (node) signals. To this end, some works investigate the potential of newly surfaced Compressive Sensing (CS) schemes for reducing transmission costs, thereby extending the energy sustainability of such nodes. Specifically, Bao et al. (2011) were the first to demonstrate the potential of employing CS techniques for recovery of the partially transmitted response information recorded on a highway bridge over the Yellow River in Binzhou of Shandong province, China. In that work, an assessment of the most suitable sparsifying basis was performed, where both a Fourier and a wavelet basis approach were compared, indicating a preference to the latter for civil engineering deployments.

In subsequent work, Bao et al. (2012) presented a framework for missing data recovery for SHM applications. Bao investigated the case where node-level pre-treated data were partially lost during the transmission process and subsequently recovered via CS at the base station level. The proposed framework was validated on actual recordings from the Jinzhou West Bridge in north-east China and the National Aquatics Center in Beijing. In further work, Bao et al. (2014) tackled the issue of potential data loss, while retraining estimation accuracy via projection of the raw recorded signal onto a random matrix, and transmission of the resulting transformed signal. The framework was validated, via a network of Imote2 wireless sensors on the Songpu Bridge in Harbin, China.

One of the first complete long-term deployments, adopting CS techniques, was carried out by O'Connor et al. (2013) on the Telegraph Road Bridge (TRB) in Monroe, Michigan. In that work, a Narada-based implementation of a random time domain sampling scheme resulted in successful signal recovery. The reconstructed signals were further used to appropriately calibrate a finite element model of the monitored system. In more recent work, O'Connor et al. (2014) employed CS to simultaneously reduce data sampling rates, on-board storage requirements, and communication data payloads once again in a field deployment on the TRB. Next, the randomly selected samples were further transmitted to an off-site server to perform recovery via a Compressive Sensing Matching Pursuit (CoSaMP) algorithm. In the final step, the recovered signals were used for extraction of the structure's mode shapes and frequencies via use of system identification tools. O'Connor concluded that the first structural natural frequencies and modal shapes can be recovered by using as little as 20% of the complete (reference) data series.

In associated recent work Zou et al. (2015) present an implementation of a CS-based framework for data loss recovery into a wireless smart sensor Imote2 platform. In the approach presented therein Random Demodulation is utilised for power-efficient reconstructions on a series of experiments, demonstrating a successful practical application of CS.

An interesting insight into the data compression problem was presented in Yang and Nagarajaiah (2014), where a novel lossy data compression scheme for monitoring seismic response data based on principled independent component analysis (PICA) is presented. It is therein demonstrated that independent component analysis is able to transform a multivariate data-set into a sparse representation space optimal for coding and compression. A potential domain of application of PICA is rapid and reliable data transfer, communication (e.g. multi-hop wireless sensor network), storage, and retrieval in online or post-disaster (e.g. earthquake) monitoring and assessment applications of civil infrastructures.

In recent work, Huang introduced a Bayesian CS framework dedicated in dealing with the trade-off between the signal reconstruction error and the amount of signal compression achieved via a CS methodology. By using the proposed method an error aware reconstruction procedure is derived (Huang et al., 2014).

The authors successfully validate this framework using both synthetic signals and real acceleration data obtained from a SHM system, deployed on a bridge. Additionally, the proposed method allows for effective quantification of reconstruction error bounds.

As explained in detail in what follows, existing works incorporating a CS approach commonly rely on the so-called Basis Pursuit De-Noising problem (Candès & Romberg, 2007). Building on those efforts, the work presented herein extends this framework by investigating a spectro-temporal approach to CS, relying on re-weighted Basis Pursuit De-Noising (Becker, 2011), resulting in enhancement of the quality of information extracted from partially transmitted data. Regarding the nature of extracted information, and particularly in what condition assessment and damage detection in civil structures is concerned, it has been pointed out in several works (Fan & Qiao, 2011; Kim et al., 2003) that natural frequency information alone may be insufficient. In ensuring a richer information content, it is of essence to additionally extract modal shape information, as this constitutes a more sensitive feature, which may convey a more informative documentation on the nature and location of potential damage/deterioration.

In previous work of the authoring team Klis and Chatzi, (2014), a framework was proposed for reducing transmission and therefore energy consumption for WSNs, relying on model-based compression on output-only data. System identification procedures, such as the Eigensystem Realization Algorithm and the Unscented Kalman Filter, were recombined in order to track the underlying states of the system and extract structural characteristics, leading to continuous condition assessment and detection of possible damage. This earlier approach necessitated availability of an underlying model of the system, possibly assembled via an early and comprehensive monitoring campaign.

In this work, a different approach is pursued, which alleviates the requirement of an existing model of the system, resulting in a model-less data compression framework. The proposed scheme, relying on CS, allows for efficient recovery of modal shapes from partially transmitted spectro-temporal information. A case study is presented demonstrating the ability of the proposed method to deliver reliable estimates of the structural modes. To this end, synthetic experiments are performed for a four-storey steel frame structure, set as a benchmark by the American Society of Civil Engineers (ASCEs) Johnson et al. (2004) resulting in a synthetic vibration-based testbed.

As demonstrated in this work, the system characteristics, i.e. natural frequencies and mode shapes, can be extracted using a WSN enhanced via the proposed spectro-temporal CS scheme. Availability of such information enables formation of indicators tied to structural performance and condition, which can in turn facilitate life-cycle planning (management & maintenance). Such a goal however requires the monitoring system to provide this information in continuous or densely spaced intervals. Within this context, this work provides a comparison of the proposed WSN solution to tethered and other WSN alternatives in terms of both accuracy and cost efficiency. The obtained results render this scheme an attractive alternative to wired deployments.

2. Monitoring of structural vibrations: general framework

2.1. Structural systems as signal filters

Structural systems operating in the linear range are typically governed by the following equation:

$$\mathbf{M}\ddot{\mathbf{z}}(t) + \mathbf{C}\dot{\mathbf{z}}(t) + \mathbf{K}\mathbf{z}(t) = \mathbf{f}(t) \tag{1}$$

where $\mathbf{M}, \mathbf{C}, \mathbf{K} \in \mathbf{R}^{n \times n}$ designate the mass, damping and stiffness matrix respectively and $\mathbf{z} = [\mathbf{z}_1, \mathbf{z}_2, \ldots, \mathbf{z}_n]^T$ denotes the system response vector when the system is driven by an external force $\mathbf{f} = [\mathbf{f}_1, \mathbf{f}_2, \ldots, \mathbf{f}_n]^T$. The structural response $\mathbf{z}$ may be calculated via use of modal expansion as:

$$\mathbf{z}(t) = \mathbf{\Phi}q(t) = \sum_{i=1}^{n} \phi_i \mathbf{q}_i(t) \tag{2}$$

where $\mathbf{q}(t) = [\mathbf{q}_1(t), \mathbf{q}_2(t), \ldots, \mathbf{q}_n(t)]^T$ denotes the vector of modal coordinates and $\mathbf{\Phi} = [\phi_1^T, \ldots, \phi_n^T]$ denotes the matrix of mass-normalised modal shapes , for which $\phi_i^T \mathbf{M} \phi_i = \delta_{ij}$. One may then recast the original equation of motion into the following form:

$$\mathbf{I}\ddot{q}(t) + \mathbf{\Gamma}\dot{\mathbf{q}}(t) + \mathbf{\Omega}q(t) = \mathbf{\Phi}^T \mathbf{f}(t) \tag{3}$$

$$\mathbf{\Omega} = diag([\omega_1^2, \ldots, \omega_n^2]) \tag{4}$$

$$\mathbf{\Gamma} = diag([2\zeta_1\omega_1, \ldots, 2\zeta_n\omega_n]) \tag{5}$$

where $\mathbf{\Omega}$ is referred to as the *generalised stiffness matrix* and $\mathbf{\Gamma}$ stands for the *modal damping matrix*. When mass-normalised eigen-vectors are employed for the transformation, as implied above, then the transformed system degenerates into a set of decoupled equations, the solution of which is attained via use of the Duhamel Integral also known as Convolution Integral:

$$q_i(t) = \frac{1}{\omega_i} \int_0^t f_i(t) h_i(t - \tau) \mathrm{d}\tau = (f_i * h_i)(t) \tag{6}$$

where $h_i(t)$ is the impulse response of the system and the $(\cdot * \cdot)$ operator indicates the convolution operator. The displacement response on the ith degree of freedom may then be obtained as:

$$y_i(t) = z_i(t) = \sum_{j=1}^{n} \mathbf{\Phi}_{ij} q_j(t) = \sum_{j=1}^{n} \mathbf{\Phi}_{ij}(f_j * h_j)(t) \tag{7}$$

By employing the Fourier transform on Equation (7) and via subsequent application of the convolution theorem, the following is rendered:

$$Y_i(\omega) = \sum_{j=1}^{n} \mathbf{\Phi}_{ij} F_j(\omega) H_j(\omega) \tag{8}$$

This result indicates that the trigonometrical components of the response can be grouped into two categories. The first category, $F_j(\omega)$, corresponds to the contribution of the forcing term in the response $q_j(t)$, which is present as a result of the particular solution of the differential equation. The second component, $H_j(\omega)$,

is linked to the inherent characteristics of the vibrating system (natural frequencies, damping and modal shapes), resulting as the homogeneous part of the solution of the differential equation. By observing the spectral representation of the signal one may define the so-called support of a vector, as the subset of components where the vector is not zero-valued:

$$supp(Y_i) = \{\omega \in \mathbf{\Omega} \ : \ Y_i(\omega) \neq 0\} \tag{9}$$

In practice, the support is selected in terms of exceedance of a user-specified threshold ϵ, as opposed to its strict theoretical definition in terms of zero entries, as:

$$supp(Y_i, \epsilon) = \{\omega \in \mathbf{\Omega} \ : \ |Y_i(\omega)| > \epsilon\} \tag{10}$$

The selected support allows for reconstruction of the signal from sparse observations, as illustrated in the next section. It should additionally be noted that the support $supp(Y_i, \epsilon)$ is bounded, implying that the spectral representation of the response is sparse. The statement that $y_i(t)$ is in turn K-sparse, or compressible, denotes that it can be exactly or accurately reconstructed using $K << N$ non-zero coefficients, or otherwise K salient harmonic components.

2.2. The missing data estimation problem

The formulation of the missing data problem is adopted from the work of Selesnick (2012) and additionally draws influence from the works of Becker (2011), Candès et al. (2006), Candès and Romberg (2007), Candès and Plan (2011). Let $x_i = [x_{1i}, x_{2i}, \ldots, x_{Ni}]^T \in \mathbb{R}^N$ designate the complete ith signal recorded by the ith sensor of length of N. Then, the missing data estimation problem can be expressed as:

$$\mathbf{y}_i = \mathbf{S}\mathbf{x}_i \tag{11}$$

where $\mathbf{y}_i = [y_{1i}, y_{2i}, \ldots, y_{Mi}]^T \in \mathbb{R}^M$ is the observed sample vector of length of M, which comprises a sub-vector of $\mathbf{x}_i$, and $\mathbf{S} \in \mathbb{R}^{M \times N}$ is a zero–one selection matrix, which is assumed known a-priori. Therefore the problem may be expressed as the task of inferring the full response time-series $\mathbf{x}_i$ given the incomplete observations $\mathbf{y}_i$ and the selection matrix $\mathbf{S}$, so that (11) holds.

The Discrete Fourier Transform (DFT) orthonormal basis matrix $\mathbf{A}$ is firstly defined as:

$$\mathbf{A}_{i,l} = \frac{1}{\sqrt{N}} e^{-j2\pi i \frac{l}{N}} \tag{12}$$

Based on the observation made in the prevision section regarding the sparse spectral representation of the response signal, it is safe to assume that the response signal $\mathbf{x}$ can be sparsely represented via a DFT orthonormal basis $\mathbf{A}$

$$\mathbf{x}_i = \mathbf{A}\mathbf{c}_i \tag{13}$$

where $\mathbf{c}_i = [c_{1i}, c_{2i}, \ldots, c_{Ni}]^T \in \mathbb{R}^N$ is a vector of coefficients. By plugging Equations (11) into (13) the observed vector $\mathbf{y}_i$ may be reconstructed via use of only few c_i sparse coefficients:

$$\mathbf{y}_i = \mathbf{SAc}_i \tag{14}$$

It should be noted that if the dimension of the observed vector $\mathbf{y}$ is smaller than the dimension of $\mathbf{c}$, the problem (14) is ill-conditioned, i.e. any set of coefficients satisfying Equation (14) can be treated as a valid solution. However, for the purposes of the problem investigated herein, the sought solution is obtained as the most sparse vector $\mathbf{c}$ fulfilling the given conditions. This may be obtained via the following optimisation problem:

$$\underset{\mathbf{c}_i}{\operatorname{argmin}} \ \|\mathbf{c}_i\|_0 \ \text{subject to} \ \mathbf{y}_i = \mathbf{SAc}_i \tag{15}$$

The major hindrance in solving the problem of Equation (15) is that this comprises a Nondeterministic Polynomial time-Hard (NP-Hard) combinatorial problem, which may not be solved efficiently via existing state-of-the-art approaches. Candés addressed this issue by investigating an equivalent problem under relaxed $\|\cdot\|_1$ constrains and conditions which would lead to a unique solution of (14) Candès and Romberg (2007). He concluded that solving a relaxed $\|\cdot\|_1$ convex optimisation problem known as a *basis pursuit problem*

$$\underset{\mathbf{c}}{\operatorname{argmin}} \ _i\|\mathbf{c}_i\|_1 \ \text{subject to} \ \mathbf{y}_i = \mathbf{SAc}_i \tag{16}$$

is equivalent to solving the $\|\cdot\|_0$ NP-Hard combinatorial problem once the so-called *Restricted Isometry Property* (RIP) condition holds (Candès & Romberg, 2007). This implies that both the combinatorial selection of coefficients $\mathbf{c}$ and the result of the convex optimisation under the RIP condition would lead to same unique result. The second approach however is significantly simpler in terms of computational toll, and at the same time efficient, since it guarantees that the original response signal $\mathbf{x}$ is fully recovered provided that the matrix product $\mathbf{SA}$ obeys the RIP condition.

Proof (Definition - RIP): For each integer $K = 1, 2, \ldots$ let us define the isometry constant δ_K of matrix $\mathbf{SA}$ as the smallest integer such that,

$$(1 - \delta_K)\|\mathbf{c}_i\|_2^2 \le \|\mathbf{SAc}_i\|_2^2 \le (1 + \delta_K)\|\mathbf{c}_i\|_2^2 \tag{17}$$

holds for all K-sparse vectors. A vector is characterised as K-sparse if it contains at most K non-zero entries. Then, the matrix $\mathbf{SA}$ is said to satisfy the K-RIP with a corresponding restricted isometry constant δ_K. The K-RIP indicates how close a subset of $\mathbf{SA}$ is to an orthonormal system when restricted to sparse linear measurements. As Candès et al.(2006) demonstrate, the RIP yields the condition $\delta_{2K} < \sqrt{2} - 1$ for ensuring that a solution to (16) also solves (15), while maintaining a good quality of the approximation solution. □

An issue pertinent to the case of SHM, lies in the quality of the recorded signals, which are typically contaminated with some level of instrument noise. Therefore, in the more realistic case where noise contamination of the recorded signals is allowed, i.e. $\mathbf{y}_i = \mathbf{SAc}_i + \mathbf{z}_i$, $\mathbf{z}_i \propto \mathcal{N}(0, \sigma^2)$, a convex optimisation problem known as the *Basis Pursuit De-Noising problem* (BPDN) is formulated:

$$\underset{\mathbf{c}_i}{\operatorname{argmin}} \ \|\mathbf{c}\|_1 \ \text{subject to} \ \|\mathbf{y}_i - \mathbf{SAc}_i\|_2 \le \epsilon \tag{18}$$

The *BPDN* is the approach utilised so far in CS implementations for WSN networks within the civil infrastructure domain O'Connor et al. (2013), O'Connor et al. (2014), Wang (2007), Wang et al. (2015), Wang and Hao (2015), Zou et al. (2015). In enhancing the current stat-of-the-art, this work examines its re-weighted variant, known as the *re-weighted Basis Pursuit De-Noising problem* (rwBPDN), also referred to as l_1 *analysis* Becker et al. (2011), defined as:

$$\underset{\mathbf{c}_i}{\operatorname{argmin}} \ \|\mathbf{Wc}_i\|_1 \ \text{subject to} \ \|\mathbf{y}_i - \mathbf{SAc}_i\|_2 \le \epsilon \tag{19}$$

where $\mathbf{W} = diag([w_1, w_2, \ldots, w_N])$ is a weighting matrix whose role is to orient the solution towards a desired $\mathbf{c}_i$ structure. The usage of the weighting coefficents is the main feature behind the spectro-temporal CS presented in the next section. It should be additionally noted that the formulations (18) and (19) are central to the signal recovery problem and can be solved given that the $\mathbf{SA}$ product holds the condition RIP.

2.3. Spectro-temporal CS

Figure 1 provides an overview of the work-flow and actions performed by each component of the framework. The right half of the figure represents the actions performed at the base station level (global), whereas the left half represents the actions executed on sensor node level (local). The execution starts by obtaining the ϵ_l support selection threshold at each sensing node, typically selected at level $\epsilon_l \approx 1$. Next, the wireless nodes execute a time-synchronised measurement and perform a windowing action, namely the recorded signal $\mathbf{x}_i$ of dimension N is divided into R frames of dimension N_R. This results in a set of signals $\mathbf{x}_{ij}$, where $j = 1 \ldots R$. The support selection is carried out for every jth data frame. The formulated support $\mathbf{U}_{ij}$ as well as its complementary set $\mathbf{U}_{ij}^c$ form key components of the re-weighted framework. Specifically, the spectral domain representation of the signal, Y_{ij}, is decomposed into a 'noisy' and 'clean' component as follows:

$$\begin{aligned}
\mathbf{c}_{ij} &= \mathbf{U}_{ij}\mathbf{c}_{ij} + \mathbf{U}_{ij}^c \mathbf{c}_{ij} \\
\mathbf{U}_{ij} &= supp(Y_{ij}, \epsilon_{ij}) \\
\mathbf{U}_{ij}^c &= \sim supp(Y_{ij}, \epsilon_{ij})
\end{aligned} \tag{20}$$

where the support selection thresholds $\epsilon_{ij} = \epsilon_l \frac{\|\mathbf{c}_{ij}\|_1}{N_R}$ are calculated for every data frame on the basis of the preselected prior to the measurement ϵ_l threshold value. This results in a diagonal binary selection thresholding matrix $\mathbf{U}_{ij}$, where ones are assigned to those coefficients of $\mathbf{c}_{ij}$ which are larger than ϵ_{ij}. Next, the wireless sensors transmit the diagonal elements of the thresholding matrix $\mathbf{U}_{ij}$ back to the server. The server then calculates the K-sparsity of the signal, i.e. $K_{ij} = \sum (\mathbf{U}_{ij})$ and forms the associated diagonal weighting matrix $\mathbf{W}_{ij}$ as:

$$\mathbf{W}_{ij} = (\mathbf{U}_{ij} + a\mathbf{I})^{-1} \tag{21}$$

In a next step, the ith wireless node estimates and transmits the jth data frame noise variance using the *off-support elements*, included in $\mathbf{U}_{ij}^c$:

$$\sigma_{u,ij}^2 \approx var(\mathbf{A}\mathbf{U}_{ij}^c\mathbf{c}_{ij}) \qquad (22)$$

where $a << 1$. The server then estimates the number of time domain samples required for signal reconstruction, according to the guideline provided in Mansour and Saab (2014). In this work it is stated that l_1 analysis requires:

$$M_{ij} \geq K_{ij} + E_{ij} \, \log \frac{N_R}{E_{ij}} \qquad (23)$$

time domain samples. The variable E_{ij} indicates the extent of the error in the support estimation, i.e. the number of erroneous sparse support elements for the ith node's jth data frame. These errors inevitably arise as a result of the applied support selection threshold, ϵ_{ij}. When the latter is set to a level that is too high, it may effectively result in omission of important coefficients from the sparse projection coefficient vector $\mathbf{c}_{ij}$. In order to address this shortcoming, in this work a simplification is made, namely E_{ij} is assumed tied to the size of the estimated sparsity of the signal as $E_{ij} = c_K K_{ij}$. The value of coefficient c_K depends on the level of desired accuracy. A value of $c_K = 0.1$ is herein adopted to allow for sufficient reconstruction of the structural modes, as demonstrated in the next section.

Based on the number of selected samples M_{ij}, the server randomly selects from the jth data frame of the ith sensor $\mathbf{y}_{ij}$ sub-vectors for transmission using a uniform distribution. In doing so, a random selection matrix $\mathbf{S}_{ij} \in \mathbb{R}^{M_{ij} \times N_R}$ is utilised, which comprises binary elements under the constraint that each column contains a unique entry equal to 1. The unit entries correspond to the indices of the utilised sparse samples. M_{ij} samples are then drawn from $\mathbf{y}_{ij}$ and are subsequently transmitted back to the server. Once all the necessary time domain samples are transmitted to the server from the jth data frame of the ith sensor, along with the corresponding $\mathbf{W}_{ij}$ weighting matrix, the jth data frame is reconstructed using the *rwBPDN* problem:

$$\hat{\mathbf{c}}_{ij} = \underset{\mathbf{c}}{\text{argmin}} \, \|\mathbf{W}_{ij}\mathbf{c}\|_1 \text{ subject to } \|\mathbf{y}_{ij} - \mathbf{S}_{ij}\mathbf{A}\mathbf{c}\|_2 \leq \sigma_{u,ij} \quad (24)$$

Lastly, the spectral data from the jth data frame are projected back to the time domain:

$$\hat{\mathbf{x}}_{ij} = \mathbf{A}\hat{\mathbf{c}}_{ij} \qquad (25)$$

Once this process has been executed for each data frame, the estimate of the full time domain signal is attained as:

$$\hat{\mathbf{x}}_i = \left[\hat{\mathbf{x}}_{i1}, \hat{\mathbf{x}}_{i2}, \dots, \hat{\mathbf{x}}_{iR}\right]^T \qquad (26)$$

Naturally, in this process an appropriate solver should be employed for solution of the *rwBPDN* problem. The algorithm adopted herein for signal recovery is described in the next section.

2.4. The NESTA algorithm

Although the CS methodology is relatively recent, the number of available solvers capable of dealing with the formulations

of Equations (16),(18),(19) is already quite vast. Most of the methods derived for solving Equation (18) fall into the category of convex problem relaxation, i.e. replacement of the l_1 norm with an equivalent formulation of the problem, or via greedy pursuit for obtaining a sparse and approximate solution. The most popular relaxation-based solvers include Gradient Projection for Sparse Reconstruction Figueiredo et al. (2007), Sparse reconstruction by separable approximation Wright et al. (2009), l_1 regularised least squares problems (l_1 ls) Kim et al. (2007), Spectral projected gradient (SPGL1) Birgin et al. (1999), Fast Iterative Soft-Thresholding Algorithm Beck and Teboulle (2009) and the Nesterov Algorithm (NESTA) Becker et al. (2011). Among the greedy-based algorithms the most popular ones involve CoSaMP Needell and Tropp (2009), Orthogonal Matching Pursuit Tropp (2004), and regularised Orthogonal Matching Pursuit Needell and Vershynin (2009).

For solving the problem stated in Equation (19) the NESTA solver is herein adopted. The NESTA is based on the work of Nesterov (1983), who initially introduced an algorithm which allows for minimisation of a convex smooth function $f(\mathbf{c})$, specified on a convex bounded set $\mathbb{Q}_p$, referred to as the primal feasible set.

$$\min_{\mathbf{c} \in \mathbb{Q}_p} f(\mathbf{c}) \qquad (27)$$

The function f is assumed to be differentiable and its gradient ∇f as Lipschitz, with L specifying an upper bound on the Lipschitz constant. Nesterov's algorithm minimises f over $\mathbb{Q}_p$ by iteratively estimating three sequences $\{\mathbf{c}_i\}, \{\bar{\mathbf{y}}_i\}, \{\bar{\mathbf{z}}_i\}$. The outline of the NESTA per the work of Becker et al. (2011) is provided below:

Proof (Algorithm NESTA): Initialise $\mathbf{c}_0$. For $i > 0$

(1) Compute $\nabla f_\mu(\mathbf{c}_i)$
(2) Compute $\bar{\mathbf{y}}_i$:
$\bar{\mathbf{y}}_i = \underset{\mathbf{c} \in \mathbf{Q}_p}{\text{argmin}} \, \frac{L_\mu}{2}\|\mathbf{c}_i - \mathbf{c}\|_2^2 + \langle \nabla f_\mu(\mathbf{c}_i), \mathbf{c} - \mathbf{c}_i \rangle$
(3) Compute $\bar{\mathbf{z}}_i$:
$\bar{\mathbf{z}}_i = \underset{\mathbf{c} \in \mathbf{Q}_p}{\text{argmin}} \, \frac{L_\mu}{\sigma_p}p_p(\mathbf{c}) + \langle \sum_{i \leq j} \alpha_j \nabla f_\mu(\mathbf{c}_i), \mathbf{c} - \mathbf{c}_i \rangle$
(4) Update $\mathbf{c}_i$:
$\mathbf{c}_i = \beta_i\bar{\mathbf{z}}_k + (1 - \beta_i)\bar{\mathbf{y}}_i$

Stop when $\|\mathbf{c}_i - \mathbf{c}_{i-1}\| < threshold$ □

The scalar sequences α_i, β_i appearing in the above description are defined as $\alpha_i = \frac{1}{2(i+1)}$ and $\beta_i = \frac{1}{2(i+1)}$, as specified in the Nesterov (2005). Moreover, $p_p(\mathbf{c})$ is the so-called prox-function for the primal feasible set $\mathbb{Q}_p$. This function is strongly convex with parameter σ_p, set to $\sigma_p = 1$ herein. In more recent work Nesterov (2005) extended this method for the minimisation of non-smooth convex functions f, assuming this can be expressed as:

$$f(\mathbf{c}) = \|\mathbf{c}\|_1 = \max_{\mathbf{u} \in \mathbb{Q}_d} \langle \mathbf{u}, \mathbf{W}\mathbf{c} \rangle \qquad (28)$$

where $\mathbf{c} \in \mathbb{R}^n, \mathbf{u} \in \mathbb{R}^p$ and $\mathbf{W} \in \mathbb{R}^{p \times n}$. $\mathbb{Q}_d$ is referred to as the dual feasible set and is assumed to be convex. A smooth

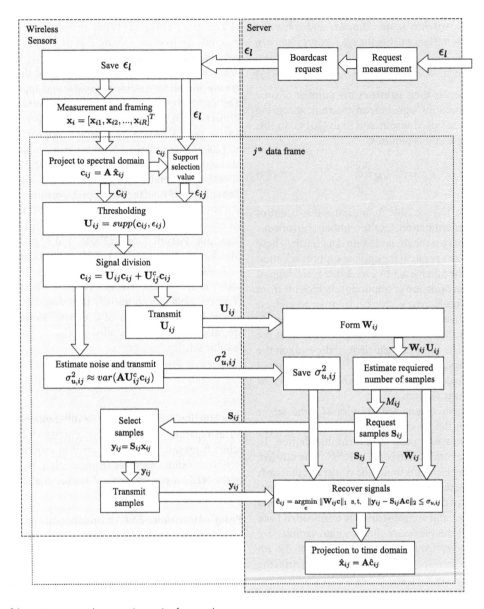

Figure 1. Overview of the spectro-temporal compressive sensing framework.

approximation of $f_\mu(\mathbf{c})$ of the original function $f(\mathbf{c})$ is then employed as:

$$f_\mu(\mathbf{c}) = \max_{\mathbf{u} \in \mathbb{Q}_d} \langle \mathbf{u}, \mathbf{Wc} \rangle - \mu p_d(\mathbf{u}) \qquad (29)$$

where $p_d(\mathbf{u})$ is a prox-function for $\mathbb{Q}_d$ and $\mu \to 0$ is a smoothing approximation parameter. The detailed description of the algorithm is provided in the work of Becker et al. (2011), who implemented the algorithm within the general CS framework, for solving Equation (19). Following the NESTA formulation, Becker et al. assumed that the l_1 norm of the coefficient vector $\mathbf{c}$ may be written as:

$$||\mathbf{c}||_1 = \max_{\mathbf{u} \in \mathbb{Q}_d} \langle \mathbf{u}, \mathbf{Wc} \rangle \qquad (30)$$

where $\mathbf{W}$ is defined in the Equation (19) and the dual feasible set is defined as and l_{inf}-ball, i.e. $\mathbb{Q}_d = \mathbf{u} : ||\mathbf{u}||_{inf} \leq 1$. Since the function $||(\mathbf{c})||_1$ is convex, but non-smooth, an approximation per Equation (32) is adopted. The aim is to solve the following optimisation problem:

$$\underset{\mathbf{c}}{\mathrm{argmin}} \, f_\mu(\mathbf{Wc}) \text{ subject to } ||\mathbf{y} - \mathbf{SAc}||_2 \leq \epsilon \qquad (31)$$

According to Nesterov (1983), the latter is equivalent to implementing the previously outlined NESTA algorithm on the smooth constrained optimisation problem:

$$\min_{\mathbf{u} \in \mathbb{Q}_p} f_\mu(\mathbf{Wc}) \text{ where } \mathbb{Q}_p = \{\mathbf{c} : ||\mathbf{y} - \mathbf{SAc}||_2 \leq \epsilon\} \qquad (32)$$

The ability of the algorithm to efficiently solve the *rwBPDN* problem largely relies on the use of the weighting matrix $\mathbf{W}$,

(a)

(b)

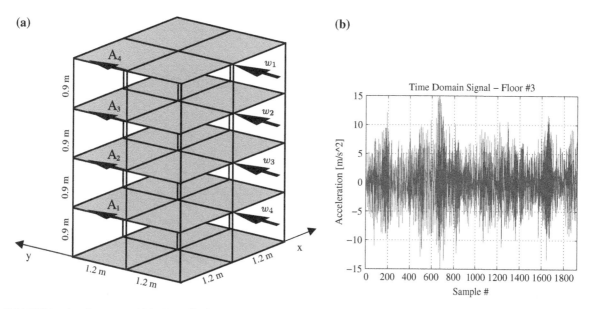

Figure 2. (a) ASCE benchmark structure with horizontal load sources w_1, w_2, w_3, w_4 and acceleration measurement locations A_1, A_2, A_3, A_4; (b) Generated acceleration time series for location A_3.

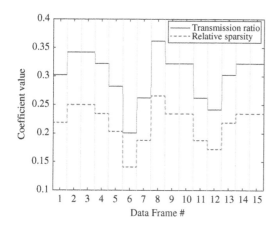

Figure 3. Relative sparsity and transmission ratio per data frame, floor #3. $\epsilon_l = 1.2$.

whose purpose lies in emphasising a desired structure for the vector of the sparse domain coefficients **c**. The calculation of the re-weighting product $||\mathbf{Wc}||_1$ results in penalising the coefficients c_j corresponding to a high weighting value w_j. This forces the solution of Equation (19) to concentrate on the elements of low corresponding weights w_i. It is reminded that matrix **W** is constructed via spectral data delivered by the sensing nodes, according to Equation (21).

3. Validation of the proposed framework on synthetic data

3.1. Description of the test structure

The validation of the proposed framework is carried out using synthetically generated data-sets obtained from the Finite Element model of a four-storey, two-bay by two-bay steel-frame quarter-scale ASCE benchmark structure, described in Johnson et al. (2004). The structural model has been developed in the Earthquake Engineering Research Laboratory at

the University of British Columbia and is used for simulating introduced structural damages and corresponding acceleration sensor readings. The data generation process utilises a simplified model variant of the structure in which the model is constructed using a 12-DOF shear-building that allows only for two horizontal translations and one rotation per floor, as indicated in Figure 2(a).

Synthetic response data series are generated via use of the ASCE-Datagen software, which is designed to solve for a pre-selected excitation case, referred to as *Case ID*, and a prese-lected damage scenario, referred to as *Damage ID*, via use of a specific integration scheme, referred to as *Method ID*. For the analysis carried out herein, the *Case ID* is set to 1, which corresponds to white noise excitation applied independently at all stories in the y direction, as illustrated in the Figure 2(a). These excitations are modelled as filtered Gaussian white noise processes, passed through a sixth-order low-pass Butterworth filter with a 100 Hz cut-off frequency Johnson et al. (2004). In the carried out simulation, the *force intensity* parameter, *damping ratio*, *simulation step* parameter and *measurement noise level* parameter are maintained as the program default values. The duration time of the simulation is set to 60 sec. The selected integration scheme uses the defaults *MATLAB's* integrator for linear systems, i.e. *lsim*. Observations in the form of acceleration are assumed available in the locations denoted as A_1, A_2, A_3, A_4 in Figure 2(a). In a post-processing phase, the obtained data are re-sampled to 140 Hz in order to simulate a realistic WSN sampling scheme. The parameter values adopted in the simulation are explicitly listed in Table 1. An example of a generated acceleration time series corresponding to location A_3 is included in Figure 2(b).

3.2. Application of spectro-temporal CS

In this section, the proposed spectro-temporal CS approach, *rwBPDN*, is cross-compared against the standard BPDN

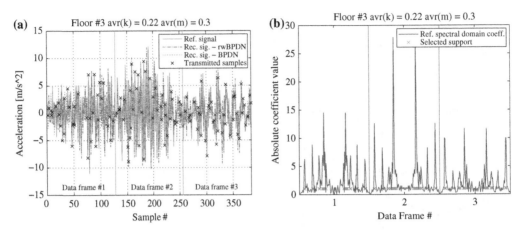

Figure 4. The data framing process for the recorded response signals, in the temporal and spectral domain, per data frame #1, #2, #3 for floor #3. (a) Time domain representation of the reference and recovered signal. Framing of the data. (b) Spectral domain representation per data frame #1, #2, #3, and selected support elements.

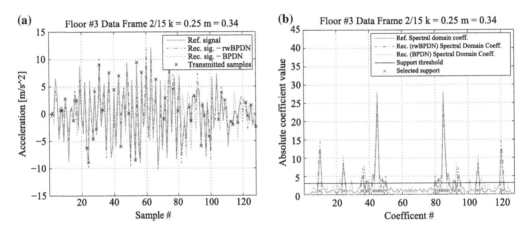

Figure 5. Data frame #2, floor #3 in the temporal and spectral domain representation. (a) Reference and recovered time domain signal. Transmitted samples in black. (b) Reference and recovered spectral domain representation. The transmitted support is noted in magenta.

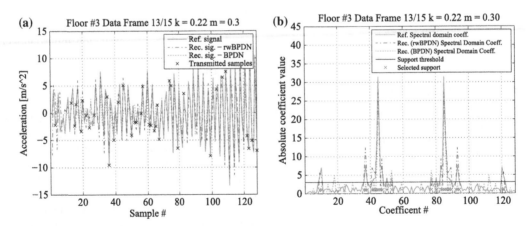

Figure 6. Data frame #13, floor #3 in the temporal and spectral domain representation. (a) Reference and recovered time domain signal. Transmitted samples in black. (b) Reference and recovered spectral domain representation. The transmitted support is noted in magenta.

approach implemented in existing literature works O'Connor et al. (2014). The acceleration signals, generated for the test-case assumed herein, define a set $\ddot{\mathbf{x}}_{A_1}, \ldots, \ddot{\mathbf{x}}_{A_4} \in \mathbb{R}^{N \times 1}$, with $N = 1920$. The length of the reconstruction data frames is selected as $N_R = 128$, which results in dividing the recorded signals into $R = 15$ individual frames. For every data frame the recovery process is performed separately. Figure 4(a) illustrates the results of the framing process for signal $\ddot{\mathbf{x}}_{A_3}$, which produces the data-frame set $\ddot{\mathbf{x}}_{A_3} = [\ddot{\mathbf{x}}_{A_31}, \ldots, \ddot{\mathbf{x}}_{A_315}]$. Next, each data frame is projected from the time domain onto the spectral

Table 1. Aggregated simulation parameters.

	Value	Units
Case ID	1	–
Damage ID	1	–
Method ID	1	–
Filtering	off	–
Simulation step	1e-3	[s]
Noise level	10	[%]
Damping ratio	0.01	[1]
Force intensity	150	[1]
Duration	60	[s]

domain as $c_{A3j} = \mathbf{A}^{-1}\ddot{\mathbf{x}}_{A3j}$, $j = 1 \ldots R$. The results of the projection process are illustrated in Figure 4(b). The frames denoted via a red background incorporate representation details in both the time domain, in Figures 5(a), 6(a), and the spectral domain in Figures 5(b), 6(b).

For every data frame the support selection is performed according to the threshold $\epsilon_{A3j} = \epsilon_l \frac{\|c_{A3j}\|_1}{N_R}$. Figures 5(b) and 6(b) indicate the assigned support selection threshold via a black solid line corresponding to $\epsilon_l = 1.2$, i.e. $\epsilon_{A32} = 3.11, \epsilon_{A313} = 3.05$. The support matrix for each frame is then formulated, comprising elements which satisfy $|c_{ij}| > \epsilon_{ij}$. The selected support indices, essentially designating significantly contributing coefficients, are indicated per data frame in Figure 4(b) via purple markers. The same notation is maintained for the spectral domain representation in Figures 5(b) and 6(b).

The selected support indices are transmitted back to the server, where the diagonal weighting matrices per frame, $\mathbf{W}_{ij}$, are formed as $\mathbf{W}_{ij} = (\mathbf{U}_{ij} + a\mathbf{I})^{-1}$. The off-support elements are used for estimating and transmitting the associated noise variance $\sigma_{u,ij}^2 \approx var(\mathbf{A}\mathbf{U}_{ij}^c c_{ij})$, in this case $\sigma_{u,A32}^2 = 1.1464, \sigma_{u,A313}^2 = 1.4652$, back to the server. The server then employs the support matrix, $\mathbf{U}_{ij}$, for estimating the number of required transmission samples according to the relation: $M_{ij} \geq K_{ij} + c_K K_{ij} \log \frac{N_R}{c_K K_{ij}}$, where a coefficient $c_K = 0.1$ is employed. The latter yields $K_{A32} = 32, K_{A313} = 28$, which in turn results in a nuber of transmitted samples $M_{A32} = 43, M_{A313} = 39$. The resulting relative sparsity, defined as $k_{ij} = K_{ij}/N_R$, and relative transmission ratios, $m_{ij} = M_{ij}/N_R$, per data frame are indicated in Figure 3.

Once the noise variance $\sigma_{u,ij}^2$ is established, the sub-vectors of the signals are randomly selected via matrix S_{ij}, which specifies the transmitted samples per sensor according to $\mathbf{y}_{ij} = S_{ij}\ddot{\mathbf{x}}_{ij}$. The transmitted samples are indicated in Figures 4(b), 5(a) and 6(a) via black markers. Once all required samples per data frame are transmitted back to the server, the $rwBPDN$ optimisation problem is solved for every data frame according to Equation (24). This procedure is illustrated in Figures 5(b) and 6(b), where the absolute value of the recovered spectral domain signal is plotted in red. As a last step of the reconstruction, the spectral domain estimate is projected back to time domain. The recovered results for frame number 2,13 are indicated in Figures 5(a) and 6(a) via red dashed lines. The $rwBPDN$ recovered results for the whole time series, i.e. via aggregating the recovered frames are shown in Figure 4(a) in red, whereas the $BPDN$ reconstruction results are marked in green. Although the goal herein lies in modal-shape extraction and not signal reconstruction, one may already note a superior performance for the proposed $rwBPDN$ scheme.

4. Reconstruction of structural modes via partial transmissions

In order to assess the performance of the proposed framework, a parametric analysis is conducted where the recovery procedure is repeated for varying support selection thresholds $\epsilon_l = \{0.8, 1.2, 1.35, 1.5\}$. Next, a modal identification procedure is carried out on the recovered time series via use of the Multi-variant Auto-Regressive identification toolbox provided in the *ARfit* Matlab package Schneider and Neumaier (2001). The support is selected with respect to the frequencies reported in Johnson et al. (2004) as 9.41 Hz (mode 1), 25.60 Hz (mode 2), 38.85 Hz (mode 3) and 48.68 Hz (mode 4). These frequencies pertain to bending modes along the y direction of the model of Figure 2(a). Figures 7(a) and (b), 8(a) and (b) illustrate the identified modal shapes compared against the identification results obtained from reference (complete) time series marked via a solid blue line. In these figures the identification results obtained using the $BPDN$ approach are marked via a green dotted line. The results are cross-checked against the mode shapes obtained using the proposed spectro-temporal CS algorithm relying on re-weighted Basis Pursuit De-Nosing ($rwBPDN$). These results are marked in red via a dash-dotted line. As may be observed, an increase in the support selection threshold ϵ_l, implies a corresponding decrease in the amount of transmitted data m_{ij}, which inevitably leads into some deterioration in the quality of the reconstructed modal shapes. However, the consistency in the recovery of these shapes via spectro-temporal CS remains quite remarkable even for very low transmission levels, as illustrated in Figure 8(b). As is further obvious, the proposed $rwBPDN$ approach delivers significant enhancement when compared to the standard $BPDN$ alternative. In order to further validate the quality of the identified eigenvectors the AutoMAC criterion is employed. This results via calculation of the Modal Assurance Criterion (MAC) values between a pair of identified $\hat{\boldsymbol{\phi}}_i$ and reference eigenvectors $\boldsymbol{\phi}_j$:

$$\text{AutoMAC}(i,j) = \frac{\|\hat{\boldsymbol{\phi}}_i^T \boldsymbol{\phi}_j\|}{\sqrt{\|\hat{\boldsymbol{\phi}}_i^T \hat{\boldsymbol{\phi}}_i\| \|\boldsymbol{\phi}_j^T \boldsymbol{\phi}_j\|}} \tag{33}$$

Based on this criterion, when the diagonal components of the resulting AutoMAC matrix are equal to one and the off-diagonal terms lie below an acceptable threshold, the set of identified eigenvalues is deemed as successful. This is illustrated in Figure 9, for the case of $\epsilon_l = 1.2$, which corresponds to an average transmission ratio of $avr(m_{ij}) = 0.31$, results indicate successful reconstruction.

5. Cost efficiency of the proposed STCS using the ASCE benchmark case study

The efficiency of the proposed scheme is evaluated based on the financial costs an infrastructure operator needs to invest in order to deploy and maintain the monitoring system. The costs are compared in two theoretical deployment scenarios of a permanent vibration-based monitoring systems, i.e. a WSN-based monitoring deployment using fully transmitted data (WSN-SHM) and finally a deployment of WSN featuring the proposed

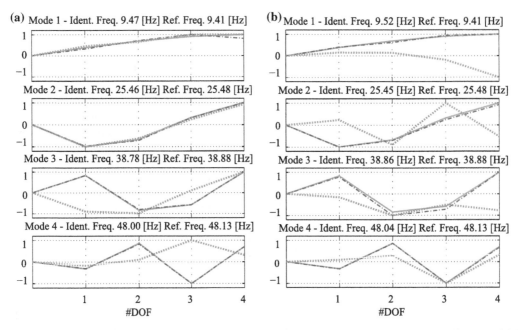

Figure 7. Mode shape identification results for support selection thresholds $\epsilon_l = 0.8$ and $\epsilon_l = 1.2$. The blue solid line designates the reference model, the green dotted line corresponds to the *BPDN*, whereas red dash-dot line denotes the *rwBPDN*. (a) $\epsilon_l = 0.8$, trans. ratio $avr(m_{ij}) = 0.46$ (b) $\epsilon_l = 1.2$, trans. ratio $avr(m_{ij}) = 0.31$.

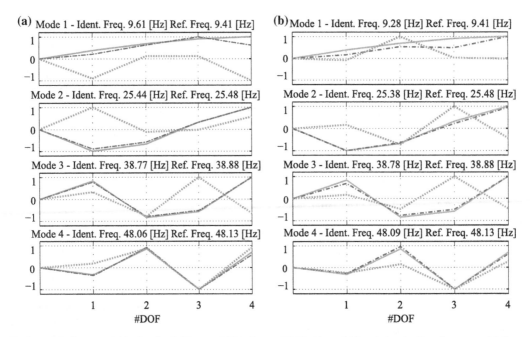

Figure 8. Identification results for support selection thresholds $\epsilon_l = 1.35$ and $\epsilon_l = 1.5$. The blue solid line designates the reference model, the green dotted line corresponds to the *BPDN*, whereas red dash-dot line denotes the *rwBPDN*. (a) $\epsilon_l = 1.35$, trans. ratio $avr(m_{ij}) = 0.28$ (b) $\epsilon_l = 1.5$, trans. ratio $avr(m_{ij}) = 0.25$.

specto-temporal data compression scheme (STCS-WSN-SHM). Both of the deployment scenarios consider the earlier presented ASCE testbed structure as the example structure.

5.1. Initial deployment costs

The typical costs of instrumentation for the communication infrastructure of a cable-based monitoring system, as reported in 2002, rise to approximately 5000$ per measurement channel, including materials and associated labour costs (Celebi, 2002; Lynch, 2006). When switching to a wireless-based solution this

cost, as reported in 1996, is expected to drop to around 1000$ per measurement channel (Lynch, 2006; Maser et al., 1996). Clearly, a simple calculation demonstrates that the communication infrastructure makes up roughly 80% of the overall cost of a tethered monitoring system (Lynch, 2006; Straser et al., 1998). In addition, additional hard to estimate costs may be identified, associated with interruption of the normal operation of the structure during the deployment of the cables. Straser et al. (1998) have demonstrated that the deployment time of WSNs, and therefore the time of distribution of infrastructure, is significantly smaller than in the case of a tethered monitoring

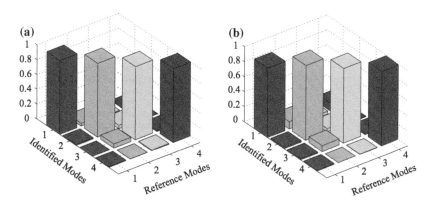

Figure 9. AutoMAC criteria plots for $\epsilon_l = 1.2$ and $\epsilon_l = 1.35$. (a) $\epsilon_l = 1.2$, trans. ratio $avr(m_{ij}) = 0.31$ (b) $\epsilon_l = 1.5$, trans. ratio $avr(m_{ij}) = 0.25$.

system. However, in this work focus is put purely on the costs covered by the infrastructure operator.

5.2. Monitoring system maintenance actions

Feltrin et al. (2011) demonstrated that the availability of property designed WSN can reach 99%. Therefore, in the simplified theoretical case scenario, failure of the measurement nodes can be completely neglected. As a result, the only remaining factor impacting the availability of the WSN monitoring system is battery replacement. In order to estimate the number of interventions required by the infrastructure operator, assumptions have to be made on the type of employed wireless sensor, the mode of its operation, the required radio transmission power for achieving a reliable connectivity within the network, the type of used network topology, the battery capacity, the length of the measurement and the adopted sampling rate.

5.2.1. Wireless node specification and assumptions

In order to obtain realistic estimates on the necessary number of interventions per year, the power consumption and data transmission characteristics of actual *WiseNode_v4* wireless nodes are adopted in this work (Novakovic et al., 2009). The corresponding specifications are reported in Table 2. The calculation performed does not account for more complex phenomena, such as ageing of batteries, communication unreliability, or the impact external conditions have on the battery lifetime. In the estimation it is assumed that the nodes of the WSN-SHM wireless system operate according to state-flow presented in Figure 10(a). The WSN featuring the proposed specto-temporal data compression scheme, presented in Figure 10(b), is enriched with a signal processing functionality. The calculated state-flows are composed out of *sensing and analog to digital conversion states*, as denoted via use of subscript S and ADC accordingly; a, *signal processing state* denoted via subscript DSP, and a *data transmission state*, denoted via subscript T, followed by a prolonged state of *idle state* denoted via subscript Z. The described state-flows correspond to a single measurement executed by the WSN.

For investigating the long-term behaviour of the monitoring system, it is assumed that such a sequence of operations is executed $M_N = 24$ per day, which corresponds to one measurement per hour. Further on, a requirement of using a radio

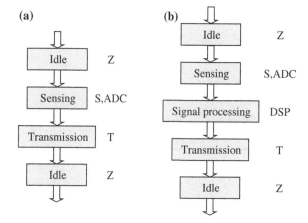

Figure 10. (a) Stateflow of the WSN monitoring system operating in a full data transmission mode (WSN-SHM). (b) Stateflow of the WSN monitoring system operating with the proposed specto-temporal data compression, i.e. in partial transmission mode (STCS-WSN-SHM).

power signal of minimum 10 dBm is assumed for obtaining a reliable *star topology* communication between the server and the wireless nodes. The sensing and analog to to digital conversion module is assumed to be using a $ADC_{bit} = 16$ bit word size and acquires signal with a rate of $F_s = 100\,\text{Hz}$ samples per second. The sensors are assumed to be able to store $N = 6144$ samples. As demonstrated in the previous section, the spectro-temporal CS scheme is capable of providing sufficiently good data operating at a transmission ratio of $m = 0.3$. Further on, a reconstruction data frame of dimension $N_R = 256$ is assumed, which corresponds to dividing the signal into $R = 24$ individual data frames. Further on, only the diagonal components of the support $\mathbf{U}_{ij}$ are assumed to be stored for transmission, in bit format. A summary of these parameters is reported in Table 3.

5.2.2. Mode of operation and associated energy expenditure

In order to assess the energy expenditure required to operate the WSN node on a daily basis, the time spent on a single sequence of operations needs to be evaluated first. In case of state S and the simultaneously executed state ADC, this time would be defined as: $t_{S,ADC} = N/f = 61.44\,\text{s}$. The state produces $I_{S,ADC} = N \times ADC_{bit} = 12288\,\text{byte}$. Next, in the case of the proposed STCS-WSN-SHM the acquired signal undergoes

Table 2. WiseNode_v4 wireless node power consumption specification Novakovic et al. (2009).

Parameter	Value	Unit	Description
P_Z	81	μW	Power consumption of Wireless Sensor Module in idle state
P_S	3.9	mW	Power consumption of the used acceleration sensor board
P_ADC	2.16	mW	Power consumption of the analog to digital conversion board, 16 bit, 100 Hz sampling rate
P_DSP	5.7	mW	Power consumption of data processing module calculating 256-point FFT
P_T	103.8	mW	Power consumption in radio transmit mode at 10 dBm
E_FWD	2106	μJ	Energy needed for transmitting one data package at 10 dBm
I_FWD	7	byte	Information transmitted per one data package
U_b	3	V	Supply voltage / average battery discharge

Table 3. Assumed WSN and environmental parameters.

Parameter	Value	Unit	Description
B_l	3×2122	mAh	Battery capacity. Three AAA size alkaline batteries
N	6144	–	Length of single measurement
N_R	256	–	Length of reconstruction frame
R	24	–	Number of reconstruction frames
ADC_bit	16	bit	Length of used analog to digital conversion word
F_s	100	Hz	Sampling frequency
M_N	24	–	Number of measurements per day
F_s	10	dBm	Assumed minimal radio signal power for obtaining reliable communication
m	0.3	–	Assumed compression ratio for used spectro-temporal compression scheme

Table 4. Estimated time spend by *WiseNode_v4* per sequence and per day.

Parameter	Per sequence (s)	Per day (s)	Description
$t_{S,ADC}$	61.44	1474.56	Time spend in sensing and analog to digital conversion states
t_{T_f}	35.12	842.88	Time spend in signal transmission state using full transmission scheme
t_{Z_f}	–	84082.56	Time spend in idle state using full transmission scheme
t_{DSP}	3	72	Time spend in signal processing state using partial transmissions scheme
$t_{T_{stcs}}$	12.74	305.76	Time spend in signal transmission state using partial transmissions scheme
$t_{Z_{stcs}}$	–	84547.68	Time spend in idle state using partial transmissions scheme

Table 5. Estimated energy spend by *WiseNode_v4* per sequence and per day.

Parameter	Value per sequence (J)	Value per day (J)	Description
$E_{S,ADC}$	0.37	8.93	Energy spend in sensing and analog to digital conversion states
E_{T_f}	3.64	87.49	Energy spend in signal transmission state using full transmission scheme
E_{Z_f}	–	6.81	Energy spend in idle state using full transmission scheme
E_{DSP}	0.017	0.41	Energy spend in signal processing state using partial transmissions scheme
$E_{T_{stcs}}$	1.32	31.73	Energy spend in signal transmission state using partial transmissions scheme
$E_{Z_{stcs}}$	–	6.84	Energy spend in idle state using partial transmissions scheme

Table 6. Summary of comparison results.

Type of deployment	Initial cost	Battery rep.	Natural freq.	Modal shapes	Damping factors
WIRE-SHM	5.000$ / channel	–	yes	yes	yes
WSN-SHM	1.000$ / channel	1.8 years	yes	yes	yes
STCS-WSN-SHM	1.000$ / channel	3.9 years	yes	yes	no

signal processing, i.e. for each $j = 1, \ldots, R$ data frame obtained from ith sensor the Fourier transformation $\mathbf{c}_{ij}$ needs to be calculated, and its absolute value is in turn calculated and thresholded in order to select the support $\mathbf{U}_{ij}$ and estimate the noise level $\sigma_{u,ij}^2$. As the most complex and computationally heavy operation lies in the Fourier transformation, this is herein selected for estimation. Other operations relating to the time expenditure of state DSP are omitted for the sake of simplicity. The detailed report of Novakovic et al. (2009) describing the *WiseNode_v4* provides both the time and energy expenditure of the *WiseNode_v4* executing the *Integer based approximation of Fourier transformation* on a 256-point window. The provided

data result in $t_{DSP_R} = 0.125\,\text{s}$ and $P_{DSP} = 5.7\,\text{mW}$. Since the processing is executed for each of the R data frames, the overall associated processing time may be calculated as $t_{DSP} = t_{DSP_R} \times R = 3\,\text{s}$. The 256-point Fourier transformation window produced a 256-bit diagonal for the thresholding matrix $\mathbf{U}_{ij}$, which translates to 32 additional bytes per data frame. As a result DSP *state* produces an additional $I_{DSP} = 32 \times 24 = 768$ byte.

The time required by the *WiseNode_v4* for transmitting a single packet of data is calculated using the node specifications as $t_{T1} = E_{FWR}/P_T \approx 20\,\text{ms}$ where E_{FWR} is the energy needed to transmit one packet of data and P_T is the power expenditure during the transmission (Novakovic et al., 2009).

According to the same specifications one transmitted packet can carry $I_{T1} = 7$ bytes of useful, i.e. no transmission protocol-related, information. Therefore the wireless sensor WSN-SHM operating in a full transmission scheme remains in transmission state for $t_{T_f} = t_{T1} \frac{I_{S,ADC}}{I_{T1}} \approx 35.12$ s, whereas the wireless sensor operating in the proposed scheme remains in the transmission state for $t_{T_{stcs}} = t_{T1} \frac{m\, I_{S,ADC} + I_{DSP}}{I_{T1}} \approx 12.74$ s.

The WSN nodes operating in full transmission mode are assumed to remain in *idle* state for the rest of day, i.e. $t_{Z_f} = t_D - M_N(t_{S,ADC} + t_{T_f}) \approx 84082.56$ s, where t_D is the number of seconds during the day. On the other hand, the WSN nodes employing the proposed scheme remain in low power *idle* mode for $t_{Z_{stcs}} = t_D - M_N(t_{S,ADC} + t_{DSP} + t_{T_{stcs}}) \approx 84547.68$ s. Next, given the time characteristics and average power consumption per state, an estimation of the energy consumption of the nodes is carried out. The results are summarised in Tables 4 and 5.

The estimated total daily energy required by a *WiseNode_v4* node operating in the WSN-SHM scheme is equal to $E_{WSN} \approx 103.23$ J, whereas a node operating in the proposed STCS-WSN-SHM (partial transmission) scheme produces $E_{STCS} \approx 47.93$ J. Assuming the wireless nodes are powered via three alkaline AAA size batteries with energy storage of $B_I = 3 \times 2122$ mAh and are discharged using the voltage $U_b = 3$ V, they render an energy reserve approximately equal to $E_b = B_I U_b = 68752.8$ J. Such an energy reserve would allow for operating the node working in the WSN-SHM scheme for $D_{WSN-SHM} \approx 1.8$ years, while for the STCS-WSN-SHM scheme a duration of $D_{STCS-WSN-SHM} \approx 3.9$ years is estimated.

5.2.3. Summary of cost efficiency analysis

The cost efficiency of the proposed scheme is validated in this work in terms of i) the expected battery replacement periods, and therefore in terms of the expected maintenance costs that the infrastructure operator needs to cover, ii) the quality of provided information (accuracy in the prediction of modal characteristics) and iii) the initial system deployment cost. The results are summarised in Table 6.

In the summary, besides the WSN-SHM and STCS-WSN-SHM schemes, a classic tethered monitoring system is considered (WIRE-SHM). The comparison of the two WSN alternatives, i.e. STCS-WSN-SHM with WSN-SHM, clearly indicates that the proposed STCS-WSN-SHM scheme renders a scheme that is less demanding in terms of maintenance, and still allows for detection of the most important modal analysis features ,i.e. natural frequencies and modal shapes. Additionally, in contrast to the standard WIRE-SHM system, the reported expected initial deployment costs of the wireless system are significantly lower (Celebi, 2002; Lynch, 2006; Maser et al., 1996). The proposed enhanced scheme theretofore opens up opportunities for safety assessment by means of cheap, cost-effective and easily deployable, wireless structural health monitoring systems.

6. Conclusion

In this work, a novel data compression framework for WSNs is proposed. A remedy to heavy transmission costs is obtained by optimally combining the spectro-temporal information that is already present in the signal with a recently surfaced CS paradigm. A robust signal reconstruction technique is attained

allowing for reliable identification of structural modal shapes. The scheme is founded upon the adoption of a re-weighted Basis Pursuit De-Noising (*rwBPDN*) problem formulation. The approach is validated on a series of synthetic simulations generated for a benchmark four-storey frame structure of the American Society of Civil Engineers. It is demonstrated that the *rwBPDN* scheme enhances the potential for modal shape extraction when compared to the *BPDN* approach, employed so far in the literature. As a result, the potential for detecting possible changes in structural condition and likely damage is also strengthened by the proposed formulation. A comparative investigation indicates that the expected number of battery replacement maintenance operations is significantly decreased once the proposed scheme is applied, rendering this approach a good candidate for cost efficient long-term WSN deployments.

Disclosure statement

No potential conflict of interest was reported by the authors.

Funding

This work was supported by the Swiss National Science Foundation under Research [grant number 200021–143212]: *Implementation of Wireless Sensor Networks for Monitoring of Large Civil Structures.*

References

Bao, Y., Beck, J. L., & Li, H. (2011). Compressive sampling for accelerometer signals in structural health monitoring. *Structural Health Monitoring, 10*, 235–246.

Bao, Y., Li, H., Sun, X., Yu, Y., & Ou, J. (2012). Compressive sampling based data loss recovery for wireless sensor networks used in civil structural health monitoring. *Structural Health Monitoring, 12*, 78–95.

Bao, Y., Zou, Z., & Li, H. (2014, March 9). Compressive sensing based wireless sensor for structural health monitoring. In *Proceedings – sensors and smart structures technologies for civil, mechanical, and aerospace systems 9061* (pp. 90611W-1–90611W-10). San Diego, CA, USA.

Beck, A., & Teboulle, M. (2009). A fast iterative shrinkage-thresholding algorithm for linear inverse problems. *SIAM Journal on Imaging Sciences, 2*, 183–202.

Becker, S., Bobin, J., & Cands, E. J. (2011). Nesta: A fast and accurate first-order method for sparse recovery. *SIAM Journal on Imaging Sciences, 4*(1), 1–39.

Becker, S.R. (2011). Practical compressed sensing: Modern data acquisition and signal processing, (PhD thesis), Citeseer, Pasadena, CA, USA.

Birgin, E. G., Martínez, J. M., & Raydan, M. (1999). Nonmonotone spectral projected gradient methods on convex sets. *SIAM Journal on Optimization, 10*, 1196–1211.

Candès, E., & Plan, Y. (2011). A probabilistic and ripless theory of compressed sensing. *Information Theory, IEEE Transactions on, 57*, 7235–7254.

Candès, E., & Romberg, J. (2007). Sparsity and incoherence in compressive sampling. *Inverse Problems, 23*, 969–985.

Candès, E., Romberg, J., & Tao, T. (2006). Robust uncertainty principles: Exact signal reconstruction from highly incomplete frequency information. *Information Theory, IEEE Transactions on, 52*, 489–509.

Celebi, M. (2002). *Seismic instrumentation of buildings (with emphasis on federal buildings). Technical report, Report No. 0-7460-68170*. Menlo Park, CA: United States Geological Survey.

Fan, W., & Qiao, P. (2011). Vibration-based damage identification methods: A review and comparative study. *Structural Health Monitoring, 10*, 83–111.

Feltrin, G., Meyer, J., Bischoff, R., & Motavalli, M. (2009, July 22–24). Modular wireless sensor network for long-term structural health monitoring. In *4th International conference on structural health monitoring of intelligent infrastructure (SHMII-4)*. Zurich: Swiss Federal Laboratories for Materials, Testing and Research (Empa).

Feltrin, G., Saukh, O., Bischoff, R., Meyer, J., & Motavalli, M. (2011). Structural monitoring with wireless sensor networks: Experiences from field deployments. In*Proceedings of first middle east conference on smart monitoring, assessment and rehabilitation of civil structures SMAR*. Dubai: International Society for Structural Health Monitoring of Intelligent Infrastructure (ISHMII) and International Institute of FRP in Construction (IIFC).

Figueiredo, M., Nowak, R., & Wright, S. (2007). Gradient projection for sparse reconstruction: Application to compressed sensing and other inverse problems. *Selected Topics in Signal Processing, IEEE Journal of, 1*, 586–597.

Hsu, T.-Y., Huang, S.-K., Lu, K.-C., Loh, C.-H., Wang, Y., & Lynch, J. P. (2011). On-line structural damage localization and quantification using wireless sensors. *Smart Materials and Structures, 20*, 105025-1–105025-11.

Huang, Y., Beck, J. L., Wu, S., & Li, H. (2014). Robust bayesian compressive sensing for signals in structural health monitoring. *Computer-Aided Civil and Infrastructure Engineering, 29*, 160–179.

Johnson, E. A., Lam, H. F., Katafygiotis, L. S., & Beck, J. L. (2004). Phase i iasc-asce structural health monitoring benchmark problem using simulated data. *Journal of Engineering Mechanics, 130*, 3–15.

Kim, J.-T., Ryu, Y.-S., Cho, H.-M., & Stubbs, N. (2003). Damage identification in beam-type structures: Frequency-based method vs mode-shape-based method. *Engineering Structures, 25*, 57–67.

Kim, S.-J., Koh, K., Lustig, M., Boyd, S., & Gorinevsky, D. (2007). An interior-point method for large-scale l1-regularized least squares. *Selected Topics in Signal Processing, IEEE Journal of, 1*, 606–617.

Klis, R., & Chatzi, E. (2013, December 9–11). Wireless sensor network powered by dye-sensitized solar cells. In *Proceedings of the 6th international conference on structural health monitoring of intelligent infrastructure Hong Kong*. Hong-Kong: International Society for Structural Health Monitoring of Intelligent Infrastructure (ISHMII)

Klis, R., & Chatzi, E. (2014). Model-based data compression for vibration monitoring using wireless sensor networks. In H. Furuta, D. M. Frangopol, & M. Akiyama, (Eds.), *Life-cycle of structural systems design, assessment, maintenance and management* (pp. 138–145). Tokyo: CRC Press.

Le Cam, V., Le Maulf, R., Lemarchand, L., Martin, W., & Le Pen, M. (2012, July 3–6). An optimized electronic device for solar power harvesting dedicated to wireless sensor networks. In *6th European workshop on structural health monitoring*. Dresden: Fraunhofer-Institut für Zerstörungsfreie Prüfverfahren (IZFP).

Lynch, J. P. (2006). A summary review of wireless sensors and sensor networks for structural health monitoring. *The Shock and Vibration Digest, 38*, 91–128.

Lynch, J. P., Law, K. H., Kiremidjian, A. S., Kenny, T. W., Carryer, E., & Partridge, A. (2001, September 12–14). The design of a wireless sensing unit for structural health monitoring. In *Proceedings of the 3rd international workshop on structural health monitoring*. Stanford, CA: Stanford University.

Mansour, H., & Saab, R. (2014). Recovery analysis for weighted l1-minimization using a null space property. *CoRR*, abs/1412.1565.

Maser, K., Egri, R., Lichtenstein, A., & Chase, S. (1996). Field evaluation of a wireless global bridge evaluation and monitoring system (wgbems). In *Proceedings of 11th engineering mechanics conference* (Vol. 2, pp. 955–958). Fort Lauderdale, FL: ASCE.

Miller, T. I., Spencer, Jr, B. F., Li, J., & Jo, H. (2010). *Solar energy harvesting and software enhancements for autonomous wireless smart sensor networks*. Technical Report, Newmark Structural Engineering Laboratory. Champaign, IL: University of Illinois at Urbana-Champaign.

Nagayama, T., Spencer, B. F., & Rice, J. A. (2009). Autonomous decentralized structural health monitoring using smart sensors. *Structural Control and Health Monitoring, 16*, 842–859.

Needell, D., & Tropp, J. (2009). Cosamp: Iterative signal recovery from incomplete and inaccurate samples. *Applied and Computational Harmonic Analysis, 26*, 301–321.

Needell, D., & Vershynin, R. (2009). Uniform uncertainty principle and signal recovery via l2 regularized orthogonal matching pursuit. *Foundations of Computational Mathematics, 9*, 317–334.

Nesterov, Y. (1983). A method of solving a convex programming problem with convergence rate o (1/k2). *Soviet Mathematics Doklady, 27*, 372–376.

Nesterov, Y. (2005). Smooth minimization of non-smooth functions. *Mathematical programming, 103*, 127–152.

Novakovic, A., Meyer, J., Bischoff, R., Feltrin, G., Motavalli, M., El-Hoiydi, A., Restrepo, A., & Decotignie, J.-D. (2009). *Low power wireless sensor network for monitoring civil infrastructure*. Technical report, Empa, Swiss Federal Laboratories for Materials Testing and Research. CSEM, Swiss Center for Electronics and Microtechnology.

O'Connor, S. M., Lynch, J. P., & Gilbert, A. C. (2013). Implementation of a compressive sampling scheme for wireless sensors to achieve energy efficiency in a structural health monitoring system. In *Proceedings of SPIE 8694, nondestructive characterization for composite materials, aerospace engineering, civil infrastructure and homeland security* (pp. 86941L-1–86941L-11). San Diego, CA, USA.

O'Connor, S. M., Lynch, J. P., & Gilbert, A. C. (2014). Compressed sensing embedded in an operational wireless sensor network to achieve energy efficiency in long-term monitoring applications. *Smart Materials and Structures, 23*, 085014-1–085014-16.

Rice, J. A., Mechitov, K. A., Sim, S. H., Spencer, B. F., & Agha, G. A. (2011). Enabling framework for structural health monitoring using smart sensors. *Structural Control and Health Monitoring, 18*, 574–587.

Schneider, T., & Neumaier, A. (2001). Algorithm 808: Arfit - a matlab package for the estimation of parameters and eigenmodes of multivariate autoregressive models. *ACM Transactions on Mathematical Software, 27*, 58–65.

Selesnick, I. (2012). *Introduction to sparsity in signal processing*. Lecture notes. Retrieve from http://eeweb.poly.edu/iselesni/lecture_notes/sparsity_intro/index.html

Sichitiu, M. L., & Dutta, R. (2007). On the lifetime of large wireless sensor networks with multiple battery levels. *Ad Hoc & Sensor Wireless Networks, 4*, 69–96.

Sim, S.-H., Carbonell-Márquez, J. F., Spencer, B., & Jo, H. (2011). Decentralized random decrement technique for efficient data aggregation and system identification in wireless smart sensor networks. *Probabilistic Engineering Mechanics, 26*, 81–91.

Spencer, B. F., Ruiz-Sandoval, M. E., & Kurata, N. (2004). Smart sensing technology: Opportunities and challenges. *Structural Control and Health Monitoring, 11*, 349–368.

Straser, E., Kiremidjian, A., Meng, T., & Redlefsen, L. (1998, February 2–5). A modular, wireless network platform for monitoring structures. In *Proceedings of the 16th international modal analysis conference* (pp. 450–456). Santa Barbara, CA: Society for Experimental Mechanics.

Tropp, J. (2004). Greed is good: Algorithmic results for sparse approximation. *Information Theory, IEEE Transactions on, 50*, 2231–2242.

Wang, W., Wang, P., Zhou, W., & Li, H. (2015, March 8). The study of damage identification based on compressive sampling. In *Proceedings of structural health monitoring and inspection of advanced materials, aerospace, and civil infrastructure*, (pp. 94371L-1–94371L-6). San Diego, CA, USA.

Wang, Y. (2007). Wireless sensing and decentralized control for civil structures: Theory and implementation, PhD thesis, Stanford University, Stanford, CA, USA.

Wang, Y., & Hao, H. (2015). Damage identification scheme based on compressive sensing. *Journal of Computing in Civil Engineering, 29*, 04014037.

Wang, Y., Lynch, J. P., & Law, K. H. (2007). A wireless structural health monitoring system with multithreaded sensing devices: Design and validation. *Structure and Infrastructure Engineering, 3*, 103–120.

Wright, S., Nowak, R., & Figueiredo, M. (2009). Sparse reconstruction by separable approximation. *Signal Processing, IEEE Transactions on, 57*, 2479–2493.

Yang, Y., & Nagarajaiah, S. (2014). Data compression of structural seismic responses via principled independent component analysis. *Journal of Structural Engineering, 140*, 04014032.

Zhu, D., Yi, X., Wang, Y., Lee, K.-M., & Guo, J. (2010). A mobile sensing system for structural health monitoring: Design and validation. *Smart Materials and Structures, 19*, 055011-1–055011-11.

Zou, Z., Bao, Y., Li, H., Spencer, B., & Ou, J. (2015). Embedding compressive sensing-based data loss recovery algorithm into wireless smart sensors for structural health monitoring. *Sensors Journal, IEEE, 15*, 797–808.

Index

Printed and bound by CPI Group (UK) Ltd, Croydon, CR0 4YY

24/10/2024

01778291-0017